Rhizosphere Biology

Series Editor

Anil Kumar Sharma, Biological Sciences, CBSH, G.B. Pant University of Agriculture & Technology, Pantnagar, Uttarakhand, India

The Series Rhizosphere Biology, emphasizes on the different aspects of Rhizosphere. Major increase in agricultural productivity, to meet growing food demands of human population is imperative, to survive in the future. Along with methods of crop improvement, an understanding of the rhizosphere biology, and the ways to manipulate it, could be an innovative strategy to deal with this demand of increasing productivity. This Series would provide comprehensive information for researchers, and encompass all aspects in field of rhizosphere biology. It would comprise of topics ranging from the classical studies to the most advanced application being done in the field. Rhizoshpere is a dynamic environment, and a series of processes take place to create a congenial environment for plant to grow and survive. There are factors which might hamper the growth of plants, resulting in productivity loss, but, the mechanisms are not very clear. Understanding the rhizosphere is needed, in order to create opportunities for researchers to come up with robust strategies to exploit the rhizosphere for sustainable agriculture.

There are titles already available in the market in the broad area of rhizosphere biology, but there is a major lack of information as to the functions and future applications of this field. These titles have not given all the up-to-date information required by the today's researchers and therefore, this Series aims to fill out those gaps.

More information about this series at http://www.springer.com/series/15861

Ramesh Namdeo Pudake • Binod Bihari Sahu •
Maya Kumari • Anil K. Sharma
Editors

Omics Science for Rhizosphere Biology

Springer

Editors
Ramesh Namdeo Pudake
Amity Institute of Nanotechnology
Amity University Uttar Pradesh
Noida, Uttar Pradesh, India

Binod Bihari Sahu
Department of Life Sciences
National Institute of Technology
Rourkela, Odisha, India

Maya Kumari
Directorate of Life Sciences
Defence Research and Development Organisation
New Delhi, Delhi, India

Anil K. Sharma
Department of Biological Sciences, C.B.S.H.
G.B. Pant University of Agriculture and Technology
Pantnagar, Uttarakhand, India

ISSN 2523-8442 ISSN 2523-8450 (electronic)
Rhizosphere Biology
ISBN 978-981-16-0888-9 ISBN 978-981-16-0889-6 (eBook)
https://doi.org/10.1007/978-981-16-0889-6

This Springer imprint is published by the registered company Springer Nature Singapore Pte Ltd.
The registered company address is: 152 Beach Road, #21-01/04 Gateway East, Singapore 189721, Singapore

Preface

Since the beginning of life on earth, rhizospheric microbial communities and plants have been living together. During these million years of coexistence, they both are being exposed to various environmental conditions and coevolved in various types of interactions. Both the partners are taking clues from each other for their survival and development. Researchers have always wanted to know more about the mechanism behind these interactions, and the recent development in the field of omics science (metagenomics, metatranscriptomics, metaproteomics, and so on) has promoted the interest in knowing more and more about how rhizospheric microbiomes impact physiology and growth of crop plants. The studies are shedding light on how the association is established between the host and surrounding microbiome, accessing different biochemical processes happening in symbiotic or antagonistic interactions, and bioprospecting of valuable products. These advancements in technologies will be useful to establish functional mechanism of host–microbe relationship. Recent research in this field has achieved some results and now we know that a specific portion of soil microbiome is always actively associated with roots of plants and contributes significantly to host physiological performance. The recent focus is now to identify and isolate this core microbiome for the betterment of plant growth under various stress conditions. However, there are still some bottleneck like the cost of analysis and lesser availability of databases, and more informed and focussed future work will be helpful in utilization of plant microbe ecosystem in fuller extent. This book has reviewed the recent studies and emphasized on specific focus of future research. The chapters contributed by prominent scientists will be a ready source of information for young researchers who want to contribute to this field.

Reviewing the current research and writing it in the chapter form demands lot of efforts and dedication, and for that we thank our authors for their valuable contribution. Also, we extend our sincere thanks to Ms. Raagai Priya Chandrasekaran of Springer for her valuable support to facilitate completion of this book.

Noida, India	Ramesh Namdeo Pudake
Rourkela, India	Binod Bihari Sahu
New Delhi, India	Maya Kumari
Pantnagar, India	Anil K. Sharma

Contents

Editors and Contributors

About the Editors

Ramesh Namdeo Pudake is an assistant professor in Amity University Uttar Pradesh—one of the top ranked private universities in India. Dr. Pudake is a PhD in Crop Genetics and Breeding from China Agricultural University, Beijing, PRC. After his PhD he is engaged in research in a range of organisms but with a focus on crop plant genomics. He has also worked in the Department of Agronomy at Iowa State University Ames, IA, USA, on host–pathogen interaction and gene mapping. Currently, he is focusing on the research on different application genomics in plant–pathogen interaction. Dr. Pudake has published more than 35 research publications, 2 books, and 15 book chapters. He also has one scholarship award from Chinese Government for pursuing his doctoral studies. He is an expert reviewer for several journals of repute.

Binod Bihari Sahu has been working in Department of Life Science at National Institute of Technology Rourkela since 2014 as an assistant professor. His current work is focused on translational research in rice blast and sheath blight resistance. Dr. Sahu worked as a postdoctoral research associate at Iowa State University for 5 years with focus on nonhost resistance in Arabidopsis against soybean fungal pathogens, *Phytophthora sojae* and *Fusarium virguliforme*. He identified few NHR genes and got a US patent. Besides biotic stress research experience, he worked with abiotic stress in his Ph.D. thesis from Institute of Life Sciences Bhubaneswar. He published the key findings of his research with 18 articles in international journals and 6 book chapters and is working as academic editor to PLoS One in addition to reviewer to several research journals.

Maya Kumari is working as Scientist-E in Directorate of Life sciences, Defence Research and Development Organization, Govt. of India, and has more than 12 years of research experience in the field of plant molecular biology. Her research work in functional and molecular breeding for detection of molecular mechanism of plant biotic stress tolerance is of high scientific and social relevance. Currently, she is coordinating various projects related to biological defence technologies and acting as interface between army, industry and DRDO Labs. She has published more than 40 research papers, and act as reviewer for scientific articles in reputed journals. Besides her published work, Dr. Maya persuaded to achieve many awards including Masters Gold Medal, Endeavour Research Scholarship, DEST, Australia, to carry out research at CLIMA, UWA, Australia, Young Women Scientist Scholarship, Department of Science and Technology, Junior/Senior Research Fellow of Council of Scientific and Industrial Research, Govt. of India, Chinese Government Scholarship, China Scholarship Council, P.R. of China.

Anil K. Sharma is a professor at Department of Biological Sciences, CBSH, and Director of Extension Education, G.B. of Pant University of Agriculture & Technology in Pantnagar. He completed his postdoctoral studies at the GSU, Louisiana (USA), and has been a visiting scientist at the University of Basel, Switzerland (2003), and University of Helsinki, Finland (2013). He has extensive research and teaching experience and served as a reviewer for various funding bodies and international research journals. He also holds two patents in plant biology and microbiology, respectively. The recipient of various prestigious grants, he has published more than 90 research articles and 30 review articles, as well as 2 books with prominent publishers.

Contributors

Rahul Arora Division of Biosciences, University College London, London, United Kingdom
The Francis Crick Institute, London, United Kingdom

Usha Balasundaram Department of Genetic Engineering, Faculty of Engineering and Technology, SRM Institute of Science and Technology, SRM Nagar, Kattankulathur, Kanchipuram, Chennai, TamilNadu, India

Annapurna Bhattacharjee Department of Biochemical Engineering and Biotechnology, Indian Institute of Technology Delhi, New Delhi, India

D. Blaise Division of Crop Production, ICAR—Central Institute for Cotton Research, Nagpur, Maharashtra, India

Jaspreet Kaur Boparai Department of Biotechnology, Sri Guru Granth Sahib World University, Fatehgarh Sahib, Punjab, India

Shazia Anwer Bukhari Department of Biochemistry, GC University—Faisalabad, Faisalabad, Pakistan

Debatri Chakraborty Department of Biotechnology and Medical Engineering, National Institute of Technology, Rourkela, Odisha, India

Puneet Singh Chauhan Microbial Technology Division, Council of Scientific and Industrial Research-National Botanical Research Institute (CSIR-NBRI), Lucknow, India
Academy of Scientific and Innovative Research (AcSIR), Ghaziabad, India

Pradeep K. Das Mohapatra Department of Microbiology, Raiganj University, Uttar Dinajpur, West Bengal, India

Yashika Dhingra Department of Plant Molecular Biology, University of Delhi, Delhi, India

Shiv Shanker Gautam Department of Microbiology, College of Basic Sciences and Humanities, GB Pant University of Agriculture and Technology (GBPUA&T), Pantnagar, Uttarakhand, India

Mansour Ghorbanpour Department of Medicinal Plants, Faculty of Agriculture and Natural Resources, Arak University, Arak, Iran

Reeta Goel Department of Microbiology, College of Basic Sciences and Humanities, GB Pant University of Agriculture and Technology (GBPUA&T), Pantnagar, Uttarakhand, India

Tanya Gupta Department of Biotechnology and Medical Engineering, National Institute of Technology, Rourkela, Odisha, India

Munazza Ijaz Department of Bioinformatics and Biotechnology, GC University—Faisalabad, Faisalabad, Pakistan

Muhammad Iqbal Department of Environmental Science and Engineering, GC University—Faisalabad, Faisalabad, Pakistan

Priya Jaiswal Microbial Technology Division, Council of Scientific and Industrial Research-National Botanical Research Institute (CSIR-NBRI), Lucknow, India
Academy of Scientific and Innovative Research (AcSIR), Ghaziabad, India

Srishti Kar Microbial Technology Division, Council of Scientific and Industrial Research National Botanical Research Institute (CSIR-NBRI), Lucknow, India

Shivani Khatri Department of Biochemical Engineering and Biotechnology, Indian Institute of Technology Delhi, New Delhi, India

Bahman Khoshru Department of Soil Biology and Biotechnology, Faculty of Agriculture, University of Tabriz, Tabriz, Iran

Lisha Khungar Department of Plant Molecular Biology, University of Delhi, Delhi, India

Behnam Asgari Lajayer Department of Soil Science, Faculty of Agriculture, University of Tabriz, Tabriz, Iran

Bhaswatimayee Mahakur Department of Botany and Biotechnology, Ravenshaw University, Cuttack, Odisha, India

Mahmood-ur-Rahman Department of Bioinformatics and Biotechnology, GC University—Faisalabad, Faisalabad, Pakistan

Sankalp Misra Microbial Technology Division, Council of Scientific and Industrial Research-National Botanical Research Institute (CSIR-NBRI), Lucknow, India

Debasis Mitra Department of Microbiology, Raiganj University, Uttar Dinajpur, West Bengal, India

Abiramavalli Moorthy Department of Genetic Engineering, Faculty of Engineering and Technology, SRM Institute of Science and Technology, SRM Nagar, Kattankulathur, Kanchipuram, Chennai, TamilNadu, India

Shokufeh Moradi Department of Soil Biology and Biotechnology, Faculty of Agriculture, University of Tabriz, Tabriz, Iran

Pallavi J.N.L. College Khagaul, Patliputra University, Patna, Bihar, India

Debasish Pati Department of Life Science, NIT, Rourkela, Odisha, India

Ramesh Namdeo Pudake Amity Institute of Nanotechnology, Amity University Uttar Pradesh, Noida, Uttar Pradesh, India

Mrinalini Singh Pundir Department of Biological Sciences, College of Basic Science and Humanities, G.B. Pant University of Agriculture and Technology, Pantnagar, U.S Nagar, UK, India

Sivasankari Ramadurai Department of Genetic Engineering, Faculty of Engineering and Technology, SRM Institute of Science and Technology, SRM Nagar, Kattankulathur, Kanchipuram, Chennai, Tamil Nadu, India

Bilal Rasool Department of Zoology, GC University—Faisalabad, Faisalabad, Pakistan

Binod Bihari Sahu Department of Life Science, NIT, Rourkela, Odisha, India

Indira P. Sarethy Department of Biotechnology, Jaypee Institute of Information Technology, Noida, India

Angana Sarkar Department of Biotechnology and Medical Engineering, National Institute of Technology, Rourkela, Odisha, India

Pushpender Kumar Sharma Department of Biotechnology, Sri Guru Granth Sahib World University, Fatehgarh Sahib, Punjab, India

Shilpi Sharma Department of Biochemical Engineering and Biotechnology, Indian Institute of Technology Delhi, New Delhi, India

Ravindra Soni Department of Agricultural Microbiology, College of Agriculture, Indira Gandhi Krishi Vishwavidyalaya, Raipur, Chhattisgarh, India

Nidhi Srivastava Department of Biotechnology, Jaypee Institute of Information Technology, Noida, India

Deep Chandra Suyal Department of Microbiology, Eternal University, Sirmour, Himachal Pradesh, India

Shikha Thakur CSIR—National Chemical Laboratory, Pune, India

Issayeva Akmaral Umirbekovna Department of Chemistry, Biology and Agriculture, Shymkent University, Karatau, Kazakhstan

Kulandaivelu Velmourougane Division of Crop Production, ICAR—Central Institute for Cotton Research, Nagpur, Maharashtra, India

Muhammad Zubair Department of Bioinformatics and Biotechnology, GC University—Faisalabad, Faisalabad, Pakistan

Chapter 1
Rhizosphere Metagenomics: Methods and Challenges

Abiramavalli Moorthy and Usha Balasundaram

Abstract "Rhizosphere" is a narrow region associated with plant roots, acting as a residing place for millions of microorganisms. The rhizosphere-associated microbes are collectively called as root microbiome or rhizobiome. These microbiomes play a vital role in plant health by manipulating their growth and development. Rhizobiomes include both beneficial communities which enhance plant growth and improve plant defense mechanisms and pathogens which are harmful to plants. Nevertheless, the beneficial communities compete with the pathogens and colonize the roots. Though the significance of rhizosphere microbial community is well acknowledged, characterization of a plenty of microbes colonizing the rhizosphere is not done. Studying the rhizobiome of a crop species is an essential factor of crop improvement. "Metagenomics" is a frontier science that deals with study of metagenomes found in an environment such as rhizosphere. In this chapter, we have reviewed the most important metagenomic approaches and attributes to study the microbial diversity in the rhizosphere. We have discussed about the methods and software programs available for metagenome assembly, binning strategies, taxonomic classification, and functional annotation of metagenomics datasets. In addition, we have briefly pointed out the bottlenecks of the metagenomics approaches in studying the rhizobiomes.

Keywords Assembly · Binning · Classification · Metagenomics · Microbiomes · Rhizosphere

---

A. Moorthy · U. Balasundaram (✉)
Department of Genetic Engineering, Faculty of Engineering and Technology, SRM Institute of Science and Technology, SRM Nagar, Kattankulathur, Kanchipuram, Chennai, TamilNadu, India

R. N. Pudake et al. (eds.), *Omics Science for Rhizosphere Biology*, Rhizosphere Biology, https://doi.org/10.1007/978-981-16-0889-6_1

1.1 Introduction

In 1904, a German agronomist, Lorenz Hiltner, coined the term "rhizosphere" from the words rhiza and sphere (rhiza-root; sphere-environment of influence). The term describes an associated area of plant roots and the surrounding soil (Hartmann et al. 2009). The rhizosphere soil has proximity of 1–2 mm to the plant root. This is the zone where the plant roots release their metabolites and colonization of a wide range of microbiome that feed on these metabolites occur. The rhizosphere region can be divided into three zones (Fig. 1.1). The endorhizospheric zone is the cortex and endodermis of the root. The second, rhizoplane zone is the surface of root where the roots release their metabolites that attracts the microbes for colonization. And ectorhizospheric zone where soil region found between the rhizoplane and the bulk soil. This is where the free-living microorganisms reside.

Apart from these three zones, there are certain well-defined specific layers in the rhizosphere where fungal association occurs and this region is termed as mycorrhizospheric zone (Odelade and Babalola 2019). However, rhizosphere of any plant does not have a defined size or shape, it just changes along the root surface.

Plant roots in general release a wide variety of chemical compounds into their rhizosphere as a result of their metabolic processes. This process is termed as rhizodeposition, and the released metabolites are called rhizodeposits or root exudates. Rhizodeposits can be broadly categorized into two categories: low molecular weight compounds and high molecular weight compounds. Among these, organic acids, amino acids, sugars, phenolics, and flavonoids are low molecular weight compounds released from the roots which act as a readily available major carbon

Fig. 1.1 Schematic diagram showing rhizosphere soil zones with associated microbes

source for the soil-borne microbial community (Hinsinger et al. 2009). Proteins, polysaccharides, and mucilage's are high molecular weight compounds exuded by the roots (Hayat et al. 2017). Basically, these plants derived compounds act as key signals and attract various microorganisms towards them. The attracted microbes colonize the root surface found in the rhizoplane and consume the released metabolites for their energy. In turn, these colonized microbes produce certain secondary metabolites, antibiotics, phytohormones, and a few other compounds which help the host plant in their growth, development, and in defense mechanisms. This leads to a better productivity of the plants. In some cases, pathogenic bacterial community will also get attracted towards the plant-derived metabolites. In such cases, a negative interaction develops between the host plant and the pathogens, the invading pathogenic organisms secrete toxic metabolites, leading to detrimental effects of the host plants.

However, plants are smart enough to protect themselves from these pathogens by modifying their root exudates composition which selectively attracts beneficial bacteria that outcompete the soil-borne pathogens (Pérez-Jaramillo et al. 2016). For instance, Rudrappa et al. (2008) have demonstrated this selective recruitment of growth-promoting bacterial community in the rhizosphere of Arabidopsis upon pathogen infection, by predominantly releasing malic acid. Similarly, under nutrient deprivation especially iron, plants release a different set of metabolites that attract microbes which produce siderophores to make the soil nutrient (iron) available for the plants (Carvalhais et al. 2013). In case of phosphate and nitrogen deprivation, plants exude strigolactones and flavonoids which specifically recruit mycorrhizas and nitrogen-fixing bacteria, respectively (Bertin et al. 2003; Akiyama et al. 2005; Hassan and Mathesius 2012). These observations elucidate the significance of rhizodeposits in plant–microbe interactions and in sustainable maintenance of the rhizosphere. Additionally, even in the nonexistence of a plant, the root exudates when blended with the soil precisely alter the bacterial community of the soil by selective attraction. This has been proven through 16S rRNA pyrosequencing (Badri et al. 2013). This clearly demonstrates the significance of root exudates in microbe–microbe interactions.

In addition to Plant Growth-Promoting Rhizobacteria (PGPRs), there are fungal partners that have a symbiotic relationship with plants by feeding on root exudates. Arbuscular Mycorrhizal Fungi (AMF) and a few non-mycorrhizal endophytic strains provide benefits to host plants by increasing the availability of soil nutrients, fighting against pathogen attack and enhancing the overall yield of the host plants (Parniske 2008). The ability of a plant to alter its root microbiome for its own benefits (Germida and Siciliano 2001) lead to diversity in rhizobiome of each plant species. Not only species variations but also the genotype of the host plant determines the rhizobiome diversity. Plant diversity is directly related to the belowground microbial diversity. In addition, belowground microbial diversity is instigated also depending on the following factors such as the soil type, physico-chemical properties of the soil, climatic conditions, nutritional status of the host plant, host root exudate composition, and developmental stage of the host plant (Berg 2009). Therefore, it is important to clearly understand the biogeochemical processes that occur in the rhizosphere soil for substantial maintenance of the soil environment and the organisms colonizing it (Morrissey et al. 2004). Also, to utilize the root colonizing microbiome as an efficient tool for improved agricultural productivity, it is vital to understand the microbial diversity in terms of their structure, composition, ecology, and activities in various environments (Pinton et al. 2007). This would provide a source of information to prefer a selective microbiome that helps in crop improvement (Panke-Buisse et al. 2015).

1.2 Rhizosphere Metagenomics

Soil-borne microorganisms are the key players of biogeochemical processes of the ecosystem. Millions of prokaryotic and eukaryotic taxa are found in the soil such as bacteria, archaea, viruses, phages, fungi, and protozoa. A report suggests that, generally soil possess around 4×10^6 different microbial taxa (Curtis et al. 2002). In addition to this, recent reports suggest that every 1 g soil contains 1000–10,000 bacterial taxa (Wagg et al. 2014; French et al. 2017). The diversity in the soil-borne microorganisms is huge, influenced by various important factors (Berg 2009). So, it is important to unveil the complexity of belowground microbial diversity using suitable methods. There have been various methods to study the microbial diversity since 1880s till date. The methods that are followed to identify the taxonomic diversity and phylogenetic relationships of microorganisms can be divided into two types: culture-dependent and independent methods. Culture-dependent approaches include (1) plate count method, (2) BIOLOG plate method (Biolog Inc., Hayward, CA, USA), where only microbes which can grow under laboratory conditions can be identified and studied. This method was not suitable for the unculturable microbes found in the soil.

Culture-independent approaches are of four types: (1) microbial lipid based (Phospholipid fatty acid analysis-PLFA, fatty acid methyl esters-FAME), (2) non-PCR-based (DNA re-association, GC content-based), (3) PCR-based (RFLP, RAPD, DGGE, FISH, and so on), and (4) sequencing-based (clonal library sequencing, marker-based, and whole metagenome shotgun sequencing). Among these, sequencing techniques such as marker gene based and shotgun sequencing (Fig. 1.2) are being widely used recently because of the ease of the techniques, its high efficiency, and low cost.

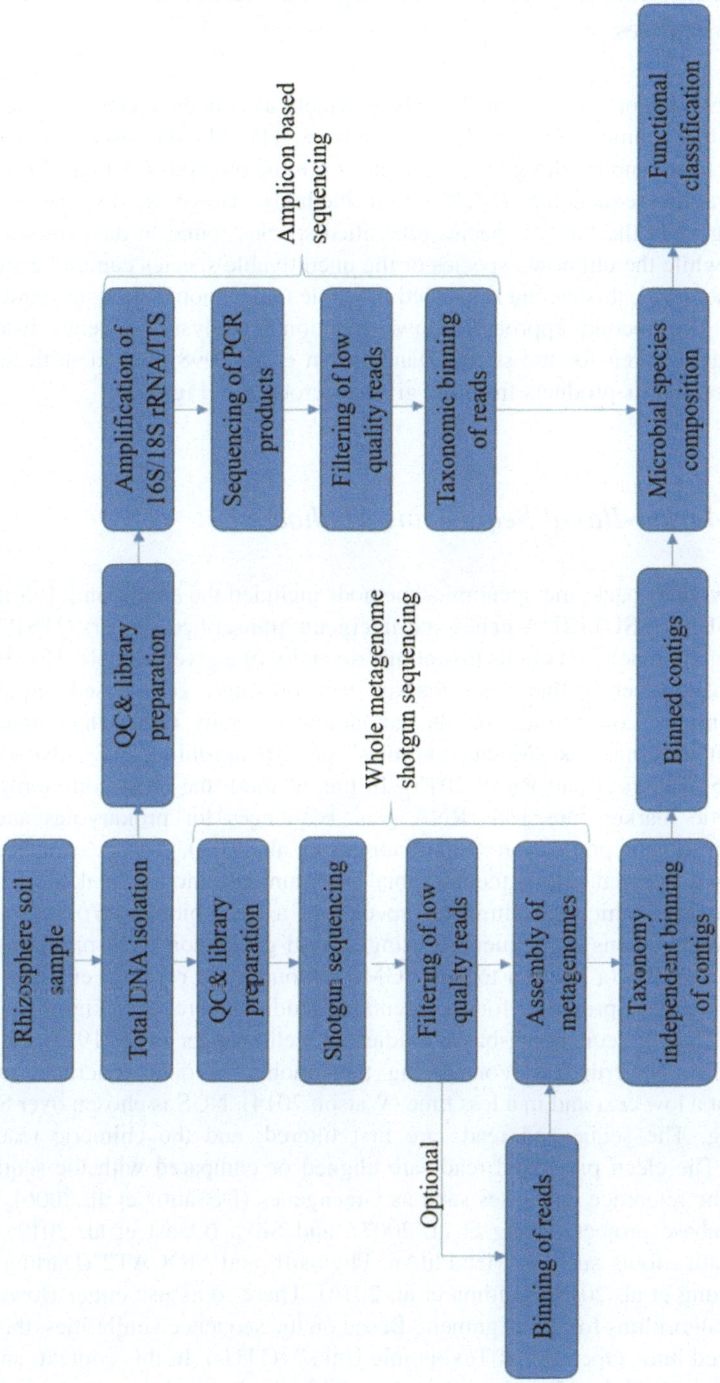

Fig. 1.2 Schematic diagram showing the workflow of the rhizosphere metagenomics

1.3 Metagenomics Methods to Explore Microbial Diversities

There are two major methods that have been widely used in the recent times to study the microbial community in the rhizosphere. (1) Marker-based sequencing, (2) Whole metagenome shotgun sequencing. Both the methods are high-throughput next-generation sequencing (NGS)-based methods. However, in marker gene sequencing, only the known species (classified species found in databases) can be identified while the unknown species or the uncultivable species cannot be studied. On the other hand, this lacuna is bridged in whole metagenome shotgun sequencing approach. The second approach allows functional analysis of genes from the genomes undertaken for the study (Handelsman et al. 1998) and identification of novel genes and its products from the given microbiome (Fig. 1.2).

1.3.1 Marker-Based Sequencing Method

Until a few years back, metagenomics methods included the amplicon [(16S rRNA/ Small-SubUnit (SSU) rRNA genes and intergenic transcribed spacers (ITS)]-based sequencing approach that elicits taxonomic diversity of a given sample. However, it was proposed recently that since marker gene or single gene-based sequencing approach highly concentrates on the taxonomic diversity rather than functional diversity it is termed as "Metataxonomics" or "Metaprofiling" (Escobar-Zepeda et al. 2015; Marchesi and Ravel 2015). In this method, the most commonly used phylogenetic markers are 16S rRNA gene sequences for prokaryotes and ITS regions for fungal population (Eloe-Fadrosh et al. 2016). These sequences are universally conserved among the microbial communities; and are flanked by hyper-variable regions which discriminates species of a microbiome (Woese and Fox 1977). The amplicons are sequenced using second-generation NGS platforms such as Illumina MiSeq or the Ion torrent PGM. Although, the recently emerged NGS platform is widely preferred for metagenomic studies, there were significant progresses in Sanger sequencing-based studies as well (Sanger et al. 1977; Gillespie et al. 2002). Nevertheless, considering the amount of data generated (almost 100 GB) at a low cost and in a less time (Watson 2014), NGS is chosen over Sanger sequencing. The sequenced reads are first filtered, and the chimeric reads are removed. The clean processed reads are aligned or compared with the sequences found in the reference databases such as Greengenes (DeSantis et al. 2006), Ribosomal database project (Wang et al. 2007), and Silva (Quast et al. 2012) using bioinformatics tools such as MetaPhlAn, Phylosift, and MOCAT2 (Darling et al. 2014; Truong et al. 2015; Kultima et al. 2016). These tools use either Bowtie2 or HMMER algorithms for the alignment. Based on the sequence similarities, the reads are clustered into "Operational Taxonomic Units" (OTUs). In this context, an OTU is a cluster or a group of reads which share 97% similarity and are expected to be from the same species. This method can be called as "direct taxonomic classification" (Breitwieser et al. 2019) since it matches each read directly to the reference database, and it is particularly helpful if the research question is quantifying the known community. Over the last 40 years, 16S rRNA genes have been used as a standard marker for taxonomic annotation of soil microbiome. The major drawbacks of this approach is that, it relies exclusively on a single marker gene region (16S rRNA for prokaryotes and ITS for fungi) to interpret the species diversity. If any horizontal gene transfer events have occurred in the target soil region, then this marker-assisted sequencing approach would become uninformative. If there is no closely related species found in the database, the software cannot classify those reads. In addition, this method cannot detect viruses (if any) in the sample. Although this method is fast and relatively cheap, classification of unknown species is not possible (Breitwieser et al. 2019). Therefore, to overcome these limitations, scientists prefer whole metagenome shotgun sequencing method, particularly if their research purpose is to identify and quantify microorganisms from all domains.

1.3.2 Shotgun Metagenome Sequencing Method

Shotgun sequencing of the complete collection of genomes (metagenome) available in a given space at a given time is referred to as "Metagenomics" (Scholz et al. 2016). In the field of ecology, "Metagenomics" is analogous to culture-free genomic approaches to study the microbiomes (complete collection of microbes found in a particular ecosystem) in its natural environment (Chen and Pachter 2005). It is a frontier science that combines microbiology, molecular biology, and biotechnology to answer the two popular ecological queries such as: Who are they? What are they doing in that vicinity? (Singh et al. 2010). This approach can reveal both coding and non-coding part of the genomes which can be used further to identify new markers, genes and its products (Escobar-Zepeda et al. 2015). Shotgun metagenome sequencing for microbial diversity studies have been attempted so far, through various NGS platforms such as 454 pyrosequencing, Ion torrent PGM, Illumina, and PacBio (Glenn 2014). At present, the second- and third-generation technologies (Illumina, PacBio and Oxford Nanopore Technology) are leading in metagenomics studies. In Illumina, millions of short reads will be generated (150–400 bp in length), whereas in PacBio and ONT longer reads (6–20 kb) but fewer in number will be generated. In the latter, there is a high raw error rate but are correctable. Ease of library preparation, depth of sequencing (generates almost 100 GB data), time and cost reduction favors these two technologies. Also, such deep sequencing would reveal even the least abundant species found in the given sample. The major benefits of this method are that researchers can achieve deep sequencing of a sample with a lesser starting material and generate a bigger sequence datasets (around 100 GB at a time). In addition, the datasets can be analyzed more precisely by measuring both structural diversity and functional diversity. Furthermore, if the sample comprises of unexplored genomes, a de novo assembly of the dataset followed by downstream analysis is possible with this method. However, this approach requires high end computational resources and softwares to assemble and analyze bigger datasets. Also, prior knowledge on the UNIX operating system is essential for a researcher to analyze the big data.

1.3.3 Attributes to Quantify Microbial Diversity

Before stepping into quantification or classification of metagenomic datasets, one must know the attributes or measures that classify their structure. The key element of a biological community is “species diversity”. So far researchers have proposed a few concepts and metrics to quantify a microbial community. The two important attributes followed by researchers till date to determine microbial diversity are species richness (refers to number of species in a particular area) and species differential abundance (which quantifies the uneven distribution of communities). The latter can be tested against a theoretical community which has equal distribution of all species. So, when two or many communities that share equal species richness, but different abundances are compared, the community, or the group which has the lowest abundance value will be considered as more diverse. In addition to this, different metrics have been used in metagenomics approaches to quantify the diversity of one or many communities. To quantify the diversity within one community (Operational Taxonomic Units-OTUs), alpha diversity is being used. Regional diversity between many communities is being measured through gamma diversity.

Linking the two metric data, there is a third metric named beta diversity which opens up about the diversity of many communities in a given area (Krebs et al. 2014). However, these metrics could not reach a maximum sample size. So, statistical non-parametric estimators were designed for quantifying microbial diversity: Simpson’s index and Shannon-Weaver’s index. Simpson index gives weightage to most common species whereas Shannon index gives higher weightage to rare species. Both the estimators are good for species richness concept; however, meticulous sampling is required when the sample has more rare species in it. There came an invention of a new index, tail statistic, which measures the rare species in a complex sample precisely lending more weightage to the less abundant taxa (Li et al. 2012). These metrics are useful in comparing communities between samples. But the comparison should be vigilant that no data is biased to provide an informative result (Bonilla-Rosso et al. 2012).

1.3.4 Classification of Metagenomic Datasets

As mentioned earlier, the classification method is chosen depending on the research purpose. If the objective is targeted to a particular known species or to just detect the presence or absence of a known community, then marker-based metagenomic sequencing followed by direct taxonomic classification of the sequenced reads is preferred. But when the research question is on identifying and quantifying the microbiome from an unexplored environment such as rhizosphere soil, it requires de novo assembly-based classification since soil environments might possess a wide range of uncultivable microbial domains. Classification of datasets generated from shotgun metagenomics method can be done in two aspects: (1) structural classification (2) functional classification (de Fátima Alves et al. 2018). In this context, structure refers to the population abundance and richness in an environment. Here, one can elaborately study the connections between individual members of a community, which is mandatory to decipher their biological functions (Tringe et al. 2005; Vieites et al. 2008). This approach brings out the phylogenetic diversity among the microbiomes. In functional classification, sequenced reads are assembled into longer contigs and scaffolds, and are compared against gene databases and pathway databases to decipher the functional significance of the given sample.

1.3.5 Structural Classification

To structurally analyze the datasets generated from sequencing, a process called "binning" is being applied. Clustering of sequencing reads into groups or compartments or bins are referred to as "binning." Binning is the subsequent step of both marker gene sequencing approach and shotgun sequencing approach. Depending on the sequencing approach, we choose (amplicon-based/shotgun) binning methods, which can be sorted into various categories.

1.3.5.1 Binning Strategies Based on Taxonomy

In taxonomy-based binning, the softwares work by assigning the individual reads obtained from marker gene sequencing/shotgun sequencing against a reference database which then group the reads into bins based on their taxonomic similarity. However, this taxonomy-dependent strategy is not suitable for microbes whose genomes are not found in the reference databases analyzed. Software simply groups those reads that are not mapped to any reference as "unassigned." Therefore, it can be put forth that taxonomy-dependent binning could decipher the taxonomic diversity of only known microorganisms. Furthermore, this strategy can be subdivided into three: (1) alignment based, (2) composition based, and (3) hybrid based.

1.3.5.2 Alignment-Based Classification

In alignment-based analysis, BLAST and BLAT were the preferred algorithms to compare the reads to reference genomes initially. However, in recent times, because of the bigger datasets, Hidden Markov Models (HMMs) and Bowtie2 algorithms are used by the bioinformatics tools, MetaPhlAn, Phylosift, MOCAT2, and GOTTCHA (Darling et al. 2014; Freitas et al. 2015; Truong et al. 2015; Kultima et al. 2016). This method is faster when compared to other two methods. It simply aligns the reads against the databases which contains millions of genes and genomic signatures from various genomes of microorganisms. The publicly available reference databases are NCBI, EMBL, PFAM, UniProt, Genbank, DDBJ, NCBI Refseq, and Ensembl. Reads are grouped based on their alignment quality and the topmost hits obtained from the comparison. Publicly available softwares that adopt this alignment-based strategy are MG-RAST server and CAMERA. This strategy can predict the species relative abundance and species overall composition found in the given sample. However, these softwares can predict the similarities of only known species. While the datasets from a not well-explored niche such as soil generally will have a lot of unknown or completely new species which is not present in the reference databases. In such cases, specialized methods that works based on sequence composition are implemented to address this issue.

1.3.5.3 Composition-Based Classification

This is an alignment free and a reliable method for taxonomic diversity prediction. Sequence composition is the key attribute which is used to quantify the conservation among species. Instead of alignment, this method uses algorithms that rely on matching of k-mers. A k-mer is defined as subsequences of a nucleotide sequence of length k. In this strategy, the bioinformatics tools such as Kraken and CLARK (Wood and Salzberg 2014; Ounit et al. 2015) first create a database with k-mers and an identifier for every k-mer. The k-mers are basically generated from the reads by clustering each read into short k-mers. K-mers can be of any size but it can neither be too long nor be too short which may lead to nonspecific matches or doesn't match at all. The ideal length of the k-mers, $k = 20–31$, is followed by the familiar tools. So, the database created by the tools will ideally have k-mers of every genome and their taxonomy IDs, that are acting as references for the query k-mers generated from the metagenomic dataset. When a k-mer matches with more than one taxa, the Least Common Ancestor (LCA) of those taxa will be considered. In addition to k-mers of fixed length, there are certain tools (extension of Kraken) (Wood and Salzberg 2014) which construct k-mers or seeds with variable length. These seeds are considered as genomic signatures and are used for matching metagenomic datasets with the reference sequences. In this case, an exact match of the bases in the signature is trivial whereas it requires only a part of it to match perfectly. The accuracy of classification for these tools is diverse. For instance, Kraken works well up to taxa level classification. While Bracken (an extension of Kraken), (Lu et al. 2017) and CLARK (Ounit and Lonardi 2016) can precisely classify even at species or genus levels.

1.3.5.4 Hybrid-Based Classification

Hybrid-based binning combines the two strategies (alignment based and composition based). First phase will be aligning the clean reads against reference sequences for similarities among the species using a local aligner such as BLAST (Altschul et al. 1990; Camacho et al. 2009). Second phase will be comparison of the results obtained from alignment with the reference sequences. That is, tools like MEGAN utilizes the LCA of the alignment generated in the first phase for further classification. In other tools such as Taxator-tk, the sequence overlaps of the query (resulted from local alignment) will be merged into longer sequences. These longer sequences will then be matched against reference genome sequences. The pairwise distance between the query and the reference genome is calculated and used for further classification. This method ensures accuracy and secures time. The programs that work on this hybrid-based binning are PHINX (Mohammed et al. 2011), PhymmBL (Brady and Salzberg 2009), Amphora2 (Wu and Scott 2012), and MaxBin (Wu et al. 2014).

1.3.5.5 Amino Acid-Based Taxonomic Classification

Compared to DNA sequences, amino acid sequences are evolutionarily highly conserved among microbial population. Hence, classifying the microbiome based on their amino acid sequences instead of DNA sequences would be more sensitive form of classification. In this method, the sequenced reads are first translated into amino acid sequences and aligned against reference protein databases. Then based on the alignment results, species composition will be quantified through LCA approach. The most widely used bioinformatics tools that apply these strategies are DIAMOND for translation of datasets and MEGAN for composition prediction. A few other classifiers such as Kaiju uses Burrows-Wheeler transform (BWT) algorithm to index the reference databases and stores the pre-constructed index in FM-index table. So, when a dataset is assigned to this tool, it first translates all the reads in all the six frames. Then compare the amino acid fragments against the pre-indexed reference databases for perfect matches.

1.3.5.6 Taxonomy-Independent Binning

When a sample for metagenomics study is chosen from a least explored environmental niche such as rhizosphere, comprising uncharacterized and unculturable microbes, the downstream analysis of the datasets such as binning, assembly, and classification highly varies. In this scenario, the process of binning is not based on taxonomy but based on sequence features. The principle behind taxonomy-independent binning is that each species will have a unique sequence composition. So, the reads which share the same tetra nucleotide frequencies or k-mer frequencies are assumed to be originated from the genome of the same species. In general, binning and assembly of metagenomes are interrelated processes which can be performed one after the other or in an integrated platform. However, in this scenario, binning is always performed before a metagenomic assembly. Because binning of reads into different taxa will reduce the ambiguity in mixing of genomic information. Theoretically, each bin is considered as separate taxon and assembled separately. There are several methods such as Likelybin is available for binning. Likelybin (Kislyuk et al. 2009) is a simplified approach where the reads are clustered in an assumption that tetra nucleotide frequency of all the reads are similar within a microbial taxon. This model cannot be used when there is an occurrence of horizontal gene transfers in the community. So, more complex models such as PHYSCIMM (Kelley and Salzberg 2010), MetaWatt (Strous et al. 2012), CONCOCT (Alneberg et al. 2014), and Latent Strain Analysis (LSA) (Cleary et al. 2015) are available for clustering of complicated reads. The outputs or bins generated from these methods are further used for de novo assembly and taxonomic classification.

1.3.5.7 De Novo Assembly of Metagenomic Datasets

An assembly is a process of reconstruction of original genome sequence in silico using the reads generated from sequencing the genomic fragments. A de novo metagenomic assembly is nothing but a parallel reconstruction of all the genomes present in a metagenomic sample without a reference genome. Generally, de novo assemblers use either of the following specialized algorithms for assembling the reads (1) Overlap Layout Consensus (OLC), (2) de Bruijn graph based. OLC assemblers perform a pair-wise alignment between the reads. Significantly overlapping reads are connected to form a graph which is then used for construction of longer contigs. While, in de Bruijn assemblers, instead of pair-wise comparison, the reads are split into k-mers (k can be any length) and the graph is generated by reading the successive k-mers of every read. Most of the recently developed genome assemblers adopt de Bruijn algorithm because it requires less computational resources when compared to OLC assemblers. However, de Bruijn assemblers are highly prone to sequence errors and false joining of sequences because of the short k-mers. MetaVelvet and RayMeta are the well-accepted tools for assembling of metagenomics datasets. These tools follow de Bruijn algorithm to construct contigs. After constructing the longer contigs, scaffolds are generated and validated statistically. N50 values (length of the 50% of scaffolds assembled) and length of the contigs generated are significant attributes in validating an assembly, which should be equal and longer, respectively (Mäkinen et al. 2012). Online tools such as MetaQUAST are available to do the validation of assemblies (Mikheenko et al. 2016). The validated assemblies are then compared with the reference genomes for taxonomic classification.

1.3.5.8 Contig Binning

Assembled contigs and scaffolds are clustered into bins through binning algorithms discussed in the earlier sections. Either contigs will be aligned against the reference databases or binning will be performed based on composition of the species such as GC content, genomic signatures, and nucleotide frequency distribution between the species. Furthermore, after clustering the contigs into bins, these bins are validated by softwares for incompleteness or contamination. Each bin is considered as one taxon or one species. So, if there is repetition of marker genes, it is considered as contamination and if there is absence of marker genes in a bin then it is considered as incomplete assembly. In both the cases, bins can be reclustered. CheckM and BUSCO are two such tools which carry out the validation of contig binning (Parks et al. 2015; Simão et al. 2015).

1.3.6 Functional Classification

The functional classification part comprises of gene and promoter predictions and gene function annotations by correlating the datasets with reference protein databases. The term "metagenomics" does not stop only with analyzing the microbial diversity, but also deals with predicting the significant genes of the metagenomes and predicting their probable functions. Once the assemblers generate contigs and scaffolds, they will be subjected to gene prediction analysis through bioinformatics tools. The gene finder softwares such as MetaGene Annotator (Noguchi et al. 2008), Glimmer (Delcher et al. 2007), Glimmer-MG (Kelley et al. 2012) critically search even shorter contigs and scaffolds for the presence of translation initiation sites, start and stop codons and Open Reading Frames (ORFs) based on the most potential and reliable features of gene prediction and classification, di-codon usage, and GC composition. Importantly, the softwares predict the gene length and their location in the genome. These popular software programs work through statistical models such as Hidden Markov Model (HMM) and Ribosomal Binding Site (RBS) model. In addition to this, tools such as FragGeneScan (Rho et al. 2010) can predict genes straightaway from the short sequenced reads itself without requiring an assembly, using HMMs. It is reported that FragGeneScan performs well when compared to other gene prediction tools and classifiers (Roumpeka et al. 2017). Adding to this, there are integrated and interactive programs such as MetAMOS (Treangen et al. 2013) and Anvi'o (Eren et al. 2015) that performs metagenomic assembling and gene annotation as well. Furthermore, once predicting and classifying the probable protein coding genes is done, they will be searched against protein databases such as Interpro (Hunter et al. 2012), Kyoto Encyclopedia of Genes and Genomes (KEGG) (Kanehisa et al. 2007), and MetaCyc (Caspi et al. 2016) to predict their likely functions. Last, but not the least, a metagenomic assembly and annotation generate a large file in a format which is not easily accessible on desktops and laptops. In such cases, there are web applications such as Meta4 (Richardson et al. 2013) and online portals such as MG-RAST (Glass et al. 2010), and recent one EDGE (Philipson et al. 2017) for accessing these information.

1.4 Milestones of Rhizosphere Metagenomics

The well-known rhizosphere microbiomes identified and studied so far are plant growth-promoting rhizobacteria, nitrogen-fixing bacteria, biocontrol agents, mycorrhizas, protozoa, mycoparasites, a few deleterious organisms such as nematodes and pathogens (Mendes et al. 2013). Till date, the interactions between the plant roots and the associated microbiomes ("root microbiome" or "rhizobiome") have been well documented in various rhizosphere soils. In general, microbial diversity analysis of various different soils indicated that there are more than one million different bacterial taxa found in a single gram of soil. The prominent communities in the soils analyzed are *Bacteroidetes, Betaproteobacteria, and Alphaproteobacteria* (Gans et al. 2005; Roesch et al. 2007). A lot of reports on rhizobiomes through metagenomics approach have been documented. For instance, marker-based sequencing of rhizospheres of 14 different plant species depicted the presence of approximately 1200 distinct taxa in it, with the *Proteobacteria* as the predominant phylum (Hawkes et al. 2007). Similarly, marker-based pyrosequencing (SSU-rRNA library and nifH cluster I library) in *Erica andevalensis* in a metal-rich acidic soil and maize, respectively, demonstrated the abundance of *Actinobacteria, Acidobacteria, Azospirillum, Bradyrhizobium,* and *Ideonella.* Almost 27% of the clones analyzed belonged to uncharacterized taxa (Mirete et al. 2007; Roesch et al. 2007). PhyloChip investigation or 16S microarray of oats microcosms revealed that there were 1917 distinguishable taxa in it (DeAngelis et al. 2009). Similar analysis on potato rhizosphere revealed the presence of almost 2432 OTUs (Weinert et al. 2011). The rhizosphere of sugar-beet was predicted to have 33,346 OTUs in total (Mendes et al. 2011). Over all, it is estimated that, approximately 55,000 OTUs are found in rhizospheres of various plant species (Mendes et al. 2013). Furthermore, functional classification of metagenomics datasets generated from various rhizospheric soils, predicted many significant beneficial genes such as genes that confer metal tolerance and salt tolerance (Mirete et al. 2007, 2015; Guazzaroni et al. 2013). These kinds of research and inventions in the rhizosphere metagenomics studies are trending recently. Structure and composition of rhizobiomes of many important plant species such as *Oryza sativa, Zea mays, Setaria italica, Gossypium hirsutum L., Solanum lycopersicum, Allium cepa L., and Nicotiana tabacum* have been documented recently (Table 1.1).

Table 1.1 The dominant microbial taxa associated with the rhizosphere of various important plants detected through metagenomics methods (Only recent reports are listed in this table)

S. no.	Plant	Rhizosphere-associated microbes	References
1	Rice (*Oryza sativa*)	*Proteobacteria, Gemmatimonadetes, Verrucomicrobia, Geobacter Anaeromyxobacter, Clostridiaceae, Opitutaceae and Herbaspirillum*	Breidenbach et al. (2016)
2	Maize (Zea mays)	*Acidobacteria, Proteobacteria, Actinobacteria, Bacteroidetes, Chloroflexi, Firmicutes, Gemmatimonadetes, Chitinophaga, Nitrospira, Flavabacterium, Nonomuraea, Thiobacillus, Phenylbacterium, Bradyrhizobium, Erwinia, Inquitinus, Nitrosovibrio, and Rickettsia*	Yang et al. (2017)
3	Cotton (*Gossypium hirsutum*)	*Proteobacteria, Verrucomicrobia, Planctomycetes, Acidobacteria, Actinobacteria, Bacteroidetes, Firmicutes, Chloroflexi*	Qiao et al. (2017)
4	Tomato (*Solanum lycopersicum*)	*Proteobacteria, Acidobacteria, Actinobacteria, Bacteroidetes, Firmicutes, other unclassified bacteria.*	Li et al. (2014)
5	Fox tail millet (*Setaria italica*)	*Proteobacteria, Acidobacteria, Actinobacteria, Bacteroidetes, Firmicutes, Thaumarchaeota*	Jin et al. (2017)
6	Onion (*Allium cepa*)	*Proteobacteria, Firmicutes, Actinobacteria*	Ikeda et al. (2014)
7	Tobacco (*Nicotiana tabacum*)	*Proteobacteria, Verrucomicrobia, Acidobacteria, Actinobacteria, Firmicutes, Gemmatimonadetes, Planctomycetes, Bacteroidetes*	Saleem et al. (2016)

1.5 Challenges of Rhizosphere Metagenomics

There are a few practical difficulties in a successful metagenomics study of rhizosphere soils. First and foremost is the quality of starting genetic material. A successful metagenomics study requires a high quality, intact, pure DNA from the rhizosphere soil. Presence of polyphenolics in the DNA will highly hinder the PCR and subsequent sequencing steps (Ranjan et al. 2005; Sharma et al. 2007). To address this issue, there are a lot of modified protocols and suitable isolation kits available (Felczykowska et al. 2015; Tanveer et al. 2016). However, the rhizobiome of different plants will be different and hence demand different extraction procedures which were proven through PCR-DGGE profiles (Niemi et al. 2001). The second challenge is to choose a suitable sequencing platform as there are various NGS platforms in the market, choosing a sequencing technique that is appropriate to the research question is a consideration. The third concern is assembly of metagenomes. As single genome assembly itself is a complex process because of the presence of repetitive elements in the reads, reconstruction of multiple genomes present in a single sample is definitely a difficult process requiring specialized assemblers, algorithms, and softwares. As rhizosphere comprises of a wide variety of microbes which are not always clonal and there are chances for horizontal gene transfers, the assembly and classification becomes even harder. Assemblers might predict repeat variations as species variations. The next challenge is the identification of rare species, as rhizosphere is a niche where a lot of uncharacterized and unculturable microbes reside, covering a specific species with any depth of sequencing is not feasible. Even with large size datasets, unless the sample itself contains only a few communities, it is not easy to pull out the rare species. The last issue is wrong matches in taxonomic classification, when the reads are directly compared against the reference databases, there is a possibility that the match could point even noncoding regions and show hits from an outlying organism. This is based on the quality of the database (Carr and Borenstein 2014). This would definitely lead to a wrong taxonomic grouping. Furthermore, in functional annotations, gene prediction errors are common as the species richness in the sample increase, softwares can predict false genes by detecting the structural level homology instead at the sequence level. But there are specific substitution parameters for the softwares to overcome this error (Yooseph et al. 2008).

1.6 Concluding Remarks

In this chapter, we have briefed about the significance of rhizosphere microbiome and reviewed the various metagenomics methods available for analyzing the microbial diversity of the rhizosphere and also about the probable pitfalls of a rhizosphere metagenomics study. Application of metagenomics approaches to study the rhizosphere microbiome would clearly contribute to characterization of a wide range of unknown and unculturable microbes and to discover novel candidate genes and gene products. Next-generation sequencing is a promising approach to study the microbial diversity in a rarely explored niche such as rhizosphere. A wide range of computational softwares are being developed for analysis of metagenomics datasets. However, an expertise in handling these large datasets and bioinformatics tools is a pre-requirement in metagenomics study.

Acknowledgments The authors are thankful to the reviewer for his/her cautious evaluation of our manuscript and his/her valuable comments and suggestions. A special thanks to Dr. Ashok Kumar M, Senior Research Fellow, Department of Genetic Engineering, SRM Institute of Science and Technology, Kattankulathur, for his assistance in drawing the figures for the manuscript.

References

Akiyama K, Matsuzaki K, Hayashi H (2005) Plant sesquiterpenes induce hyphal branching in arbuscular mycorrhizal fungi. Nature 435:824–827

Alneberg J, Bjarnason BS, De Bruijn I et al (2014) Binning metagenomic contigs by coverage and composition. Nat Methods 11:1144–1146

Altschul SF, Gish W, Miller W et al (1990) Basic local alignment search tool. J Mol Biol 215:403–410

Badri DV, Chaparro JM, Zhang R et al (2013) Application of natural blends of phytochemicals derived from the root exudates of Arabidopsis to the soil reveal that phenolic-related compounds predominantly modulate the soil microbiome. J Biol Chem 288:4502–4512
Berg G (2009) Plant–microbe interactions promoting plant growth and health: perspectives for controlled use of microorganisms in agriculture. Appl Microbiol Biotechnol 84:11–18
Bertin C, Yang X, Weston LA (2003) The role of root exudates and allelochemicals in the rhizosphere. Plant Soil 256:67–83
Bonilla-Rosso G, Peimbert M, Alcaraz LD et al (2012) Comparative metagenomics of two microbial mats at Cuatro Ciénegas Basin II: community structure and composition in oligotrophic environments. Astrobiology 12:659–673
Brady A, Salzberg SL (2009) Phymm and PhymmBL: metagenomic phylogenetic classification with interpolated Markov models. Nat Methods 6:673
Breidenbach B, Pump J, Dumont MG (2016) Microbial community structure in the rhizosphere of rice plants. Front Microbiol 6:1537
Breitwieser FP, Lu J, Salzberg SL (2019) A review of methods and databases for metagenomic classification and assembly. Brief Bioinform 20:1125–1136
Camacho C, Coulouris G, Avagyan V et al (2009) BLAST+: architecture and applications. BMC Bioinformatics 10:421
Carr R, Borenstein E (2014) Comparative analysis of functional metagenomic annotation and the mappability of short reads. PLoS One 9(8):e105776
Carvalhais LC, Dennis PG, Fan B et al (2013) Linking plant nutritional status to plant-microbe interactions. PLoS One 8(7):e68555
Caspi R, Billington R, Ferrer L et al (2016) The MetaCyc database of metabolic pathways 600 and enzymes and the BioCyc collection of pathway/genome databases. Nucleic Acids Res 44: D471–D480
Chen K, Pachter L (2005) Bioinformatics for whole-genome shotgun sequencing of microbial communities. PLoS Comput Biol 1(2):e24
Cleary B, Brito IL, Huang K et al (2015) Detection of low-abundance bacterial strains in metagenomic datasets by eigengenome partitioning. Nat Biotechnol 33:1053
Curtis TP, Sloan WT, Scannell JW (2002) Estimating prokaryotic diversity and its limits. Proc Natl Acad Sci U S A 99:10494–10499
Darling AE, Jospin G, Lowe E et al (2014) PhyloSift: phylogenetic analysis of genomes and metagenomes. PeerJ 2:e243
de Fátima Alves L, Westmann CA, Lovate GL et al (2018) Metagenomic approaches for understanding new concepts in microbial science. Int J Genomics 2018:2312987
DeAngelis KM, Brodie EL, DeSantis TZ et al (2009) Selective progressive response of soil microbial community to wild oat roots. ISME J 3:168–178
Delcher AL, Bratke KA, Powers EC, Salzberg SL (2007) Identifying bacterial genes and endosymbiont DNA with glimmer. Bioinformatics 23:673–679
DeSantis TZ, Hugenholtz P, Larsen N et al (2006) Greengenes, a chimera-checked 16S rRNA gene database and workbench compatible with ARB. Appl Environ Microbiol 72:5069–5072
Eloe-Fadrosh EA, Ivanova NN, Woyke T, Kyrpides NC (2016) Metagenomics uncovers gaps in amplicon-based detection of microbial diversity. Nat Microbiol 1:1–4
Eren AM, Esen ÖC, Quince C et al (2015) Anvi'o: an advanced analysis and visualization platform for 'omics data. PeerJ 3:e1319
Escobar-Zepeda A, Vera-Ponce de Leon A, Sanchez-Flores A (2015) The road to metagenomics: from microbiology to DNA sequencing technologies and bioinformatics. Front Genet 6:348
Felczykowska A, Krajewska A, Zielińska S, Łoś JM (2015) Sampling, metadata and DNA extraction-important steps in metagenomic studies. Acta Biochim Pol 62(1):151–160
Freitas T, Chain P, Lo C-C, Li P-E (2015) GOTTCHA database, version 1. Los Alamos National Laboratory. Nucleic Acids Res 43(10):e69
French KE, Tkacz A, Turnbull LA (2017) Conversion of grassland to arable decreases microbial diversity and alters community composition. Appl Soil Ecol 110:43–52

Gans J, Wolinsky M, Dunbar J (2005) Computational improvements reveal great bacterial diversity and high metal toxicity in soil. Science 309(5739):1387–1390
Germida J, Siciliano S (2001) Taxonomic diversity of bacteria associated with the roots of modern, recent and ancient wheat cultivars. Biol Fertil Soils 33:410–415
Gillespie DE, Brady SF, Bettermann AD et al (2002) Isolation of antibiotics turbomycin A and B from a metagenomic library of soil microbial DNA. Appl Environ Microbiol 68:4301–4306
Glass EM, Wilkening J, Wilke A et al (2010) Using the metagenomics RAST server (MG-RAST) for analyzing shotgun metagenomes. Cold Spring Harb Protoc 2010(1):pdb.prot5368
Glenn TC (2014) NGS field guide: overview. Mol Ecol Resour 11:759–769
Guazzaroni M, Morgante V, Mirete S, González-Pastor JE (2013) Novel acid resistance genes from the metagenome of the T into R iver, an extremely acidic environment. Environ Microbiol 15:1088–1102
Handelsman J, Rondon MR, Brady SF et al (1998) Molecular biological access to the chemistry of unknown soil microbes: a new frontier for natural products. Chem Biol 5:R245–R249
Hartmann A, Schmid M, Van Tuinen D, Berg G (2009) Plant-driven selection of microbes. Plant Soil 321:235–257
Hassan S, Mathesius U (2012) The role of flavonoids in root–rhizosphere signalling: opportunities and challenges for improving plant–microbe interactions. J Exp Bot 63:3429–3444
Hawkes CV, DeAngelis KM, Firestone MK (2007) Root interactions with soil microbial communities and processes. In: Cardon Z, Whitbeck J (eds) The rhizosphere. Academic, Cambridge, pp 1–29
Hayat S, Faraz A, Faizan M (2017) Root exudates: composition and impact on plant–microbe interaction. In: Biofilms in plant soil and health, pp 179–193
Hinsinger P, Bengough AG, Vetterlein D, Young IM (2009) Rhizosphere: biophysics, biogeochemistry and ecological relevance. Plant Soil 321:117–152
Hunter S, Jones P, Mitchell A et al (2012) InterPro in 2011: new developments in the family and domain prediction database. Nucleic Acids Res 40(10):4725
Ikeda S, Suzuki K, Kawahara M et al (2014) An assessment of urea-formaldehyde fertilizer on the diversity of bacterial communities in onion and sugar beet. Microbes Environ 29:231–234
Jin T, Wang Y, Huang Y et al (2017) Taxonomic structure and functional association of foxtail millet root microbiome. Gigascience 6:gix089
Kanehisa M, Araki M, Goto S et al (2007) KEGG for linking genomes to life and the environment. Nucleic Acids Res 36:D480–D484
Kelley DR, Salzberg SL (2010) Clustering metagenomic sequences with interpolated Markov models. BMC Bioinformatics 11:544
Kelley DR, Liu B, Delcher AL et al (2012) Gene prediction with glimmer for metagenomic sequences augmented by classification and clustering. Nucleic Acids Res 40:e9–e9
Kislyuk A, Bhatnagar S, Dushoff J, Weitz JS (2009) Unsupervised statistical clustering of environmental shotgun sequences. BMC Bioinformatics 10:316
Krebs JE, Vaishampayan P, Probst AJ et al (2014) Microbial community structures of novel Icelandic hot spring systems revealed by PhyloChip G3 analysis. Astrobiology 14:229–240
Kultima JR, Coelho LP, Forslund K et al (2016) MOCAT2: a metagenomic assembly, annotation and profiling framework. Bioinformatics 32:2520–2523
Li K, Bihan M, Yooseph S, Methe BA (2012) Analyses of the microbial diversity across the human microbiome. PLoS One 7(6):e32118
Li J-G, Ren G-D, Jia Z-J, Dong Y-H (2014) Composition and activity of rhizosphere microbial communities associated with healthy and diseased greenhouse tomatoes. Plant Soil 380:337–347
Lu H, Sun J, Zhu L (2017) The role of artificial root exudate components in facilitating the degradation of pyrene in soil. Sci Rep 7:7130
Mäkinen V, Salmela L, Ylinen J (2012) Normalized N50 assembly metric using gap-restricted co-linear chaining. BMC Bioinformatics 13:255
Marchesi JR, Ravel J (2015) The vocabulary of microbiome research: a proposal. Microbiome 3:31

Mendes R, Kruijt M, De Bruijn I et al (2011) Deciphering the rhizosphere microbiome for disease-suppressive bacteria. Science 332(6033):1097–1100

Mendes R, Garbeva P, Raaijmakers JM (2013) The rhizosphere microbiome: significance of plant beneficial, plant pathogenic, and human pathogenic microorganisms. FEMS Microbiol Rev 37:634–663

Mikheenko A, Saveliev V, Gurevich A (2016) MetaQUAST: evaluation of metagenome assemblies. Bioinformatics 32:1088–1090

Mirete S, De Figueras CG, González-Pastor JE (2007) Novel nickel resistance genes from the rhizosphere metagenome of plants adapted to acid mine drainage. Appl Environ Microbiol 73:6001–6011

Mirete S, Mora-Ruiz MR, Lamprecht-Grandío M et al (2015) Salt resistance genes revealed by functional metagenomics from brines and moderate-salinity rhizosphere within a hypersaline environment. Front Microbiol 6:1121

Mohammed MH, Ghosh TS, Singh NK, Mande SS (2011) SPHINX—an algorithm for taxonomic binning of metagenomic sequences. Bioinformatics 27:22–30

Morrissey JP, Dow JM, Mark GL, O'Gara F (2004) Are microbes at the root of a solution to world food production? EMBO Rep 5:922–926

Niemi RM, Heiskanen I, Wallenius K, Lindström K (2001) Extraction and purification of DNA in rhizosphere soil samples for PCR-DGGE analysis of bacterial consortia. J Microbiol Methods 45:155–165

Noguchi H, Taniguchi T, Itoh T (2008) MetaGeneAnnotator: detecting species-specific patterns of ribosomal binding site for precise gene prediction in anonymous prokaryotic and phage genomes. DNA Res 15:387–396

Odelade KA, Babalola OO (2019) Bacteria, fungi and archaea domains in rhizospheric soil and their effects in enhancing agricultural productivity. Int J Environ Res Public Health 16:3873

Ounit R, Lonardi S (2016) Higher classification sensitivity of short metagenomic reads with CLARK-S. Bioinformatics 32:3823–3825

Ounit R, Wanamaker S, Close TJ, Lonardi S (2015) CLARK: fast and accurate classification of metagenomic and genomic sequences using discriminative k-mers. BMC Genomics 16:236

Panke-Buisse K, Poole AC, Goodrich JK et al (2015) Selection on soil microbiomes reveals reproducible impacts on plant function. ISME J 9:980–989

Parks DH, Imelfort M, Skennerton CT et al (2015) CheckM: assessing the quality of microbial genomes recovered from isolates, single cells, and metagenomes. Genome Res 25:1043–1055

Parniske M (2008) Arbuscular mycorrhiza: the mother of plant root endosymbioses. Nat Rev Microbiol 6:763–775

Pérez-Jaramillo JE, Mendes R, Raaijmakers JM (2016) Impact of plant domestication on rhizosphere microbiome assembly and functions. Plant Mol Biol 90:635–644

Philipson CW, Davenport K, Voegtly L et al (2017) Brief protocol for EDGE bioinformatics: analyzing microbial and metagenomic NGS data. Bioprotocol 7:e2622

Pinton R, Varanini Z, Nannipieri P (2007) The rhizosphere as a site of biochemical interactions among soil components, plants, and microorganisms. In: Pinton R, Varanini Z, Nannipieri P (eds) The rhizosphere: biochemistry and organic substances at the soil-plant interface. Marcel Dekker Inc, NewYork, p 424

Qiao Q, Wang F, Zhang J et al (2017) The variation in the rhizosphere microbiome of cotton with soil type, genotype and developmental stage. Sci Rep 7:1–10

Quast C, Pruesse E, Yilmaz P et al (2012) The SILVA ribosomal RNA gene database project: improved data processing and web-based tools. Nucleic Acids Res 41:D590–D596

Ranjan R, Grover A, Kapardar RK, Sharma R (2005) Isolation of novel lipolytic genes from uncultured bacteria of pond water. Biochem Biophys Res Commun 335:57–65

Rho M, Tang H, Ye Y (2010) FragGeneScan: predicting genes in short and error-prone reads. Nucleic Acids Res 38:e191

Richardson EJ, Escalettes F, Fotheringham I et al (2013) Meta4: a web application for sharing and annotating metagenomic gene predictions using web services. Front Genet 4:168

Roesch LFW, Fulthorpe RR, Riva A et al (2007) Pyrosequencing enumerates and contrasts soil microbial diversity. ISME J 1:283–290
Roumpeka DD, Wallace RJ, Escalettes F et al (2017) A review of bioinformatics tools for bio-prospecting from metagenomic sequence data. Front Genet 8:23
Rudrappa T, Czymmek KJ, Paré PW, Bais HP (2008) Root-secreted malic acid recruits beneficial soil bacteria. Plant Physiol 148:1547–1556
Saleem M, Law AD, Moe LA (2016) Nicotiana roots recruit rare rhizosphere taxa as major root-inhabiting microbes. Microb Ecol 71:469–472
Sanger F, Nicklen S, Coulson AR (1977) DNA sequencing with chain-terminating inhibitors. Proc Natl Acad Sci U S A 74:5463–5467
Scholz M, Ward DV, Pasolli E et al (2016) Strain-level microbial epidemiology and population genomics from shotgun metagenomics. Nat Methods 13:435–438
Sharma PK, Capalash N, Kaur J (2007) An improved method for single step purification of metagenomic DNA. Mol Biotechnol 36:61–63
Simão FA, Waterhouse RM, Ioannidis P et al (2015) BUSCO: assessing genome assembly and annotation completeness with single-copy orthologs. Bioinformatics 31:3210–3212
Singh BK, Bardgett RD, Smith P, Reay DS (2010) Microorganisms and climate change: terrestrial feedbacks and mitigation options. Nat Rev Microbiol 8:779–790
Strous M, Kraft B, Bisdorf R, Tegetmeyer H (2012) The binning of metagenomic contigs for microbial physiology of mixed cultures. Front Microbiol 3:410
Tanveer A, Yadav S, Yadav D (2016) Comparative assessment of methods for metagenomic DNA isolation from soils of different crop growing fields. 3 Biotech 6(2):220
Treangen TJ, Koren S, Sommer DD et al (2013) MetAMOS: a modular and open source metagenomic assembly and analysis pipeline. Genome Biol 14:R2
Tringe SG, Von Mering C, Kobayashi A et al (2005) Comparative metagenomics of microbial communities. Science 308(5721):554–557
Truong DT, Franzosa EA, Tickle TL et al (2015) MetaPhlAn2 for enhanced metagenomic taxonomic profiling. Nat Methods 12:902–903
Vieites JM, Guazzaroni M-E, Beloqui A et al (2008) Metagenomics approaches in systems microbiology. FEMS Microbiol Rev 33:236–255
Wagg C, Bender SF, Widmer F, van der Heijden MGA (2014) Soil biodiversity and soil community composition determine ecosystem multifunctionality. Proc Natl Acad Sci U S A 111:5266–5270
Wang Q, Garrity GM, Tiedje JM, Cole JR (2007) Naive Bayesian classifier for rapid assignment of rRNA sequences into the new bacterial taxonomy. Appl Environ Microbiol 73:5261–5267
Watson M (2014) Illuminating the future of DNA sequencing. Genome Biol 15:108
Weinert N, Piceno Y, Ding G-C et al (2011) PhyloChip hybridization uncovered an enormous bacterial diversity in the rhizosphere of different potato cultivars: many common and few cultivar-dependent taxa. FEMS Microbiol Ecol 75:497–506
Woese CR, Fox GE (1977) Phylogenetic structure of the prokaryotic domain: the primary kingdoms. Proc Natl Acad Sci U S A 74:5088–5090
Wood DE, Salzberg SL (2014) Kraken: ultrafast metagenomic sequence classification using exact alignments. Genome Biol 15:R46
Wu M, Scott AJ (2012) Phylogenomic analysis of bacterial and archaeal sequences with AMPHORA2. Bioinformatics 28:1033–1034
Wu Y-W, Tang Y-H, Tringe SG et al (2014) MaxBin: an automated binning method to recover individual genomes from metagenomes using an expectation-maximization algorithm. Microbiome 2:26
Yang Y, Wang N, Guo X et al (2017) Comparative analysis of bacterial community structure in the rhizosphere of maize by high-throughput pyrosequencing. PLoS One 12:e0178425
Yooseph S, Li W, Sutton G (2008) Gene identification and protein classification in microbial metagenomic sequence data via incremental clustering. BMC Bioinformatics 9:182

Chapter 2
Metagenomic Approach in Relation to Plant–Microbe and Microbe–Microbe Interactions

Sivasankari Ramadurai, Abiramavalli Moorthy, and Usha Balasundaram

Abstract Metagenomics is the study of collective microbial genomes of environmental samples. Microorganisms are essential components of every part of life on the planet. Microorganisms do not exist as individuals, but they form complex communities (microbiomes) in the above-ground and below-ground parts of the plants, contributing to the plant's growth, health, and performance in many ways. There are several factors that regulate the structure and composition of the microbes interacting with plants. However, it has been calculated that only 0.1% of the microbes found in the environment are culturable, and remaining are uncultivable and untapped. The unculturable microbial communities are referred to as "microbial dark matter" which almost covers a high percentage of the planet's biomass and biodiversity. However, only very little is known about these microbiomes and their interactions with a host. Every microbial genome contains a unique set of genes encoding novel enzymes used for biotechnological applications. Metagenomics methods using advanced sequencing technologies enable tracking of these novel biological molecules that are available in the natural systems in a high-throughput manner. Metagenomics approach also unravels the microbial interactions between themselves and between plants and the microbes as well. This chapter describes the significance of microbial interactions and how plant–microbiome and microbe–microbe interactions are being studied by metagenomics approaches.

Keywords Metagenomics · Next-generation Sequencing Technologies · Microbiomes · Microbial interactions · Plant Microbe interaction

---

S. Ramadurai · A. Moorthy · U. Balasundaram (✉)
Department of Genetic Engineering, Faculty of Engineering and Technology, SRM Institute of Science and Technology, SRM Nagar, Kattankulathur, Kanchipuram, Chennai, Tamil Nadu, India

R. N. Pudake et al. (eds.), *Omics Science for Rhizosphere Biology*, Rhizosphere Biology, https://doi.org/10.1007/978-981-16-0889-6_2

2.1 Introduction

Microorganisms influence several important plant characteristics. A few microbes can delay growth or destroy a plant, whereas beneficial ones can enhance the plant's immunity by directly destroying pathogenic invaders (Jackson and Taylor 1996) or by inhibiting the growth of pathogens (Enebe and Babalola 2019). These microbes also play a key role in plant's hormonal levels and acquisition of nutrition from the soil. Interestingly, these microbes in their natural ecosystems exist as complex communities made of a myriad of culturable and unculturable ones to uphold their environmental stability (Kennedy and Gewin 1997). Interactions in plants can be grouped into plant–microbe and microbe–microbe interactions (Fig. 2.1).

The imperceptible participants of a microbial community vary greatly in their biochemical actions and interactions among species and within species as well. Although, phylotyping provides consistent information about the individual

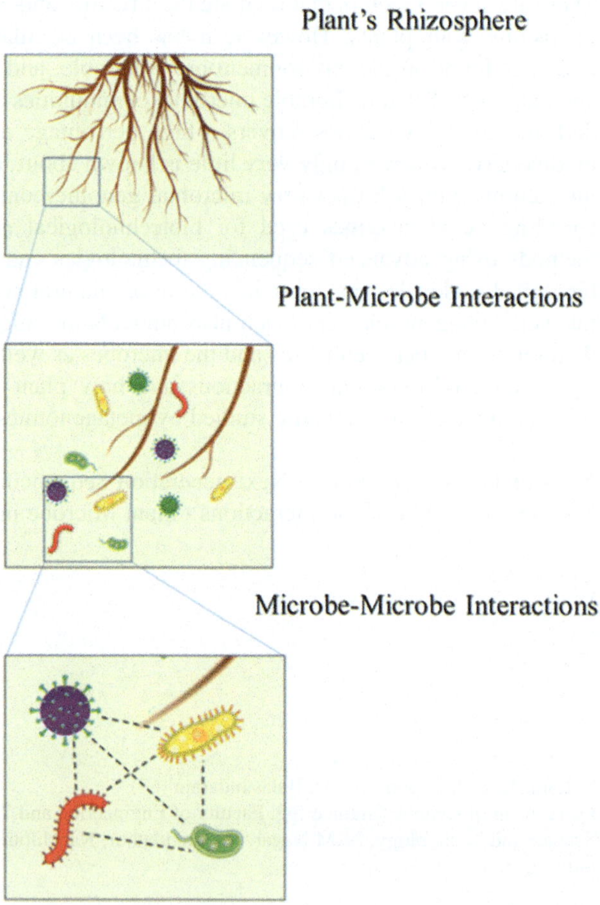

Fig. 2.1 Representation of plant–microbe and microbe–microbe interactions in rhizosphere

members of a community, because of high genomic diversity within a species makes it difficult to predict the functions of the community members. Existing culture-dependent techniques and single genome-sequencing approaches have their own drawbacks such as only lab-cultivable organisms can be detected. Thus, a new modern culture-independent technique, "metagenome sequencing" has evolved. Metagenomics took over the existing traditional culture-based techniques by identifying the hitherto unknown proteins and related sequences in microbial genomes. Moreover, this approach allows studying the microbes in their natural ecosystems (Ghosh et al. 2018).

Metagenomics is a research field that includes a collection of research strategies and computational methods for the exploration of a particular environment. Furthermore, metagenomics methods circumvent the major hurdles of environmental microbiology, the uncultivability, and the diversity of microbiomes. This technology enables detailed prediction of the numbers and diversities of the microbial communities and their compositions found in a complex environment. Nevertheless, metagenomics methods demand isolation of any environmental DNA with high purity and deep sequencing. Only then, it is possible to do a detailed taxonomic classification of the microbes present in the sample analyzed (National Research Council 2007). In earlier days, metagenomics referred to the isolation of DNA, cloning, screening of cloned libraries and sequencing the unique clones, and documenting their origin and structure based on their phylogenetic information. However, the evolution of novel, efficient and simpler methods such as direct sequencing of any environmental sample without the need for culturing it in the laboratory has progressed metagenomics into a new higher level.

Modern metagenomics techniques emphasize on three major directions. (1) Correlation of phylogeny to its possible functions, where the large segments of the DNA that possess phylogenetically important genes is screened and the flanking segments which propound a possible function for the DNA being analysed is also screened; (2) Identification of novel genes which can be exploited for industrial applications; and (3) Bulk sequencing of any environmental sample which would ultimately reveal the number, nature, and complete genetic diversity of the communities found in that sample (Leveau 2007). This approach provides a global view of the community being studied. In addition, it offers clarity about the complex interactions that happen between the communities being analyzed. Adopting metagenomics methods enables researchers to view environments themselves as biological entities possessing their own genetic repertoires and to avoid considering individual species.

2.2 Unravelling Microbiome Interactions Through Metagenomics Approach

Metagenomics methods in studying the microbial communities in a specific environment are divided into two ways such as structural metagenomics and functional metagenomics (Rabausch et al. 2013). However, a combinatorial approach of these two strategies permits deeper insights on individual units of the communities found in an ecosystem and their functions as well. Consequently, microbial interactions in complex habitats regulated by microbes can easily be studied now with the help of metagenomics. Initially, Sanger sequencing was the key metagenomic sequencing method followed by researchers, which indeed showed good progress. Nevertheless, the emergence of Next-Generation Sequencing technologies (NGS) and its massive capability to sequence millions of DNAs at the same time and at a less expense highly reinforced the application of this technology for metagenomics studies (Alves et al. 2018). With this advanced technology, any ecosystem can now be unveiled just by direct sequencing of the environmental DNA. The intense effect of NGS technologies on generating a large amount of metagenomics sequence data has made researchers to redefine metagenomics as "Random shotgun sequencing of any environmental DNA generating no less than 50 Mbp sequence data"(Kunin et al. 2008). This sequencing-based metagenomics analysis can be performed either by sequencing clones amplified with a marker (16S rRNA) gene designating the possible taxonomic source of the DNAs or by random shotgun-sequencing of the whole metagenomes until our genes of interests and taxonomic markers are found. Furthermore, these capabilities allow the identification of novel genes, their roles in conferring stress resistance, antibiotic resistance, and their taxonomic origins without even isolating and culturing of microbes. Nevertheless, these strategies have their own pros and cons. In case of 16S sequencing method, it is only possible to identify and quantify the structure and composition of the communities while the functions of those communities cannot be predicted (Soni et al. 2012). Similarly, in whole metagenomes sequencing, assembly and taxonomic classification can be done well through binning strategies. However, in this method, it is only possible to predict species differentiation, but it is very difficult to differentiate the strains of a species (Sharon and Banfield 2013).

2.2.1 Microbe–Microbe Interactions

In natural environments such as soil, microorganisms hardly occur in isolation. They are often associated with some hosts and develop interactions with the host organisms. Additionally, there will be interactions between the microbial species found in an environment in which one species outcompete the other. These interactions highly influence the plant–microbe interactions. Although plants provide resources, the interactions between the microbes which share the same habitat are critical for their community dynamics, structure, and function (Kinkel et al. 2011). The first kind of interaction between microbe–microbe is "antagonism." Some microbes produce cell wall degrading enzymes such as cellulases and chitinases which completely degrade the cell wall of the competing pathogenic microorganisms. One good example of such an antagonism is Plant Growth-Promoting Rhizobacteria (PGPR), which suppress the fungal infection caused by *Rhizoctonia solani* in several plants. This recommends that certain interactions between the microbes invading the same ecosystem are essential to overcome several detrimental effects. Besides bacteria, fungi such as Arbuscular Mycorrhizal Fungi (AMF) are unique, beneficial, and environmentally important organisms in soil. They interact with various microbes found in the soil ecosystem such as mycorrhiza helper bacteria for their mycorrhizal formation by inhibiting other fungal organisms. AMF and PGPR are indeed two principal components of the soil microbiota. They are known for their ability to induce systemic resistance against pathogenic microbes. The interactions between these two communities are synergistic and reflective on growth, health, and overall performance of the host plants in agriculture environments. These kinds of microbe–microbe interactions have a high impact on host plant's nutrition and defence mechanisms. For instance, AMF inhibits other fungal pathogen attack and PGPR such as *Rhizobium* and *Bacillus* produce siderophores which dispossess pathogens from the environment. The next kind of interaction between microbe–microbe is "amensalism". Chemical compounds released by one organism destroy the other organism. This is also known as "antibiosis" where one microbe produces antimicrobial substance causing detrimental effects to another microbe. The occurrence of this interaction is most common in beneficial bacteria while fighting against host pathogens. For example, *Bacillus*, *Streptomyces*, and *Pseudomonas* synthesize bioactive molecules such as lipopeptides which destroys the pathogenic organisms such as oomycetes and trophozoites. The third kind of interaction between microbes is "parasitism", where one species survives at the cost of the other. For instance, the bacteria *Pasteuria penetrans* suppress the root-knot nematode infection by reproducing themselves within the infected nematodes leading to detrimental effects of the nematodes and also the inhibition of penetration of new nematodes (Igiehon and Babalola 2018). On the other hand, interactions between the microbes which share the same habitat not necessarily be detrimental always except for the pathogens. The microbes indeed collaborate, co-operate with each other by sharing some genes among themselves for communal stability. This is nothing but the horizontal gene transfer events. This way they have the possibility for both the partners to influence each other's fitness (Bakker et al. 2014). These interactions were documented through cytology-based systems, microplate reader-based systems and microfluidics-based systems so far (Hennessy et al. 2017; Massalha et al. 2017).

Furthermore, microbes in general, produce various kinds of signalling molecules, termed as auto-inducers which basically observe the community density in a native ecosystem and control their behavioural pattern in a mutual way (Fuqua et al. 1994). This mechanism is termed as quorum sensing (QS). The auto-inducers produced by microbes are further categorized into three types (a) auto-inducer synthase, (b) acylhomoserine lactones (AHLs), (c) auto-inducing peptides (AIPs) (Huang et al. 2016). AHL signals are detected majorly in Gram-negative bacteria, whereas cyclic peptides are found only in Gram-positive bacteria (Hartmann et al. 2014). Microbes use this method to control the formation of biofilm, DNA transfer, pathogenic infections, to produce extracellular polysaccharides, and several other processes (Galloway et al. 2011). On the other hand, in certain cases, microbes themselves interrupt the process of QS by producing receptor molecules that inhibit QS or by producing degrading enzymes which degrades the auto-inducers. This mechanism is known as quorum quenching (QQ). Microbes use this mechanism as a competitive strategy for food and space. These are the major ways by which microbes interconnect themselves and interact with each other.

2.2.2 *Plant–Microbe Associations*

Plants are associated with a surfeit of microbes having potential roles for plant health and performance. Plants employ microbes from their associated vicinities such as the phyllosphere (leaves in contact with the external environment), the anthosphere (flowers), the carposphere (fruits), the spermosphere (germinated seeds), and the rhizosphere soil, a part that is closely associated with the roots (Hardoim et al. 2015). However, the communities that are laterally transferred from the rhizosphere are highly assorted encompassing various groups such as Acidobacteria, Verrucomicrobia, Bacteroidetes, Proteobacteria, Planctomycetes, and Actinobacteria (Fierer 2017). The spermosphere microbes are vertically transferred and thrive in the developing roots (Liu et al. 2012; Hardoim et al. 2012). Similarly, the phyllosphere of many plants includes bacterial species mostly when compared to other species of microbes such as fungi, archaea, protists, and pathogenic microbes which are found to be less in abundance in the phyllosphere (Dong et al. 2019).

A substantial amount of information on the structure, dynamics, and functional capabilities of the plant holobiont (A collection of a host plant and a myriad of living organisms in and around it form a separate ecological unit) and the secluded community members are available now. The significance of the plant holobiont has recently been widely studied because of the functional potentials of the associations. Concerning the plant–microbiome association, the associated microbial taxa can be beneficial or pathogenic to plants. Certain plant–microbe interactions secure the host plants against invading harmful organisms and help plants in their nutrition whereas a few interactions can be detrimental to the host plants (Igiehon and Babalola 2018). Hotspots of microbial interactions in a plant are shown in Fig. 2.2.

2.2.2.1 Significance of Above-Ground Microbiomes

As mentioned earlier, plant–microbiome associations can be both above-ground and below-ground. The above-ground plant parts such as leaves, stems, and other reproductive organs comprises of a collection of microbes associated with it. Specifically,

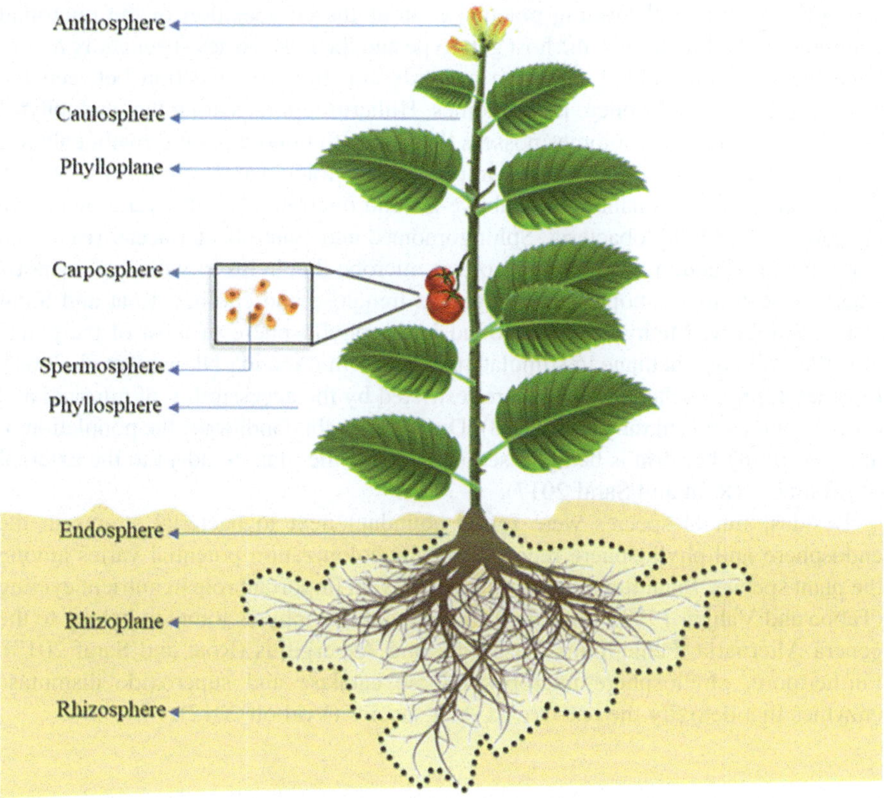

Fig. 2.2 Schematic showing hotspots of microbial interactions with the plant

microbes that colonize the outer surface of the plant parts are epiphytes and the endophytic microbes colonize the interior parts of the plants by penetration. The endophytes basically enter the plant from the soil environment through root cracks or through lateral root developing spots regulated actively (Compant et al. 2005). The endophytes are then transmitted to various parts of the plants for colonization which is controlled by the distribution of plant resources and the ability of the endophytes to colonize. Though endophytes can penetrate through aerial parts such as flowers and fruits and enter the plant, they mostly prefer entering through the xylem pathway to various parts of the plant (Compant et al. 2010). It has been stated that, in some plants, phyllosphere microbes arise from the soil ecosystems with the help of roots and are regulated by environmental factors (Vorholt 2012; Zarraonaindia et al. 2015; Wallace et al. 2018). Subsequently, the endosphere and phyllosphere microorganisms are found to be highly diverse at the genus and species level (Zarraonaindia et al. 2015).

The above-ground plant microbes in general arise from soil, kernels, or air and acclimatize themselves on or inside the plant organs. While the soil type and ecological factors and farming practice contour the composition of the microbial community. Adding to this, the host genotype and the host tissues specifically recruit microbiomes which clearly indicate a sturdy functional connection between the plants and their phyllosphere microbiomes. Hitherto reports stating that endophytes and above-ground microbiomes possess the ability to uphold plant growth, enhance immunity for disease resistance, and improve stress tolerance (Hardoim et al. 2015). For instance, the dominant taxa of above-ground microbiomes of several plants are Pseudomonas, Methylobacteria, Sphingomonas, and Enterobacteriaceae (Roat and Saraf 2017). Pseudomonas is an important microbe that helps to withstand osmotic stress by secreting osmoprotectants such as trehalose and choline (Roat and Saraf 2017). Similarly, Methylobacteria found in the phyllosphere of most of the plants uses the available methane to stimulate plant growth (Abanda-Nkpwatt et al. 2006). In general, phyllosphere microbes are restricted by the accessibility of nitrogen and carbon sources (Fürnkranz et al. 2008). During drought conditions, the population of nitrogen-fixing bacteria is being raised up to assist the plant to adapt to the external environment (Roat and Saraf 2017).

Besides, fungal species were found abundant next to bacterial strains in the endosphere and phyllosphere regions, but the colonization potential varies among the plant species. Such fungal communities play an important role in nutrient cycling (Tebbe and Vahjen 1993). Some common fungal endophytes found to belong to the genera Alternaria, Penicillium, Fusarium, and Aspergillus (Roat and Saraf 2017). Furthermore, phyllosphere microbes release catalase and superoxide dismutase enzymes that detoxify the reactive oxygen species (Vorholt 2012).

2.2.2.2 Significance of Below-Ground Microbiomes

Generally, below-ground microbial associations occur between the rhizosphere soil and the associated roots which form the rhizosphere-rhizobiome. Rhizosphere soil, a niche that lies closer to the roots, is rich in nutrients due to the deposition of mucilage and root exudates. Root exudates alter soil properties, reduce the spread of competing plant species, and influence microbial communities association. Exudate compounds include various low molecular weight and high molecular weight compounds which stimulate plant–microbe interactions, especially associations involving rhizobacteria and Arbuscular Mycorrhizal Fungi (AMF) (Monther and Kamaruzaman 2012). Root exudates are the sole energy source of soil-borne microbes and thus, they consume exudates for their survival and in turn promote plant's overall health. Consequently, there is a mutual connection with each other, resulting in suppression of the growth of pathogens, acquisition of nutrients from the soil, and promotion of the host plant's growth (Liu et al. 2019).

Rhizosphere microbes release a variety of chemicals that have specialized functions such as fixing of atmospheric nitrogen, nutrient solubilization, production of phytohormones, improving plant's defence mechanisms against pathogens and stress alleviation (Ali et al. 2009). Moreover, rhizosphere microbial communities compete internally with each other, and with invading pathogenic microbes, which affect the plant crops (Alina et al. 2015). However, the root exudates and their composition of the host plant decide the structure of the community to be colonized. For instance, in a few medicinal plants, elevated flavonoid exudation was correlated to the colonization of bacterial communities such as Bacillus and Paenibacillus (Köberl et al. 2013). Furthermore, several metagenomic analysis reports reveal that rhizosphere regions of most of the plants comprises of Proteobacteria, Firmicutes, Actinobacteria, and Bacteriodetes as their predominant genera (Akinsanya et al. 2015).

Nevertheless, the community constantly changes in the rhizosphere based on the key factors such as different developmental stages of roots of the host plant, type and texture of the soil, pH of the soil, moisture content, organic deposition, and root exudates (Rehman et al. 2019). In addition, a few other ecological factors such as climate change, farming practices, and contaminated soil also affect the microbiome structure, composition, abundance, and interactions (Fierer 2017; Hartmann et al. 2009). Besides, plant species also reflects specificity in microbiomes, each species have their own set of microbes in their rhizosphere and associated roots (Hacquard 2016). For instance, metagenomics analysis of root microbiomes of several barley plants grown in the same soil and environment depicted different root microbiome structures and compositions because of the innate immunity of each host and their respective exudate profiles (Bulgarelli et al. 2015).

2.2.2.3 Benefits of Rhizosphere Microbiomes

Plant Growth-Promoting Rhizobacteria (PGPR) are a group of rhizosphere soil-borne microbes which highly improve soil fertility and helps plants in various ways for their growth and development. PGPRs are involved in nitrogen fixation, microbial antagonism, phosphate, and potassium solubilization, and absorption of heavy metals polluting the soil. In addition, rhizobacteria help in improving the photosynthesis process, chlorophyll enhancement, and carbon assimilation. Phytohormones such as auxins or gibberellins are produced by certain root endophytes which influence the host plant's growth. Bacterium helping in nitrogen fixation includes species of *Achromobacter, Arthrobacter, Acetobacter, Azomonas, Beijerinckia, and Bacillus* (Pindi et al. 2014). *Bacillus* sp. produces organic acids such as gluconic, citric, and fumaric acids under phosphorous-limiting conditions which increases the solubility of poorly soluble phosphorus (Pindi et al. 2014). An increase in the germination of orchid seeds was observed when inoculated with *Spingomonas* and *Mycobacterium* sp. *Bacillus cepacia* and *Pseudomonas aeruginosa* were able to reduce infections caused by the pathogen *Sclerotinia sclerotiorum* in sunflower (Ambrosini et al. 2012). PGPR produces siderophores that chelate metals and thus play an important role in rhizoremediation. For instance, *Burkholderia* strains isolated from the rhizosphere and rhizoplane of tomato plants exhibited activities involved in bioremediation and acts as a potential fertilizer (Ambrosini et al. 2012).

In addition, PGPRs help plants with their defence mechanisms against invading pathogens by stimulating plants to produce proteolytic enzymes against the invading pathogens and by producing antibiotics which inhibit cell wall synthesis in invading pathogens. Additionally, PGPRs greatly support plants for their survival under stress conditions. For instance, *Achromobacter* isolates were found to enhance the growth of seedlings under water-deficient conditions. Furthermore, reports have suggested that 50% of the rhizosphere-associated bacterial communities belong to the genus *Rhizobium*, which promotes plant growth (Ambrosini et al. 2012). Besides, plant hormones were synthesized by the plant growth-promoting rhizobacteria that conquer the rhizosphere of the plant. Several characteristics of the rhizobacteria aid in promoting plant growth. One such plant growth-promoting activity of rhizobacteria is the production of phytohormones. For example, fluorescent *Pseudomonas* sp., were found to exclusively produce gibberellic acid that regulates the growth of the plant. Production of gibberellic acid was observed in *Bacillus pumilus, Bacillus licheniformis*, and *Bacillus siamensis*, BE 76 (Ambawade and Pathade 2013). Similarly, *Rhizobium* sp. was observed to enhance cytokinin production in plants by regulating the expression of the signalling pathways which triggers cortical cells of the plants to divide (Farag et al. 2013). Cytokinin-producing bacteria has been shown to enhance shoot growth in droughted plants. *Agrobacterium tumefaciens* produced trans zeatin and boosted the growth in Arabidopsis. The bacterial strains of *Bacillus, Pseudomonas, Escherichia, Micrococcus, and Staphylococcus* genera were reported to improve the endogenous IAA content of the plants especially in wheat (Ali et al. 2009). Cyanobacterial species including *Anabaena, Anabaenopsis, Chlorogloeopsis, Gloeothece, Nostoc, and Synechocystis* were found to raise the IAA levels in the plants (Hussain et al. 2010).

2.3 Sampling and Experimental Design to Study Plant–Microbiomes

The rhizosphere is one of the complex ecosystems and a potential hotspot of countless microbial interactions that occur among the microbes and between the microbes and a host plant. One gram of rhizosphere soil possesses more than a million microbes in it (Gans et al. 2005). Interestingly, the population of rhizobiomes is highly diverse than the microbiomes found in the bulk soil. The main reason behind this is root exudates (Nannipieri et al. 2008). Hence, predicting the number and diversity of the microbial communities in the rhizosphere and identifying their functions require a special technique. Metagenomics analysis of the rhizosphere is a perfect solution for this issue. In general, the experimental procedures start with sampling, isolation of high quality pure DNA from any rhizosphere soil, metagenomic library construction, followed by sequencing and analysis of the metagenomics data using softwares.

2.3.1 Soil Sampling

Sampling is an essential step for obtaining good quality metagenomic data and its interpretations, especially when a highly complex habitat such as rhizosphere is being taken for study. The sample size should be sufficient enough to represent the whole population present in the habitat analysed (Wooley et al. 2010). So, it is significant to preselect sample preparation methods which will include the entire target group of microbes. It is also very important to know the sampling strategies and various effective experimental methods (Felczykowska et al. 2015). One should try asking and answering a few important questions before stepping into sampling of a habitat. For instance, when and where the samples are collected? Similarly, how the samples are transported to the laboratories? How long has the sample been stored before the isolation of DNA? The year, month, day, and time of collection should be noted. Samples should be transported properly without getting contaminated. The samples should not be stored for long term and isolation to be done as soon as possible after sample collection (Paul and Clark 1988). The "metadata" of the sample states the physical, chemical, and environmental features of the sample and can be submitted in the bio project databases at National Center for Biotechnology Information (NCBI) for future access.

In plant–microbe interaction studies, as phyllobiomes and rhizobiomes vary in their structure and abundance according to the host's developmental stage, it is important to have enough replicates of the sample. Additionally, species arrival timing influences their distribution on phyllobiomes and rhizobiome (Kennedy et al. 2009). Thus, it is advisable to have at least five replicate samples per plant organ or sample type to resolve this issue. In addition, if the aim of the study is to analyse the root microbiomes, it is recommended to analyse bulk soil samples (i.e. soil in distance ≥2 cm to roots) as well for comparing and predicting the origin of rhizobiomes (Kuzyakov and Razavi 2019). Furthermore, samples should always be collected at their own sites (e.g. field, greenhouse) to avoid ecological factors that affect the microbiome composition. Adding to this, proper processing, storing, and transporting of the samples should be assured such as snap freezing, storing in RNA stabilization solution, and transporting with liquid nitrogen (Reid 2003; Salter et al. 2014).

2.3.2 Isolation of DNA from Rhizosphere

Isolation of DNA from any environmental sample is very complex and challenging as it may get affected by contamination. In case of a highly complex ecosystem such as rhizosphere soil, there will be various impurities, viz., humic, fulvic acids, DNases mixed in the soil which hinder other downstream processes of a metagenomic workflow (Tebbe and Vahjen 1993). Therefore, it is always important to isolate any environmental DNA in a good quality as well as quantity for a successful metagenomics analysis. Several rapid and efficient protocols that give good yield have been developed and tested in many samples (Sagar et al. 2014). Besides, there are several commercially available environmental DNA isolation kits that extract pure as well as enough DNA. PowerSoil, UltraClean, and RNA PowerSoil (Mo Bio Laboratories, California, USA); FastDNA Spin kit for soil and FastRNA Pro soil-direct kit (MP Biomedicals, Solon, Ohio, USA); and SoilMaster (Epicentre Biotechnologies, Madison, Wisconsin, USA) are some of the recent widely used kits.

2.3.3 Amplicon Sequencing

Amplicon-based sequencing approach in general demand a high quality DNA, which will be amplified by normal Polymerase Chain Reaction (PCR) with phylogenetic marker genes (16S rRNA, 18S rRNA, and internal transcribed spacer-ITS) and the resulting PCR products will be sequenced, followed by classifying and characterizing the sequenced reads using bioinformatics methods and tools (Sharpton 2014). This approach has been one of the successful metagenomics technique which unveiled earth's microbial diversity remarkably, ranging from ecosystems such as the human gut, plant roots, thermal vents of oceans, hot springs and volcano soil (Lozupone and Knight 2007), and human microbiome consortium (De León et al. 2013).

Amplification of marker genes and high-throughput sequencing of the amplicons is being widely used in recent times to reveal the organization, structure, composition, and distribution of microbiomes in the environment and is progressively used in plant–microbiome studies (Knief 2014). High specificity of amplicon sequencing in targeting individual microbial taxa and also their related genes favours this technique (Herbold et al. 2016). This technique can even predict the rare species found in the environment that is being analysed. Nevertheless, this technique is extremely sensitive; hence, it is always mandatory to include positive (known mock communities) and negative controls (reagent and extraction blanks) (Glassing et al. 2016).

Metagenomic analyses of rhizosphere microbiomes generally depend mainly on the 16S rRNA gene that codes for the small RNA subunit of the ribosome. Thus far, the most commonly used next-generation sequencing (NGS) technology for amplicon sequencing is Illumina MiSeq which allows for high coverage at a less expense (Kozich et al. 2013). There are various 16S rRNA primers available based on the hypervariable regions of the 16S gene; however, a few of them have restricted coverage. Similarly, 18S rRNA gene primers and internal transcribed spacer (ITS) have been used for studying the fungal populations (Gilbert et al. 2014). In addition, certain functional genes encoding enzymes are being used as phylogenetic markers in amplicon sequencing of rhizosphere samples for classifying microbial taxa. For instance, *pmoA* gene for methanotrophs (Suddaby and Sourbeer 1990), *amoA* gene for ammonia oxidizers (Pester et al. 2012), *nxrB* gene for nitrite oxidizers (Pester et al. 2014). A list of such functional markers is available at the FunGene database (http://fungene.cme.msu.edu/) (Fish et al. 2013).

Once primers are decided, the libraries can be constructed by adding the adapters to the primers through a one-step PCR. The barcodes present in the primer pairs help in the simultaneous sequencing of several samples. However, if the primers selected do not contain barcodes, a two-step process is followed in which the DNA is first amplified and the barcodes are then added to the amplicons (Caporaso et al. 2011; Herbold et al. 2016). Another method for amplicon sequencing also exists, where the whole 16S gene or larger fragments (up to 30 kb) can be sequenced using PacBio sequencing (Armanhi et al. 2016). The resulting data are a valuable source for the researchers, and therefore it is recommended to submit such data to the Sequence Read Archive (SRA) of International Nucleotide Sequence Database Collaboration (INSDC) (Cochrane et al. 2016).

Nevertheless, this approach also has certain limitations. First and foremost is that amplification of the sample using universal primers may not be efficient to cover the genomes that contain longer introns in their rRNA genes (Hong et al. 2009; Sharpton et al. 2011). Second, if the same genome has many rRNA gene clusters then it will directly affect the relative richness of individual microbial taxa (Větrovský and Baldrian 2013). Third, in some cases, the 16S genomic loci between distantly related groups are transferred among themselves (i.e. horizontal gene transfer); and this have a direct impact on diversity estimation (Acinas et al. 2004). In addition, error prone sequencing and chimeras also hinder the quality of the reads (Wylie et al. 2012). Furthermore, with amplicon sequencing it is only possible to understand the structural aspects of the communities; however, resolving their functions is difficult. In such circumstances, rebuilding phylogenic information in order to predict their functions is a suitable solution (Langille et al. 2013). Depending on the structural diversity of the genomes available in databases, precision of function prediction varies. Lastly, this approach is restricted to the analysis of known taxa. Studying novel unknown microbial taxa such as viruses is a challenging task with this approach.

2.3.4 Shotgun Metagenomics

Shotgun metagenomics sequencing is a comparatively advanced and powerful approach that offers a detailed understanding of microbial communities concerning their structural and functional aspects as well. However, due to the large amount of data generated from this technique, the sequence analysis part is a bit tedious. Yet, the availability of new tools and softwares sidestep this complexity and allow the scientific community to unravel the structural and functional diversities of the microbiomes. Shotgun sequencing offers a better understanding of phylogenetic information and allows identification of the functions. This approach is very efficient in predicting rare species found in a sample. However, the species richness measured through this approach is unaccountable (Poretsky et al. 2014).

Nevertheless, to influence the richness of the sample data, it is significant to think through the depth and coverage of the sequencing. Regrettably, this is not an upfront chore. Especially, in analysing a complex environmental sample with a high level of genomic diversity such as rhizobiomes, depth of sequencing should be decided thoughtfully, as such sample may reflect in poor assembly with lower coverage (Sczyrba et al. 2017). So, to decide on the depth of the sequencing and the required amount of reads, in order to uncover the entire population present in the sample, it is recommended to follow a combinatorial approach of amplicon sequencing followed by shotgun metagenomics based on the results of the 16S sequencing data (Ni et al. 2013; Rodriguez-R et al. 2018). It is true that a small amount of good quality metagenomic data is adequate to predict the taxonomic abundance but such data might lead to wrong scientific conclusions (Kwak and Park 2018; Zaheer et al. 2018). Hence, it is vital to cautiously decide the required sequencing depth along with the type of analysis that will be performed.

Once the sequencing is done, the metagenomics data generated must be analysed computationally using various tools and softwares. The workflow of analysing metagenomics data includes four steps: (1) taxonomic binning, (2) taxonomic profiling, (3) target–gene reassembly, and (4) genome binning. These steps have been briefed in Chap. 1 of this book. These allow researchers to relate taxonomic distinctiveness to their key functions. For instance, in plant–microbe interactions, symbiont functions such as nitrogen fixation, effector secretions can be predicted just by linking their taxonomic identity (Eichinger et al. 2016).

In spite of various advantages over amplicon-sequencing technique, this approach also has its own limitations. First, as the resulting data from a deep sequencing of metagenomic sample is very large and complex, it is very challenging to bin the reads with precision. In addition, in the case of rhizosphere microbiomes, the sample is highly diverse and hence not all the genomes will be entirely sequenced. Consequently, overlapping of reads from the same gene may not be possible which can result in poor assembly (Sharpton et al. 2011). Even if the reads overlap, it is not necessary that they belong to the same or different genomes, which can impact sequence assembly (Mavromatis et al. 2007; Mende et al. 2012). Second, interference of host plant's DNA in the metagenomics sample highly influences the sequencing efficiency and subsequent analysis. Thus, enrichment of microbial DNA is mandatory prior to sequencing. Several molecular methods are available for enrichment and also, several computational tools are available to filter out the host reads present if any (Schmieder and Edwards 2011a; Garcia-Garcerà et al. 2013). Luckily, progressively developing bioinformatics tools and softwares resolve these challenges and ease the analysis process (Schmieder and Edwards 2011b).

2.3.5 Bioinformatics Analysis

Bioinformatics tools and softwares have significantly contributed to have insights on microbial interactions in complex ecosystems such as rhizosphere (Spence et al.

2014; Cha et al. 2016). For instance, metagenomics unveiled the causative agent of *Rhizoctonia solani* infection as *Pseudomonas* spp. in sugar beet grown in a suppressive soil (Mendes et al. 2011). Nevertheless, it is always important to assess the computational extrapolations under regulated conditions in the lab or in the field (Vorholt et al. 2017). For classifying reads generated through 16S amplicon sequencing, a few efficient computational tools are available, e.g. Mothur (Schloss et al. 2009), QIIME (Caporaso et al. 2010), and UPARSE (Edgar 2013). Quality-wise, all three pipelines perform equally. Similarly, for classification of ITS reads, PIPITS (Gweon et al. 2015) denotes a set of commands that are prerequisites of a few software, such as VSEARCH (Rognes et al. 2016) and USEARCH (Edgar and Flyvbjerg 2015; Nilsson et al. 2019). An assessment of the available tools and software for metagenomic analysis is beyond the scope of this chapter but is available in other reports (Kopylova et al. 2016).

2.4 Concluding Remarks

The advent of molecular techniques has substantially upgraded in unravelling the plant-associated microbiomes, e.g. (1) A detailed insight of *A. thaliana* roots microbiome (Bulgarelli et al. 2012; Engelbrektson et al. 2012) and (2) prediction of key microbial taxa and genes intricate in the destruction of a fungal root pathogen (Mendes et al. 2011). Metagenomics is a promising approach to recognize the mutual effects of plants, and their associated microbiomes (Vorholt et al. 2017; Durán et al. 2018). It offers a broader view of a complex environment such as rhizosphere and their interactions when compared to the other culture-dependent techniques. Metagenomics has its own power and potential to resolve innumerable challenges in various fields such as earth sciences, life sciences, biomedicine, bioenergy, bioremediation, biotechnology, agriculture, biodefense, and microbial forensics. So, with such progressive technique in hand, it is possible to predict the interactions between natural environments and human actions which collectively contour the future of our planet.

Acknowledgments The authors are thankful to the reviewer for his/her cautious evaluation of our manuscript and his/her valuable comments and suggestions.

Conflict of Interest The authors declare that they have no conflict of interests.

References

Abanda-Nkpwatt D, Müsch M, Tschiersch J et al (2006) Molecular interaction between Methylobacterium extorquens and seedlings: growth promotion, methanol consumption, and localization of the methanol emission site. J Exp Bot 57:4025–4032

Acinas SG, Marcelino LA, Klepac-Ceraj V, Polz MF (2004) Divergence and redundancy of 16S rRNA sequences in genomes with multiple rrn operons. J Bacteriol 186:2629–2635
Akinsanya MA, Goh JK, Lim SP, Ting ASY (2015) Metagenomics study of endophytic bacteria in Aloe vera using next-generation technology. Genomics Data 6:159–163
Ali B, Sabri AN, Ljung K, Hasnain S (2009) Auxin production by plant associated bacteria: impact on endogenous IAA content and growth of Triticum aestivum L. Lett Appl Microbiol 48:542–547
Alina SO, Constantinscu F, Petruţa CC (2015) Biodiversity of Bacillus subtilis group and beneficial traits of Bacillus species useful in plant protection. Rom Biotechnol Lett 20:10737–10750
Alves LDF, Westmann CA, Lovate GL et al (2018) Metagenomic approaches for understanding new concepts in microbial science. Int J Genomics 2:1–15
Ambawade MS, Pathade GR (2013) Production of gibberellic acid by Bacillus siamensis BE 76 isolated from banana plant (musa spp). Int J Sci Res 4:2319–7064
Ambrosini A, Beneduzi A, Stefanski T et al (2012) Screening of plant growth promoting rhizobacteria isolated from sunflower (Helianthus annuus L.). Plant Soil 356:245–264
Armanhi JSL, De Souza RSC, De Araújo LM et al (2016) Multiplex amplicon sequencing for microbe identification in community-based culture collections. Sci Rep 6:29543
Bakker MG, Schlatter DC, Otto-Hanson L, Kinkel LL (2014) Diffuse symbioses: roles of plant-plant, plant-microbe and microbe-microbe interactions in structuring the soil microbiome. Mol Ecol 23:1571–1583
Bulgarelli D, Rott M, Schlaeppi K et al (2012) Revealing structure and assembly cues for Arabidopsis root-inhabiting bacterial microbiota. Nature 488:91–95
Bulgarelli D, Garrido-Oter R, Münch PC et al (2015) Structure and function of the bacterial root microbiota in wild and domesticated barley. Cell Host Microbe 17:392–403
Caporaso JG, Kuczynski J, Stombaugh J et al (2010) QIIME allows analysis of high-throughput community sequencing data. Nat Methods 7:335–336
Caporaso JG, Lauber CL, Walters WA et al (2011) Global patterns of 16S rRNA diversity at a depth of millions of sequences per sample. Proc Natl Acad Sci U S A 1:4516–4522
Cha JY, Han S, Hong HJ et al (2016) Microbial and biochemical basis of a Fusarium wilt-suppressive soil. ISME J 10:119–129
Cochrane G, Karsch-Mizrachi I, Takagi T (2016) The international nucleotide sequence database collaboration. Nucleic Acids Res 44:D48–D50
Compant S, Reiter B, Sessitsch A et al (2005) Endophytic colonization of Vitis vinifera L. by plant growth-promoting bacterium Burkholderia sp. strain PsJN. Appl Environ Microbiol 71:1685–1693
Compant S, Clément C, Sessitsch A (2010) Plant growth-promoting bacteria in the rhizo- and endosphere of plants: their role, colonization, mechanisms involved and prospects for utilization. Soil Biol Biochem 42:669–678
De León KB, Gerlach R, Peyton BM, Fields MW (2013) Archaeal and bacterial communities in three alkaline hot springs in heart lake geyser basin, yellowstone national park. Front Microbiol 4:330
Dong CJ, Wang LL, Li Q, Shang QM (2019) Bacterial communities in the rhizosphere, phyllosphere and endosphere of tomato plants. PLoS One 14:e0223847
Durán P, Thiergart T, Garrido-Oter R et al (2018) Microbial interkingdom interactions in roots promote Arabidopsis survival. Cell 175:973–983.e14
Edgar RC (2013) UPARSE: highly accurate OTU sequences from microbial amplicon reads. Nat Methods 10:996–998
Edgar RC, Flyvbjerg H (2015) Error filtering, pair assembly and error correction for next-generation sequencing reads. Bioinformatics 31:3476–3482
Eichinger V, Nussbaumer T, Platzer A et al (2016) EffectiveDB—updates and novel features for a better annotation of bacterial secreted proteins and type III, IV, VI secretion systems. Nucleic Acids Res 44:D669–D674

Enebe MC, Babalola OO (2019) The impact of microbes in the orchestration of plants' resistance to biotic stress: a disease management approach. Appl Microbiol Biotechnol 103:9–25
Engelbrektson A, Kunin V, Engelbrektson A et al (2012) Defining the core Arabidopsis thaliana root microbiome. Nature 488:86–90
Farag MA, Zhang H, Ryu CM (2013) Dynamic chemical communication between plants and Bacteria through airborne signals: induced resistance by bacterial volatiles. J Chem Ecol 39:1007–1018
Felczykowska A, Krajewska A, Zielińska S, Łos JM (2015) Sampling, metadata and DNA extraction—important steps in metagenomic studies. Acta Biochim Pol 62:151–160
Fierer N (2017) Embracing the unknown: disentangling the complexities of the soil microbiome. Nat Rev Microbiol 15:579–590
Fish JA, Chai B, Wang Q et al (2013) FunGene: the functional gene pipeline and repository. Front Microbiol 4:291
Fuqua WC, Winans SC, Greenberg EP (1994) Quorum sensing in bacteria: the LuxR-LuxI family of cell density—responsive transcriptional regulators. J Bacteriol 176:269–275
Fürnkranz M, Wanek W, Richter A et al (2008) Nitrogen fixation by phyllosphere bacteria associated with higher plants and their colonizing epiphytes of a tropical lowland rainforest of Costa Rica. ISME J 2:561–570
Galloway WRJD, Hodgkinson JT, Bowden SD et al (2011) Quorum sensing in gram-negative bacteria: small-molecule modulation of AHL and AI-2 quorum sensing pathways. Chem Rev 111:28–67
Gans J, Wolinsky M, Dunbar J (2005) Microbiology: computational improvements reveal great bacterial diversity and high toxicity in soil. Science 309:1387–1390
Garcia-Garcerà M, Garcia-Etxebarria K, Coscollà M et al (2013) A new method for extracting skin microbes allows metagenomic analysis of whole-deep skin. PLoS One 8:e74914
Ghosh A, Mehta A, Khan AM (2018) Metagenomic analysis and its applications. In: Ranganathan S et al (eds) Encyclopedia of bioinformatics and computational biology. Elsevier, Amsterdam, pp 184–193
Gilbert JA, Jansson JK, Knight R (2014) The earth microbiome project: successes and aspirations. BMC Biol 12:69
Glassing A, Dowd SE, Galandiuk S et al (2016) Inherent bacterial DNA contamination of extraction and sequencing reagents may affect interpretation of microbiota in low bacterial biomass samples. Gut Pathog 8:24
Gweon HS, Oliver A, Taylor J et al (2015) PIPITS: an automated pipeline for analyses of fungal internal transcribed spacer sequences from the Illumina sequencing platform. Methods Ecol Evol 6:973–980
Hacquard S (2016) Disentangling the factors shaping microbiota composition across the plant holobiont. New Phytol 209:454–457
Hardoim PR, Hardoim CCP, van Overbeek LS, van Elsas JD (2012) Dynamics of seed-borne rice endophytes on early plant growth stages. PLoS One 7:e30438
Hardoim PR, van Overbeek LS, Berg G et al (2015) The hidden world within plants: ecological and evolutionary considerations for defining functioning of microbial endophytes. Microbiol Mol Biol Rev 79:293–320
Hartmann A, Schmid M, van Tuinen D, Berg G (2009) Plant-driven selection of microbes. Plant Soil 321:235–257
Hartmann A, Rothballer M, Hense BA, Schröder P (2014) Bacterial quorum sensing compounds are important modulators of microbe-plant interactions. Front Plant Sci 5:131
Hennessy RC, Phippen CBW, Nielsen KF et al (2017) Biosynthesis of the antimicrobial cyclic lipopeptides nunamycin and nunapeptin by Pseudomonas fluorescens strain In5 is regulated by the LuxR-type transcriptional regulator NunF. Microbiology 6(6):e00516
Herbold CW, Pelikan C, Kuzyk O et al (2016) A flexible and economical barcoding approach for highly multiplexed amplicon sequencing of diverse target genes. Front Microbiol 6:870

Hong S, Bunge J, Leslin C et al (2009) Polymerase chain reaction primers miss half of rRNA microbial diversity. ISME J 3:1365–1373
Huang J, Shi Y, Zeng G et al (2016) Acyl-homoserine lactone-based quorum sensing and quorum quenching hold promise to determine the performance of biological wastewater treatments: an overview. Chemosphere 157:137–151
Hussain A, Krischke M, Roitsch T, Hasnain S (2010) Rapid determination of cytokinins and auxin in cyanobacteria. Curr Microbiol 61:361–369
Igiehon NO, Babalola OO (2018) Below-ground-above-ground plant-microbial interactions: focusing on soybean, rhizobacteria and mycorrhizal fungi. Open Microbiol J 12:261–279
Jackson AO, Taylor CB (1996) Plant-microbe interactions: life and death at the interface. Plant Cell 8(10):1651–1668
Kennedy AC, Gewin VL (1997) Soil microbial diversity: present and future considerations. Soil Sci 162:605–606
Kennedy J, Baker P, Piper C et al (2009) Isolation and analysis of bacteria with antimicrobial activities from the marine sponge haliclona simulans collected from irish waters. Mar Biotechnol 11:384–396
Kinkel LL, Bakker MG, Schlatter DC (2011) A Coevolutionary framework for managing disease-suppressive soils. Annu Rev Phytopathol 49:47–67
Knief C (2014) Analysis of plant microbe interactions in the era of next generation sequencing technologies. Front Plant Sci 5:216
Köberl M, Schmidt R, Ramadan EM et al (2013) The microbiome of medicinal plants: diversity and importance for plant growth, quality, and health. Front Microbiol 4:400
Kopylova E, Navas-Molina JA, Mercier C et al (2016) Open-source sequence clustering methods improve the state of the art. mSystems 1:e00003–e00015
Kozich JJ, Westcott SL, Baxter NT et al (2013) Development of a dual-index sequencing strategy and curation pipeline for analyzing amplicon sequence data on the miseq illumina sequencing platform. Appl Environ Microbiol 79:5112–5120
Kunin V, Copeland A, Lapidus A et al (2008) A bioinformatician's guide to metagenomics. Microbiol Mol Biol Rev 72:557–578
Kuzyakov Y, Razavi BS (2019) Rhizosphere size and shape: temporal dynamics and spatial stationarity. Soil Biol Biochem 135:343–360
Kwak J, Park J (2018) What we can see from very small size sample of metagenomic sequences. BMC Bioinformatics 19:399
Langille MGI, Zaneveld J, Caporaso JG et al (2013) Predictive functional profiling of microbial communities using 16S rRNA marker gene sequences. Nat Biotechnol 31(9):814–821
Leveau JHJ (2007) The magic and menace of metagenomics: prospects for the study of plant growth-promoting rhizobacteria. Eur J Plant Pathol 119:279–300
Liu Y, Zuo S, Xu L et al (2012) Study on diversity of endophytic bacterial communities in seeds of hybrid maize and their parental lines. Arch Microbiol 194:1001–1012
Liu F, Hewezi T, Lebeis SL et al (2019) Soil indigenous microbiome and plant genotypes cooperatively modify soybean rhizosphere microbiome assembly. BMC Microbiol 19:201
Lozupone CA, Knight R (2007) Global patterns in bacterial diversity. Proc Natl Acad Sci U S A 104:11436–11440
Massalha H, Korenblum E, Malitsky S et al (2017) Live imaging of root-bacteria interactions in a microfluidics setup. Proc Natl Acad Sci U S A 114:4549–4554
Mavromatis K, Ivanova N, Barry K et al (2007) Use of simulated data sets to evaluate the fidelity of metagenomic processing methods. Nat Methods 4:495–500
Mende DR, Waller AS, Sunagawa S et al (2012) Assessment of metagenomic assembly using simulated next generation sequencing data. PLoS One 7:e31386
Mendes R, Kruijt M, De Bruijn I et al (2011) Deciphering the rhizosphere microbiome for disease-suppressive bacteria. Science 332:1097–1100
Monther MT, Kamaruzaman S (2012) Arbuscular mycorrhizal fungi and plant root exudates bio-communications in the rhizosphere. African J Microbiol Res 6:7295–7301

Nannipieri P, Ascher J, Ceccherini MT et al (2008) Effects of root exudates in microbial diversity and activity in rhizosphere soils. In: Nautiyal CS, Dion P (eds) Molecular mechanisms of plant and microbe coexistence. Springer, Berlin, pp 340–365
National Research Council, Division on Earth and Life Studies, Board on Life Sciences, Committee on Metagenomics: Challenges and Functional Applications (2007) Designing a successful metagenomics project: best practices and future needs. In: National Research Council et al (eds) The new science of metagenomics: revealing the secrets of our microbial planet. The National Academic Press, Washington, DC, pp 60–65
Ni J, Yan Q, Yu Y (2013) How much metagenomic sequencing is enough to achieve a given goal? Sci Rep 3:1968
Nilsson RH, Anslan S, Bahram M et al (2019) Mycobiome diversity: high-throughput sequencing and identification of fungi. Nat Rev Microbiol 17(2):95–109
Paul EA, Clark FE (1988) Soil microbiology and biochemistry. In: Paul EA, Clark FE (eds) Soil microbiology and biochemistry. Academic, San Diego, pp 1–10
Pester M, Rattei T, Flechl S et al (2012) AmoA-based consensus phylogeny of ammonia-oxidizing archaea and deep sequencing of AmoA genes from soils of four different geographic regions. Environ Microbiol 14(2):525–539
Pester M, Maixner F, Berry D et al (2014) NxrB encoding the beta subunit of nitrite oxidoreductase as functional and phylogenetic marker for nitrite-oxidizing nitrospira. Environ Microbiol 16 (10):3055–3071
Pindi PK, Sultana T, Vootla PK (2014) Plant growth regulation of Bt-cotton through Bacillus species. 3 Biotech 4(3):305–315
Poretsky R, Rodriguez-R LM, Luo C et al (2014) Strengths and limitations of 16S rRNA gene amplicon sequencing in revealing temporal microbial community dynamics. PLoS One 9(4): e93827
Rabausch U, Juergensen J, Ilmberger N et al (2013) Functional screening of metagenome and genome libraries for detection of novel flavonoid-modifying enzymes. Appl Environ Microbiol 79(15):4551–4563
Rehman A, Ijaz M, Mazhar K et al (2019) Metagenomic approach in relation to microbe–microbe and plant–microbiome interactions. In: Kumar V et al (eds) Microbiome in plant health and disease. Springer Nature, Singapore, pp 507–534
Reid JRW (2003) Experimental design and data analysis for biologists. Austral Ecol 28:588–589
Roat C, Saraf M (2017) Unravelling the interaction of plant and their Phyllosphere microbiome. In: Singh RP et al (eds) Understanding host-microbiome interactions—an omics approach: omics of host-microbiome association. Springer Nature, Singapore, pp 157–172
Rodriguez-R LM, Gunturu S, Tiedje JM et al (2018) Nonpareil 3: fast estimation of metagenomic coverage and sequence diversity. mSystems 3(3):e00039–e00018
Rognes T, Flouri T, Nichols B et al (2016) VSEARCH: a versatile open source tool for metagenomics. PeerJ 4:e2584
Sagar K, Singh SP, Goutam KK, Konwar BK (2014) Assessment of five soil DNA extraction methods and a rapid laboratory-developed method for quality soil DNA extraction for 16S rDNA-based amplification and library construction. J Microbiol Methods 97:68–73
Salter SJ, Cox MJ, Turek EM et al (2014) Reagent and laboratory contamination can critically impact sequence-based microbiome analyses. BMC Biol 12:87
Schloss PD, Westcott SL, Ryabin T, Hall JA (2009) Introducing mothur: open source, platform-independent, community-supported software. Appl Environ Microbiol 75(23):7537–7541
Schmieder R, Edwards R (2011a) Quality control and preprocessing of metagenomic datasets. Bioinformatics 27(6):863–864
Schmieder R, Edwards R (2011b) Fast identification and removal of sequence contamination from genomic and metagenomic datasets. PLoS One 6(3):e17288
Sczyrba A, Hofmann P, Belmann P et al (2017) Critical assessment of metagenome interpretation—a benchmark of metagenomics software. Nat Methods 14(11):1063–1071

Sharon I, Banfield JF (2013) Microbiology. Genomes from metagenomics. Science 342 (6162):1057–1058
Sharpton TJ (2014) An introduction to the analysis of shotgun metagenomic data. Front Plant Sci 5:209
Sharpton TJ, Riesenfeld SJ, Kembel SW et al (2011) PhyLOTU: a high-throughput procedure quantifies microbial community diversity and resolves novel taxa from metagenomic data. PLoS Comput Biol 7(1):e1001061
Soni R, Acahrya C, Primalatha K et al (2012) Metagenomics technology. In: Kumar A, Pareek A, Gupta SM (eds) Biotechnology in medicine and agriculture principles and practice. IK International, New Delhi, p 835
Spence C, Alff E, Johnson C et al (2014) Natural rice rhizospheric microbes suppress rice blast infections. BMC Plant Biol 14:130
Suddaby EC, Sourbeer MO (1990) Drawing pediatric arterial blood gases. Crit Care Nurse 10 (7):28–31
Tebbe CC, Vahjen W (1993) Interference of humic acids and DNA extracted directly from soil in detection and transformation of recombinant DNA from bacteria and a yeast. Appl Environ Microbiol 59(8):2657–2665
Větrovský T, Baldrian P (2013) The variability of the 16S rRNA gene in bacterial genomes and its consequences for bacterial community analyses. PLoS One 8(2):e57923
Vorholt JA (2012) Microbial life in the phyllosphere. Nat Rev Microbiol 10(12):828–840
Vorholt JA, Vogel C, Carlström CI, Müller DB (2017) Establishing causality: opportunities of synthetic communities for plant microbiome research. Cell Host Microbe 22(2):142–155
Wallace JG, Kremling KA, Kovar LL, Buckler ES (2018) Quantitative genetics of the maize leaf microbiome. Phytobiomes J 2(4):208–224
Wooley JC, Godzik A, Friedberg I (2010) A primer on metagenomics. PLoS Comput Biol 6(2): e1000667
Wylie KM, Truty RM, Sharpton TJ et al (2012) Novel bacterial taxa in the human microbiome. PLoS One 7(6):e35294
Zaheer R, Noyes N, Ortega Polo R et al (2018) Impact of sequencing depth on the characterization of the microbiome and resistome. Sci Rep 8:5890
Zarraonaindia I, Owens SM, Weisenhorn P et al (2015) The soil microbiome influences grapevine-associated microbiota. Mol Biol 6(2):e02527–e02514

Chapter 3
Metagenomics and Metatranscriptomics Approaches in Understanding and Discovering Novel Molecules in Rhizosphere Environment

Jaspreet Kaur Boparai and Pushpender Kumar Sharma

Abstract Rhizosphere microbiome is a diverse and dynamic communities affecting both plant health and nutrition. A better understanding of these ecosystems however remain uncharted owing to lack of tools that facilitate comprehensive profiling of microbial communities and processes which are largely constituted by uncultivable subset. Recent advances in molecular biology techniques that include metagenomics and metatranscriptomics facilitate whole community level characterization of gene diversity and their expression patterns, respectively. While metagenomics focuses on meta-analysis of structural and functional gene contents of microbial communities in various environment niches, metatranscriptomics focuses on meta-analysis of expressed genes under set of environmental conditions. These omics technologies are new and comprehensive approaches to system biology that integrate multiple-datasets, both these techniques have shed novel insights about the community activity and responses. This chapter discusses the utilization of these tools and techniques in understanding microbial interaction networks, communications, and the chemical diversity in the rhizosphere.

Keywords Metagenomics · Metatranscriptomics · Rhizosphere microbiome · Novel biochemical · System biology

3.1 Introduction

Microorganisms are found everywhere in the environment ranging from soil, animals, plants, and lower eukaryotes (Girguis 2016), and constitute dynamic communities, diversities, and competitions in various environments (Vega and Gore 2018). The rhizosphere (the soil environment influenced by root exudates) harbors complex

J. K. Boparai · P. K. Sharma (✉)
Department of Biotechnology, Sri Guru Granth Sahib World University, Fatehgarh Sahib, Punjab, India

R. N. Pudake et al. (eds.), *Omics Science for Rhizosphere Biology*, Rhizosphere Biology, https://doi.org/10.1007/978-981-16-0889-6_3

microbial communities that affect both plant health and nutrition. Better understandings of these ecosystems remain uncharacterized due to a lack of tools that may facilitate complete profiling of extremely diverse microbial communities and processes (Zaccaria et al. 2017). Recent advances in molecular biology techniques that include metagenomics and metatranscriptomics will help in accessing whole community level characterization of genomes, gene expression patterns, and plant–microbe interactions in the rhizosphere. Rhizosphere microbiology has gained significant attention in recent years, as microorganisms present in its vicinity directly or indirectly play a critical role in plant defense and development (Jeyanthi and Kanimozhi 2018). Previously, rhizosphere microbiome was explored for its beneficial effects on plants that include process of nitrogen fixation, phosphate solubilization, production of plant growth hormones, biocontrol properties, and tolerance to various stresses (Singh et al. 2019). In the rhizosphere, the microbial communities' composition is mainly determined by soil type, its texture, plant genotype, cultivars, and developmental stage of the plants (Pantigoso et al. 2020; Buscot 2005). In addition to beneficial microbes, certain microbial pathogens often colonize the rhizosphere and may bypass the mechanism of innate plant protection to trigger disease (Mendes et al. 2013; Wang et al. 2018). This chapter will discuss the role of metagenomics and metatranscriptomics tools in studying the rhizosphere environment. Eukaryotic and prokaryotic life forms studied in extreme environment (Bochdansky et al. 2017), revealed reservoir of genes responsible for different activities in soil, however, isolating and cultivating these organisms under laboratory condition is difficult. Holistic approaches such as metagenomics and metatranscriptomics can classify genes and archive them as library of soil genes (Andrews et al. 2018). Microbial operational taxonomic units (OTU's) investigations of desert, grassland, rain forest has revealed that archaeal or fungal OTUs appeared to be equal or exceed the number of unique bacterial OTUs in these ecosystems (with minimal taxonomic overlap observed between the soil types (Fierer et al. 2007).

3.2 The Rhizosphere Microbiome—Who Is There?

Soil microbial communities constitute the world's greatest repository of biodiversity known to date (Bowker et al. 2010). The rhizosphere is inhabited by ~10^{11} microbial cells/gram root (Egamberdieva et al. 2008) and over 30,000 prokaryotic species (Mendes et al. 2013). It is an interface between plant root and soil characterized by microbe–microbe and plant–microbe interactions which are mediated mainly through the secondary metabolites (Arora et al. 2013). This interface is rich in variety of carbon compounds creating a special habitat for a variety of soil microorganisms, including bacteria, fungi, oomycetes, nematodes, protozoa, algae, viruses, archaea, and arthropods. At the root–microbe interface, there is exudation of chemicals that serve as microbial capital, while microbes in return enhance plant growth (Chaparro et al. 2014). In addition, rhizosphere acts as barrier to pathogen causing infection in plants (Berendsen et al. 2012; Ryan et al. 2008). Several processes in the rhizosphere are essential gateways for nutrients and water, it uses about 50% of the energy provided by photosynthesis in terrestrial ecosystems (Taylor et al. 2012), contributing 50% of the total CO_2 released from terrestrial ecosystems (Trumbore 2006).

3.3 Interaction Network, Communications, and Chemical Diversity in Rhizosphere

Rhizosphere constitutes the most complex ecosystem harboring variety of organisms including nematodes, fungi, bacteria, and arthropod (Sharma et al. 2019a, b). This zone is associated with the increased bacterial abundance and activity as compared to the bulk soil. Plants are known to cause a selective pressure on the microbial community found in the rhizosphere. Community-level research has identified diverse microbial communities in the different plant species, indicating role of plant-derived metabolites in determining the rhizosphere microbiome assembly (Mhlongo et al. 2018; O'Banion et al. 2020). Various plant growth-promoting rhizobacteria (PGPR) are *Bacillus*, *Pseudomonas*, *Enterobacter*, *Acinetobacter*, *Burkholderia*, *Arthrobacter,* and *Paenibacillus* (Kour et al. 2019). These species interact with each other and with the plant through chemical communication. Plant releases variety of metabolites (primary and secondary) in response to altered gene expression (Verma et al. 2018), availability of nutrients, microorganism accumulation, and biofilm formation (Nazir et al. 2019), as well during inhibition of soil–borne pathogens (De Boer et al. 2019). Signaling metabolites which regulate gene expression in the host plants are volatile organic compounds (VOCs), these molecules are low molecular weight lipophilic compounds synthesized during different metabolic pathways (Fincheira and Quiroz 2018). Quorum sensing (QS) is a mechanism used by bacteria to perceive their environment, the well-documented QS signals frequently produced by Gram-negative bacteria are AHLs (N-acyl homoserine lactones) (Papenfort and Bassler 2016).

3.4 Omics Approaches and Microbial Dynamics in Rhizosphere

3.4.1 *What Is Omics?*

Omics is a novel approach to systems biology that integrates multiple datasets from genomes, proteomes, transcriptomes, and metabolomes. All these datasets are derived from individual organisms that include bacteria, fungi, plants, animals, and even viruses. Data gathered pertaining to central dogma of molecular biology includes DNA, mRNA, and proteins is measured, analyzed, and integrated using multivariate statistics methods. This book chapter will briefly discuss application of metagenomics and metatranscriptomics tools in understanding the rhizosphere environment.

3.4.2 Metagenomics in Understanding and Discovery of Novel Molecules from Rhizosphere

Metagenomics refers to meta-analysis of genomes recovered directly from environmental samples, rhizosphere contains high levels of the soil matrix, interfering humic acids, as well as plant polyphenols and other macromolecules, all of which makes it difficult to extract biological molecules such as DNA, RNA, proteins, and metabolites (Cesco et al. 2012). Despite several challenges, metagenomics studies have successfully used to reconstruct the complete to nearly complete genomes from unknown and uncultivated phyla (Seitz et al. 2016: Bickhart et al. 2017: White III et al. 2017: Pop et al. 2004; Beaulaurier 2018). It has offered enormous opportunities for researchers to unravel information related to microbial diversity and the discovery of novel molecules from rhizosphere environment (Thijs et al. 2016; Gilbert and Dupont 2010). Several microbial communities have been identified and characterized from rhizosphere environment using the metagenomic approach (Boparai et al. 2019: Li and Qin 2005: Lu et al. 2018). A recent study carried out from five distinct *Sinorhizobium* spp. (*Sinorhizobium meliloti, Sinorhizobium medicae, Sinorhizobium fredii, Sinorhizobium teranga, Sinorhizobium saheli*) found that each bacterium has adopted slightly a different strategy to interact with diverse *Medicago* spp. in the soil environment. The study further highlighted that the genes involved in the biosynthesis of the Nod factor (lipochitooligosaccharides secreted by rhizobia that are important in the interplay of recognition between roots and microbes) in polysaccharides synthesis, denitrification, and secretion systems vary within and between species (Sugawara et al. 2013). Table 3.1 briefly describe detailed studies related to the discovery of novel molecules from the rhizosphere via metagenomic.

3.4.3 Metatranscriptomics in Understanding and Discovery of Novel Molecule and Metabolic Pathways from Rhizosphere

Metatranscriptomics has enabled researchers to profile the expressed genes from diverse microbial communities; it identified key genes interactions between plants and microbes (Schenk et al. 2012) and shed light on the structure and function of a soil community (Ambardar et al. 2016). In this method, total community RNA is randomly reversely transcribed into cDNA, and the sequences obtained are profiled

Table 3.1 List of few novel compounds/molecules discovered from rhizosphere using metagenomics approach

Origin	Method used	Location	Novel compounds discovered	Function of discovered compound	References
Plant rhizosphere	Metagenomic libraries	South Korea	Enzyme-esterase EstD2	Lipolytic activity	Lee et al. (2010)
Rhizosphere soil bacteria *Bacillus subtilis* strain SK.DU.4	16S rRNA gene sequencing	India	A surface-active lipopeptide and a peptide (Iturin like group of antimicrobial biosurfactants)	Antimicrobial activity	Baindara et al. (2013)
Paenibacillus ehimensis strain A3	RAST genome sequence analysis	India	Penisin	Antimicrobial activity	Baindara et al. (2016)
A rhizobacterium (Pseudomonas fluorescens In5)	Phylogenetic analysis based on multi-locus sequence typing	South Greenland	A novel non-ribosomal peptide synthase (NRPS) gene	Antifungal activity	Michelsen and Stougaard (2011)
Actinobacteria from the rhizosphere of sugar beet plants	16s rRNA sequencing	Netherlands	(VOCs mixture) Methyl 2-methylpentanoate and 1,3,5-trichloro-2-methoxy benzene	Antifungal activity	Cordovez et al. (2015)
Pseudomonas in rhizosphere soil of *Zea mays*	16S ribosomal DNAs (rDNAs) amplification, RAPD technique	Italy	2,4-diacetylphloroglucinol (DAPG)	Antimicrobial activity	Picard et al. (2000)
Pseudomonas isolate EA105	16S rRNA gene sequencing	California	2-Undecanol (**Alcohols**) Cyclopropane, 1- methyl-2-pentyl, 1,4-Octadiene, 1 Undecene, 1-Dodecene, Cyclodecene, 1-Tridecene **(Hydrocarbons)** 2-Udecanone, 2-Tridecanone **(Ketones)** Methyl thiolacetate, Dimethyl disulfide, S-methyl propanethioate, S-methyl 3-methylbutanethioate **(S-Containing compounds)**	Antifungal activity (Suppress rice blast infections)	Spence et al. (2014)

taxonomically. Metatranscriptome-based research has discovered novel genes in the rhizosphere (Kim et al. 2014). Though metatranscriptomic approaches from the rhizosphere are not frequently used due to difficulty in the extraction of mRNAs from the complex system (Simon and Daniel 2011), nonetheless several studies were carried out employing metatranscriptomic approach. A metatranscriptome analysis carried out on oxic and anoxic zone reported Cyanobacteria, Fungi, Xanthomonadales, Myxococcales, and Methylococcales in the oxic zone, while Clostridia, Actinobacteria, Geobacter, Anaeromyxobacter, Anaerolineae, and methanogenic archaea conquered the anoxic zone, and were maintained stably during the time of incubation. In addition, methane oxidation and photosynthesis were carried out by Methylococcales and Cyanobacteria, respectively, which were specific to the oxic zone (Kim and Liesack 2015). Researchers also investigated the plant pathogenic fungi *Rhizoctonia solani* and reported that it directly or indirectly increases the transcript of oxidative stress-related genes in certain rhizobacterial families by ppGpp signalling pathway. This pattern may be responsible for the change in microbial communities and can inhibit the growth of the fungal pathogens (Chapelle et al. 2016). One more study investigating de novo metatranscriptomics of plant and fungi in the symbiotic roots of sewage-cleaning *Eichhornia crassipes* reported upregulated and downregulated genes during symbiotic process, the study reported alteration in metabolic pathways of plant in the presence of fungi (Luo et al. 2015). Metatranscriptomics data combined in chronological and/or spatial degree can address questions related to functionally active transcripts, e.g., it has been successfully implicated in biogeochemical cycles (Muller et al. 2018), models of plant–microbe interactions (Verma et al. 2018), and functional diversity measurements (Dubey et al. 2020; Ramachandran 2008; Green et al. 2019). A recent study from our lab compared metatranscriptome from two different ecosystems that include agriculture soil and organic soil, sequencing and analysis of reads revealed variations in level of gene expression in two soil types, high expression of transcripts related to aromatic hydrocarbon metabolism and stress-related proteins in agriculture soil compared to organic soil (Sharma et al. 2019a, b). Another study conducted in our lab a few years back from agriculture soil having a long history of usage of chemical fertilizers and revealed the presence of several important genes and pathways related to the metabolism of aromatic hydrocarbon (Sharma and Sharma 2018).

Studies on the Gram-positive rhizobacterium *B. amyloliquefaciens* in response to root exudates from maize revealed alteration in transcript expression related to bacterial chemotaxis, motility, and antimicrobial peptides. Some genes with unknown functions were also showed changes in their expression pattern, indicating the role of root exudates in understanding the rhizobacteria (Fan et al. 2012). In another study transcriptome data was used to investigate the Ca^{2+}calmodulin-dependent protein kinase (Ishiguro et al. 2006), and it was demonstrated that transcript-based cloning is a healthy approach for cloning genes and, the construction of a genetic map is not required in this process. In another study, the method was successfully used to study plant productivity and rhizosphere dynamics (Zhang et al. 2010). It was discovered that NCRs (nodule Cys-rich antimicrobial peptides) (NCR) genes regulate the terminal differentiation, nitrogen fixation, and infection of

Table 3.2 Metatranscriptomics studies reporting important findings in plant rhizosphere

Plant rhizosphere	Main finding	References
Wheat, oat, pea and an oat mutant (*sad1*)	Profiling microbial communities with metatranscriptomics enables comparison of relative abundance, from different samples, across all domains of life, without bias in the polymerase chain reaction. This revealed profound differences in the rhizosphere microbiome, spatially at the kingdom level between plants	Turner et al. (2013)
Arabidopsis thaliana	Genes involved in streptomycin synthesis were substantially induced at bolting and flowering stages, possibly for suppression of disease. This suggested that plants secrete mixture of compounds and specific phytochemicals in the root exudates that are produced differentially at distinct stages of development to help orchestrate rhizosphere microbiome assemblage	Chaparro et al. (2014)
Willow tree (*Salix purpurea* and *Sambucus nigra*)	The root metatranscriptomes of two species has been compared and concluded that plants transcripts are primarily affected by willow species, whereas microbial transcripts mainly responded to contamination	Yergeau et al. (2018)
Wheat	In the non-suppressive rhizosphere samples, a large number of genes involved in detoxifying reactive oxygen species (ROS) and superoxide radicals (*sod, cat, ahp, bcp, gpx1, trx*) were expressed most likely in response to the infection of *Rhizoctonia solani* AG8 in wheat roots	Hayden et al. (2018)
Corn (*Zea mays*) and soybean (*Glycine max*)	This study identified 67 differentially expressed bacterial transcripts from the rhizosphere. Transcripts downregulated after glyphosate treatment included carbohydrate and amino acid metabolism, and the upregulated transcripts included protein metabolism and respiration. In addition, bacterial transcripts involving nutrients, including iron, nitrogen, phosphorus, and potassium, was also affected by long-term glyphosate application	Newman et al. (2016)
Sugar beet	rRNA-based analyses found that *Oxalobacteriaceae, Burkholderiaceae, Sphingobacteriaceae,* and *Sphingomonadace* were considerably more prevalent in the rhizosphere upon fungal invasion. Metatranscriptomics showed that stress-related genes (*ppGpp* metabolism and oxidative stress) were upregulated in above mentioned bacterial families	Chapelle et al. (2016)
Grassland rhizosphere	Functional evaluation of root mRNA indicates that the production of plant secondary metabolites was increased in the summer	Bei et al. (2019)

intracellular *S. meliloti* bacteroid by manipulating the bacterial cell cycle (Farkas et al. 2014; Penterman et al. 2014; Haag and Mergaert 2019) and provided strong evidence about the molecular mechanism by which NCR peptides control the *S. meliloti* cell cycle during symbiosis. Authors further reported that these peptides explicitly block cell division without affecting replication of DNA and disrupt the expression of genes involved in motility, cell division, and cell cycle regulation. Another study investigated non-fungal sources of industrially important enzymes in forest soils, via screening metatranscriptomic library and identified major plant cell wall, polymer degrading, and organic matter hydrolyzing enzymes (Damon et al. 2012). Table 3.2 briefly discusses novel findings related to metatranscriptomics.

3.5 Conclusions and Future Trends

The rhizosphere is a fascinating ecosystem for basic as well as applied microbiology, plant biology, and ecology studies. A major challenge in rhizosphere studies is to select the technology platforms that can better address the most pressing scientific questions. These considerations remain relevant in several areas: acquiring data that address a particular scientific hypothesis; preventing data deluge and long delays from research to publication; and allowing for significant scientific advances. Weak planning or inadequate knowledge also causes significant downstream problems. Multi-omic approaches will contribute to a greater understanding of species in the rhizosphere and their roles in plant development, crop production, and healthy ecosystems. Furthermore, metagenomics alone cannot be used to deduce whether all of the metabolic genes at the time of sampling are expressed, and thus another omics technique called metatranscriptomics is used to probe these questions. They will help in harnessing the rhizosphere as a resource for improved plant growth and quality, for sustainable crop production, and for increased soil carbon storage under various environmental stresses.

References

Ambardar S, Gupta R, Trakroo D, Lal R, Vakhlu J (2016) High throughput sequencing: an overview of sequencing chemistry. Indian J Microbiol 56(4):394–404

Andrews KR, De Barba M, Russello MA, Waits LP (2018) Advances in using non-invasive, archival, and environmental samples for population genomic studies. In: Population genomics. Springer, Cham. https://doi.org/10.1007/13836_2018_45

Arora NK, Tewari S, Singh R (2013) Multifaceted plant-associated microbes and their mechanisms diminish the concept of direct and indirect PGPRs. In: Plant microbe symbiosis: fundamentals and advances. Springer, New Delhi, pp 411–449

Baindara P, Mandal SM, Chawla N, Singh PK, Pinnaka AK, Korpole S (2013) Characterization of two antimicrobial peptides produced by a halotolerant *Bacillus subtilis* strain SK. DU. 4 isolated from a rhizosphere soil sample. AMB Express 3(1):2

Baindara P, Chaudhry V, Mittal G, Liao LM, Matos CO, Khatri N, Korpole S (2016) Characterization of the antimicrobial peptide penisin, a class Ia novel lantibiotic from *Paenibacillus* sp. strain A3. Antimicrob Agents Chemother 60(1):580–591

Beaulaurier JA (2018) Single molecule, long read-based methods for studying epigenetic heterogeneity in bacterial genomes and metagenomes. Doctoral dissertation, Icahn School of Medicine at Mount Sinai

Bei Q, Moser G, Wu X, Müller C, Liesack W (2019) Metatranscriptomics reveals climate change effects on the rhizosphere microbiomes in European grassland. Soil Biol Biochem 138:107604

Berendsen RL, Pieterse CM, Bakker PA (2012) The rhizosphere microbiome and plant health. Trends Plant Sci 17(8):478–486

Bickhart DM, Rosen BD, Koren S, Sayre BL, Hastie AR, Chan S, Burton JN (2017) Single-molecule sequencing and chromatin conformation capture enable de novo reference assembly of the domestic goat genome. Nat Genet 49(4):643–650

Bochdansky AB, Clouse MA, Herndl GJ (2017) Eukaryotic microbes, principally fungi and labyrinthulomycetes, dominate biomass on bathypelagic marine snow. ISME J 11(2):362–373

Boparai JK, Kaur R, Sharma PK (2019) Delineating bacterial community structure of rhizosphere metagenome from Kansal Forest. Ecol Genet Genom 12:100044

Bowker MA, Maestre FT, Escolar C (2010) Biological crusts as a model system for examining the biodiversity–ecosystem function relationship in soils. Soil Biol Biochem 42(3):405–417

Buscot F (2005) In: Varma A (ed) Microorganisms in soils: roles in genesis and functions. Springer, Berlin, pp 139–153

Cesco S, Mimmo T, Tonon G, Tomasi N, Pinton R, Terzano R, Nannipieri P (2012) Plant-borne flavonoids released into the rhizosphere: impact on soil bio-activities related to plant nutrition. A review. Biol Fertil Soils 48(2):123–149

Chaparro JM, Badri DV, Vivanco JM (2014) Rhizosphere microbiome assemblage is affected by plant development. ISME J 8(4):790–803

Chapelle E, Mendes R, Bakker PAH, Raaijmakers JM (2016) Fungal invasion of the rhizosphere microbiome. ISME J 10(1):265–268

Cordovez V, Carrion VJ, Etalo DW, Mumm R, Zhu H, Van Wezel GP, Raaijmakers JM (2015) Diversity and functions of volatile organic compounds produced by Streptomyces from a disease-suppressive soil. Front Microbiol 6:1081

Damon C, Lehembre F, Oger-Desfeux C, Luis P, Ranger J, Fraissinet-Tachet L, Marmeisse R (2012) Metatranscriptomics reveals the diversity of genes expressed by eukaryotes in forest soils. PLoS One 7(1):e28967

De Boer W, Li X, Meisner A, Garbeva P (2019) Pathogen suppression by microbial volatile organic compounds in soils. FEMS Microbiol Ecol 95(8):fiz105

Dubey RK, Tripathi V, Prabha R, Chaurasia R, Singh DP, Rao CS, Abhilash PC (2020) Metatranscriptomics and metaproteomics for microbial communities profiling. In: Unravelling the soil microbiome. Springer, Cham, pp 51–60

Egamberdieva D, Kamilova F, Validov S, Gafurova L, Kucharova Z, Lugtenberg B (2008) High incidence of plant growth-stimulating bacteria associated with the rhizosphere of wheat grown on salinated soil in Uzbekistan. Environ Microbiol 10(1):1–9

Fan B, Carvalhais LC, Becker A, Fedoseyenko D, von Wirén N, Borriss R (2012) Transcriptomic profiling of Bacillus amyloliquefaciens FZB42 in response to maize root exudates. BMC Microbiol 12(1):116

Farkas A, Maróti G, Dürgő H, Györgypál Z, Lima RM, Medzihradszky KF, Kondorosi É (2014) Medicago truncatula symbiotic peptide NCR247 contributes to bacteroid differentiation through multiple mechanisms. Proc Natl Acad Sci U S A 111(14):5183–5188

Fierer N, Breitbart M, Nulton J, Salamon P, Lozupone C, Jones R, Knight R (2007) Metagenomic and small-subunit rRNA analyses reveal the genetic diversity of bacteria, archaea, fungi, and viruses in soil. Appl Environ Microbiol 73(21):7059–7066

Fincheira P, Quiroz A (2018) Microbial volatiles as plant growth inducers. Microbiol Res 208:63–75

Gilbert JA, Dupont CL (2010) Microbial metagenomics: beyond the genome. Annu Rev Mar Sci 3 (1):347–371

Girguis P (2016) Microbial ecology: here, there, and everywhere. Nat Microbiol 1(8):1–2

Green RT, East AK, Karunakaran R, Downie JA, Poole PS (2019) Transcriptomic analysis of *Rhizobium leguminosarum* bacteroids in determinate and indeterminate nodules. Microb Genomics 5(2):e000254

Haag AF, Mergaert P (2019) Terminal bacteroid differentiation in the Medicago–rhizobium interaction–a tug of war between plant and bacteria. In: de Bruijn F (ed) The model legume *Medicago truncatula*. https://doi.org/10.1002/9781119409144.ch75

Hayden HL, Savin KW, Wadeson J, Gupta VV, Mele PM (2018) Comparative metatranscriptomics of wheat rhizosphere microbiomes in disease suppressive and non-suppressive soils for *Rhizoctonia solani* AG8. Front Microbiol 9:859

Ishiguro K, Green T, Rapley J, Wachtel H, Giallourakis C, Landry A, Daly MJ (2006) Ca2+/calmodulin-dependent protein kinase II is a modulator of CARMA1-mediated NF-κB activation. Mol Cell Biol 26(14):5497–5508

Jeyanthi V, Kanimozhi S (2018) Plant growth promoting Rhizobacteria (PGPR)–prospective and mechanisms: a review. J Pure Appl Microbiol 12(2):733–749

Kim Y, Liesack W (2015) Differential assemblage of functional units in paddy soil microbiomes. PLoS One 10(4):e0122221. https://doi.org/10.1371/journal.pone.0122221

Kim Y, Wegner CE, Liesack W (2014) Soil metatrascriptomic. In: Nannipieri P, Pietramellara G, Renella G (eds) Omics in soil science. Caister Academic Press, Norfolk, pp 63–93

Kour D, Rana KL, Yadav N, Yadav AN, Kumar A, Meena VS, Saxena AK (2019) Rhizospheric microbiomes: biodiversity, mechanisms of plant growth promotion, and biotechnological applications for sustainable agriculture. In: Plant growth promoting Rhizobacteria for agricultural sustainability. Springer, Singapore, pp 19–65

Lee MH, Hong KS, Malhotra S, Park JH, Hwang EC, Choi HK, Lee SW (2010) A new esterase EstD2 isolated from plant rhizosphere soil metagenome. Appl Microbiol Biotechnol 88 (5):1125–1134

Li X, Qin L (2005) Metagenomics-based drug discovery and marine microbial diversity. Trends Biotechnol 23(11):539–543

Lu GH, Hua XM, Liang L, Wen ZL, Du MH, Meng FF, Yang YH (2018) Identification of major rhizobacterial taxa affected by a glyphosate-tolerant soybean line via shotgun metagenomic approach. Genes 9(4):214

Luo B, Gu W, Zhong J, Wang Y, Zhang G (2015) Revealing crosstalk of plant and fungi in the symbiotic roots of sewage-cleaning *Eichhornia crassipes* using direct de novo metatranscriptomic analysis. Sci Rep 5:15407

Mendes R, Garbeva P, Raaijmakers JM (2013) The rhizosphere microbiome: significance of plant beneficial, plant pathogenic, and human pathogenic microorganisms. FEMS Microbiol Rev 37 (5):634–663

Mhlongo MI, Piater LA, Madala NE, Labuschagne N, Dubery IA (2018) The chemistry of plant–microbe interactions in the rhizosphere and the potential for metabolomics to reveal signaling related to defense priming and induced systemic resistance. Front Plant Sci 9:112

Michelsen CF, Stougaard P (2011) A novel antifungal Pseudomonas fluorescens isolated from potato soils in Greenland. Curr Microbiol 62(4):1185–1192

Muller EE, Faust K, Widder S, Herold M, Arbas SM, Wilmes P (2018) Using metabolic networks to resolve ecological properties of microbiomes. Curr Opin Syst Biol 8:73–80

Nazir R, Zaffar MR, Amin I (2019) Bacterial biofilms: the remarkable heterogeneous biological communities and nitrogen fixing microorganisms in lakes. In: Freshwater microbiology. Academic, Cambridge, pp 307–340

Newman MM, Lorenz N, Hoilett N, Lee NR, Dick RP, Liles MR, Kloepper JW (2016) Changes in rhizosphere bacterial gene expression following glyphosate treatment. Sci Total Environ 553:32–41

O'Banion BS, O'Neal L, Alexandre G, Lebeis SL (2020) Bridging the gap between single-strain and community-level plant-microbe chemical interactions. Mol Plant-Microbe Interact 33 (2):124–134

Pantigoso HA, Manter DK, Vivanco JM (2020) Differential effects of phosphorus fertilization on plant uptake and rhizosphere microbiome of cultivated and non-cultivated potatoes. Microb Ecol 80:169–180

Papenfort K, Bassler BL (2016) Quorum sensing signal–response systems in gram-negative bacteria. Nat Rev Microbiol 14(9):576

Penterman J, Abo RP, De Nisco NJ, Arnold MF, Longhi R, Zanda M, Walker GC (2014) Host plant peptides elicit a transcriptional response to control the Sinorhizobium meliloti cell cycle during symbiosis. Proc Natl Acad Sci U S A 111(9):3561–3566

Picard C, Di Cello F, Ventura M, Fani R, Guckert A (2000) Frequency and biodiversity of 2, 4-diacetylphloroglucinol-producing bacteria isolated from the maize rhizosphere at different stages of plant growth. Appl Environ Microbiol 66(3):948–955

Pop M, Phillippy A, Delcher AL, Salzberg SL (2004) Comparative genome assembly. Brief Bioinform 5(3):237–248

Ramachandran VK (2008) Microarray analysis of rhizobium leguminosarum bv. Viciae 3841 colonization of the rhizosphere. Doctoral dissertation, University of Reading

Ryan RP, Germaine K, Franks A, Ryan DJ, Dowling DN (2008) Bacterial endophytes: recent developments and applications. FEMS Microbiol Lett 278(1):1–9

Schenk PM, Carvalhais LC, Kazan K (2012) Unraveling plant–microbe interactions: can multi-species transcriptomics help? Trends Biotechnol 30(3):177–184

Seitz KW, Lazar CS, Hinrichs KU, Teske AP, Baker BJ (2016) Genomic reconstruction of a novel, deeply branched sediment archaeal phylum with pathways for acetogenesis and sulfur reduction. ISME J 10(7):1696–1705

Sharma R, Sharma PK (2018) Metatranscriptome sequencing and analysis of agriculture soil provided significant insights about the microbial community structure and function. Ecol Genet Genom 6:9–15

Sharma PK, Sharma V, Sharma S, Bhatia G, Singh K, Sharma R (2019a) Comparative metatranscriptome analysis revealed broad response of microbial communities in two soil types, agriculture versus organic soil. J Genet Eng Biotechnol 17(1):6

Sharma R, Bisaria VS, Sharma S (2019b) Rhizosphere: a home for human pathogens. In: Plant biotic interactions. Springer, Cham, pp 113–127

Simon C, Daniel R (2011) Metagenomics analyses: past and future trends. Appl Environ Microbiol 77:1153–1161

Singh J, Singh P, Ray S, Rajput RS, Singh HB (2019) Plant growth-promoting Rhizobacteria: benign and useful substitute for mitigation of biotic and abiotic stresses. In: Plant growth promoting Rhizobacteria for sustainable stress management. Springer, Singapore, pp 81–101

Spence C, Alff E, Johnson C, Ramos C, Donofrio N, Sundaresan V, Bais H (2014) Natural rice rhizospheric microbes suppress rice blast infections. BMC Plant Biol 14(1):130

Sugawara M, Epstein B, Badgley BD, Unno T, Xu L, Reese J, Farmer AD (2013) Comparative genomics of the core and accessory genomes of 48 Sinorhizobiumstrains comprising five genospecies. Genome Biol 14(2):R17

Taylor LL, Banwart SA, Valdes PJ, Leake JR, Beerling DJ (2012) Evaluating the effects of terrestrial ecosystems, climate and carbon dioxide on weathering over geological time: a global-scale process-based approach. Philos Trans Royal Soc B Biol Sci 367(1588):565–582

Thijs S, Sillen W, Rineau F, Weyens N, Vangronsveld J (2016) Towards an enhanced understanding of plant–microbiome interactions to improve phytoremediation: engineering the metaorganism. Front Microbiol 7:341

Trumbore S (2006) Carbon respired by terrestrial ecosystems–recent progress and challenges. Glob Chang Biol 12(2):141–153

Turner TR, Ramakrishnan K, Walshaw J, Heavens D, Alston M, Swarbreck D, Poole PS (2013) Comparative metatranscriptomics reveals kingdom level changes in the rhizosphere microbiome of plants. ISME J 7(12):2248–2258

Vega NM, Gore J (2018) Simple organizing principles in microbial communities. Curr Opin Microbiol 45:195–202

Verma A, Kumar S, Kumar G, Saini JK, Agrawal R, Satlewal A, Ansari MW (2018) Rhizosphere metabolite profiling: an opportunity to understand plant-microbe interactions for crop improvement. In: Crop improvement through microbial biotechnology. Elsevier, Amsterdam, pp 343–361

Wang R, Wang Y, Yang Q, Kang C, Li M (2018) Unraveling the characteristics of the microbial community and potential pathogens in the rhizosphere soil of Rehmannia glutinosa with root rot disease. Appl Soil Ecol 130:271–279

White RA III, Rivas-Ubach A, Borkum MI, Köberl M, Bilbao A, Colby SM, Hixson KK (2017) The state of rhizospheric science in the era of multi-omics: a practical guide to omics technologies. Rhizosphere 3:212–221

Yergeau E, Tremblay J, Joly S, Labrecque M, Maynard C, Pitre FE, Greer CW (2018) Soil contamination alters the willow root and rhizosphere metatranscriptome and the root–rhizosphere interactome. ISME J 12(3):869–884

Zaccaria M, Dedrick S, Momeni B (2017) Modeling microbial communities: a call for collaboration between experimentalists and theorists. Processes 5(4):53

Zhang F, Shen J, Zhang J, Zuo Y, Li L, Chen X (2010) Rhizosphere processes and management for improving nutrient use efficiency and crop productivity: implications for China. In: Advances in agronomy. Academic, Cambridge, pp 1–32

Chapter 4
"Omics" Approaches for Understanding Soil Suppressiveness in Agriculture

Shivani Khatri, Annapurna Bhattacharjee, and Shilpi Sharma

Abstract Disease suppressive soils are instrumental in combating phytopathogens, which cause severe diseases in plants. They provide an eco-friendly way to mitigate biotic stresses in agricultural systems worldwide. Efforts have been made to understand the mechanisms of disease suppressiveness in soil using culture-dependent and culture-independent methods. Omics approaches have provided useful insights into the key markers responsible for imparting antagonism against soil-borne plant pathogens. Specific genera such as *Pseudomonas*, *Bacillus*, *Streptomyces*, *Lysobacter*, and *Trichoderma*, along with antibiotics and siderophores, are the key constituents in disease suppressive soils. Disease suppressive potential of a soil is dependent on several factors such as soil pH and soil type, with one of the critical factors being the type of nutrient amendment applied to the soil. While the underlying mechanisms of growth inhibition of specific fungal pathogens such as *Fusarium* and *Rhizoctonia solani* in soil ecosystem have been well elucidated, there is restricted knowledge regarding "general-disease suppression". The mechanisms responsible for imparting broad range suppressiveness can help us develop economically favourable agricultural management practices. In this chapter, we have critically reviewed significant investigations related to specific- and general-disease suppression where omics-based approaches have been adopted to study the microbial community dynamics of disease suppressive soils. Based on the reported studies, we have identified the potential role of diverse markers, compost amendments, and different microbial strains (producing key metabolites) in disease suppressiveness of soil. Thus, we propose that using different molecular and microbial markers, mapping of disease suppressive soils can efficiently be done across the globe. In addition, the effectiveness of synthetic microbial communities, and possibilities of transforming conducive soil by microbiome transfer from suppressive soil may be explored in the context of disease suppressiveness in the future.

Shivani Khatri and Annapurna Bhattacharjee contributed equally to this work.

S. Khatri · A. Bhattacharjee · S. Sharma (✉)
Department of Biochemical Engineering and Biotechnology, Indian Institute of Technology Delhi, New Delhi, India
e-mail: shilpi@dbeb.iitd.ac.in

R. N. Pudake et al. (eds.), *Omics Science for Rhizosphere Biology*, Rhizosphere Biology, https://doi.org/10.1007/978-981-16-0889-6_4

Keywords Disease suppressive soils · Molecular markers · Microbial markers · Sustainable agriculture

4.1 Introduction

Soil-borne pathogens cause severe plant diseases and detrimentally impact agricultural productivity, worldwide. Interestingly, to naturally combat this challenge, various microbes present in plant's rhizosphere mediate disease suppressiveness by producing several crucial molecules like siderophores, volatiles/secondary metabolites, and antifungal agents in soil, which impart suppressiveness to specific pathogens (Weller et al. 2002; Joshi et al. 2009). Notably, soils exhibiting greater microbial diversity have been reported to have lower incidences of disease development in plants, majorly due to the competitive interactions among rhizospheric soil microbiota and plant pathogens (Weller et al. 2002; Penton et al. 2014). Several soil-borne pathogens, e.g. *Verticillium dahliae*, *Fusarium* spp., *Pythium* spp., *Sclerotium* spp., *Rhizoctonia solani*, *Pythopthora* spp., *Ralstonia solanacearum*, and *Erwinia carotovora*, are negatively impacted by microorganisms present in disease suppressive soils (Bonanomi et al. 2018). There are several examples of soils exhibiting disease suppressiveness. One of the widely studied examples is the suppression of *F. oxysporum* by *Pseudomonas* species, which produces pyoverdin, a potent secondary metabolite (Alabouvette 1999; Siegel-Hertz et al. 2018). Further, many studies have reported that organic amendments like compost has led to improvement in disease suppressive capacity of soils (Bonanomi et al. 2007; Postma et al. 2008; Vida et al. 2016; De Corato et al. 2018; Mehta et al. 2018). Interestingly, in another study, to decipher the mechanism of disease suppression by fluorescent pseudomonads, the *phlD* locus encoding the antibiotic 2,4-diacetylphloroglucinol (DAPG) was determined and quantified by qPCR. Interestingly, the expression of genes at *phlD* locus was found to be highly correlated with disease suppressive capability of pseudomonads in amended compost (Hunjan et al. 2017).

Disease suppressiveness of soils can be classified into two types, namely general-disease suppressiveness and specific-disease suppressiveness. ***General-disease suppressiveness*** of soils develop due to combinatorial action of several microorganisms present in soil, and can be enhanced by organic matter amendments (Weller et al. 2002; Bonanomi et al. 2010; Bonilla et al. 2012; Tomihama et al. 2016). So far, general-soil suppressiveness has not been reported to be transferrable from one region to another (Weller et al. 2002). Contrastingly, ***specific-disease suppressiveness*** of soils occurs as a result of activities of a particular microbial population, which affects the pathogen at a specific stage of its life cycle (Weller et al. 2002; Gómez Expósito et al. 2017). Specific-soil suppressiveness is effectively inheritable when a defined amount (1–10% w/w) of suppressive soil is mixed with conducive soil (which has high disease incidence) but the property is lost if soil is subjected to pasteurization or fumigation in most cases (Weller et al. 2002; Mendes et al. 2011).

Precisely, general- and specific-soil suppressiveness can be compared to innate and adaptive immunity, respectively, in animals (Raaijmakers and Mazzola 2016). Specific-soil suppressiveness can further be categorized into long-standing and induced soil suppressiveness. Long-standing soil suppressiveness is a trait inherently present in specific soils, having an unknown origin, and the property is retained even in pathogen's absence (Weller et al. 2002). Soils in Chateaurenard region of France have been shown to exhibit *Fusarium* wilt suppressiveness in various plants, and this remains a prime example of long-standing soil suppressiveness (Alabouvette 1986). Induced soil suppressiveness, on the other hand, develops as a result of crop monoculture (e.g. take-all decline disease suppressiveness induced by wheat or barley monoculture), the addition of pathogen as inoculum to soil, or by exposure of susceptible plants to the soil-borne pathogens (Weller et al. 2002). The effect of suppressive soil microbial community is found to be highly variable and specific to each host plant (Chet et al. 1990; Berg et al. 2000).

Numerous investigations have been carried out in the context of disease suppressive soils across the world (Weller et al. 2002; Schlatter et al. 2017; Gómez Expósito et al. 2017). However, the mechanisms responsible for general- as well as specific-disease suppression still remain to be comprehensively studied (Schlatter et al. 2017). Primarily, in an attempt to understand the basis of soil suppressiveness, traditional microbiological approaches like cultivation-dependent techniques (enumeration-based methods) and biocontrol assays have been employed (Schlatter et al. 2017; Gómez Expósito et al. 2017). However, the limitation of routinely used cultivation-dependent methods is that only about 1–10% of the microbiome is culturable, and information about majority of the unculturable microorganisms remains elusive. Consequently, it has not been possible to decipher the exact basis of soil suppressiveness, by relying only on cultivation-dependent techniques. Thus, to circumvent this problem, in recent decades, molecular techniques like restriction fragment length polymorphism (RFLP) (Gu and Mazzola 2003), denaturing gradient gel electrophoresis (DGGE) (Cretoiu et al. 2013; Chng et al. 2015), quantitative PCR (qPCR) (Hunjan et al. 2017), and omics-based approaches like 16S rRNA amplicon sequencing (Rosenzweig et al. 2012; Yin et al. 2013), metatranscriptomics (Chapelle et al. 2016), metabolomics (Cha et al. 2016), etc. have been used to identify key microbial members responsible for disease suppression, and to gain deeper insights into the mechanisms of soil disease suppressiveness (Gómez Expósito et al. 2017; Carrión et al. 2018; Hayden et al. 2019; Ossowicki et al. 2020).

In this chapter, we focus exclusively on different omics approach-based studies, which have endeavoured to understand the mechanisms of general- and specific-soil suppressiveness in the last two decades. Interestingly, during this period, people have increasingly adopted omics-based approaches to study soil suppressiveness (Fig. 4.1). Based on the reported investigations, we also provide a global overview of the incidences of disease suppressiveness across the world (Fig. 4.2). Further, within the purview of this chapter, we specifically review the findings of some interesting case studies, and enlist the details of the research studies (conducted during 2000–2020) related to general- and specific-soil suppressiveness (Table 4.1).

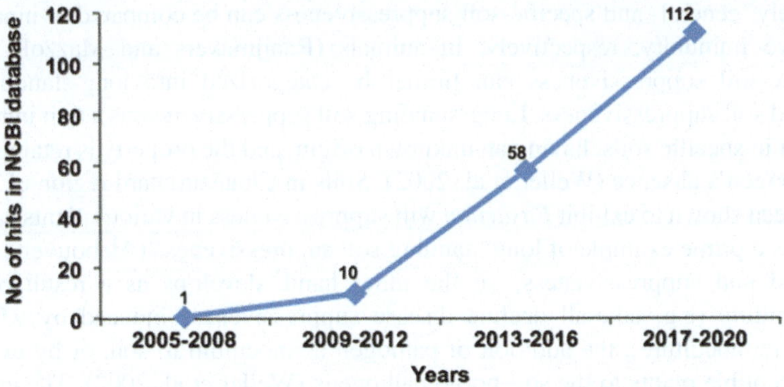

Fig. 4.1 "Omics"-based approaches for understanding soil disease suppressiveness. The number of hits (articles) is the results of searches with keywords "omics" and "disease suppressive soils" in the NCBI database

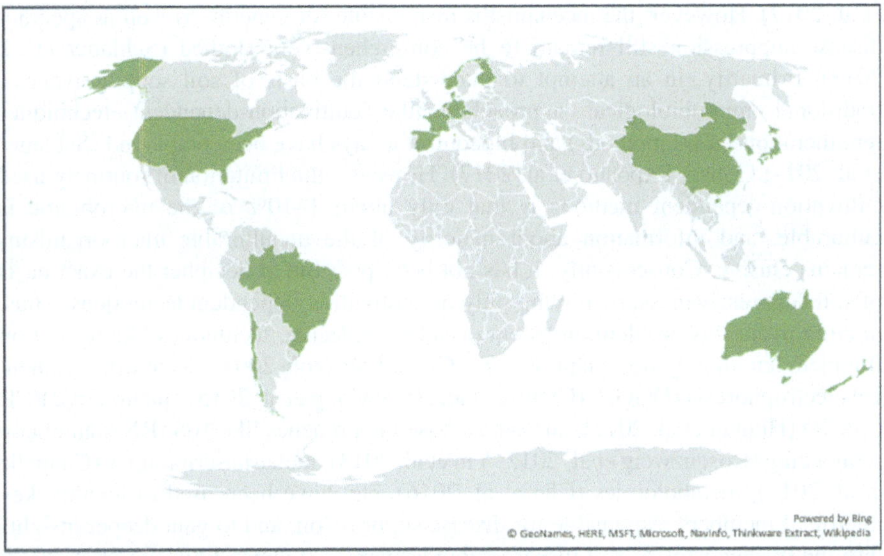

Fig. 4.2 Global overview of disease suppressive soils identified across the world. The green shaded regions represent the countries that have reported evidence of soil suppressiveness against different pathogens. The data is based on literature published in NCBI till 2019

Eventually, we explore further possibilities which can be vital for executing sustainable disease suppression programmes in agricultural systems, based on our current understanding of the subject.

Table 4.1 Overview of disease suppressiveness reported in agriculture by employing various approaches

S. no.	Pathogen(s)	Host plant	Remarks including type of approach	References
A: General suppressiveness				
			Non-omics-based approach	
1	*Pythium ultimum, Rhizoctonia solani, Phytophthora* spp., *F. oxysporum and Verticillium dahliae*	Turf grass	Bioassays Compost-based amendment induced disease suppressive ability	Noble and Roberts (2004)
2	*Rhizoctonia solani, Verticillium longisporum, Streptomyces scabies*	Sugarbeet	Bioassays Growth suppression of pathogens was the result of specific- and general-disease suppression caused by characteristic antagonistic microbial community Isolated two species, *Lysobacter* sp. and *Streptomyces* sp. exhibiting disease suppression	Postma et al. (2008)
3	*Fusarium oxysporum, Pythium aphanidermatum, Rhizoctonia solani, Pythium debaryanum*	Tomato	Plant pathogenicity assay Significant reduction of pathogen infestation was observed in compost-treated pot experiments	Dukare et al. (2011)
4	*Rhizoctonia solani, Sclerotinia minor, Fusarium oxysporum, Phytophthora cinnamomi, Pythium irregulare*	Cucumber, tomato, bean, melon	In vitro and in vivo assays Significant reduction in the growth of mycelia of all pathogenic fungi under study Characteristic antagonistic microbial members associated with compost-treated soil belonged to *Trichoderma, Aspergillus, Pseudomonas,* and *Actinomycetes*	De Corato et al. (2018)
5	*Pythium ultimum, Sclerotinia minor, Rhizoctonia minor*	Cress	Bioassay for damping-off disease, and quantitative estimation of antifungal enzymes Observed higher fungal and bacterial diversity with increased activity of enzymes (glucanases, chitobiosidase and N-acetyl-glucosaminidase) reported to exhibit antagonistic activity against wide range of fungal pathogens in compost-treated soil compared to conventional farming.	Pane et al. (2011)

(continued)

Table 4.1 (continued)

S. no.	Pathogen(s)	Host plant	Remarks including type of approach	References
			Omics-based approach	
6	*Rhizoctonia solani AG3*, *Plasmodiophora brassicae*, *Fusarium*	Different plants	METACONTROL-metagenomics project (Trans-European project) Metagenomic analysis of diverse suppressive soil samples revealed the presence of antibiotic biosynthetic clusters in microbial community prevailing in different soils	van Elsas et al. (2008)
7	*Rhizoctonia solani, Pythium ultimum, Fusarium oxysporum, Aspergillus niger*	Sugarbeet and cauliflower	16S rRNA-based sequencing of all strains and in vitro antifungal assay Isolated several *Lysobacter* strains from disease suppressive soil and their in vitro antagonistic activities were confirmed against all selected pathogens	Gómez Expósito et al. (2015)
B: Specific suppressiveness				
1	*Rhizoctonia solani*		**Non-omics-based approach**	
		French bean	Culture-based quantification of biocontrol genera form compost amended soil *Trichoderma* spp. and fluorescent pseudomonads were responsible for growth suppression of the pathogen	Joshi et al. (2009)
			Omics-based approach	
		Apple and wheat	RFLP analysis of 16S rRNA gene Change in population of fluorescent pseudomonads by changing the type of amendments and intercropping pattern	Gu and Mazzola (2003)
		Pine	In vitro pot study, terminal RFLP (T-RFLP) analysis of 16S and 18S rRNA genes Type of compost determined the structural and functional diversity of soil microflora with disease suppressive ability	Pérez-Piqueres et al. (2006)
		Sugarbeet	Using PhyloChip-based metagenomic, crucial	Mendes et al. (2011)

(continued)

Table 4.1 (continued)

S. no.	Pathogen(s)	Host plant	Remarks including type of approach	References
			bacterial taxa were identified from sugarbeet rhizospheric microbiome, which can cause disease suppression of *Rhizoctonia solani*, the fungal root pathogen	
			Metagenomics and metatranscriptomics Analysis indicated that specific bacterial family members like *Oxalobacteriaceae*, *Burkholderiaceae,* etc. from sugarbeet rhizospheric suppress *R. solani* due to upregulation of stress-related genes	Chapelle et al. (2016)
			16S rRNA PhyloChip-based analysis Heat perturbations negatively impacted disease suppressiveness of soil by altering the rhizobacterial community composition	van der Voort et al. (2016)
			Comparative genomics and metabolomics analysis upon isolation of *Burkholderiaceae* strains, and their genome sequencing, LC/MS, and in vitro and in vivo antifungal assays Studied the antagonistic activity of members belonging to family *Burkholderiaceae* producing sulphurous volatile molecules with antifungal activities in soil, and deciphered their role in disease suppressiveness	Carrión et al. (2018)
		Wheat	454 pyrosequencing of 28S LSU gene sequencing revealed differences in fungal community structures of suppressive and non-suppressive soils About 40 genera possessing pathogen suppressiveness were identified in suppressive soils	Penton et al. (2014)

(continued)

Table 4.1 (continued)

S. no.	Pathogen(s)	Host plant	Remarks including type of approach	References
			Metabolomics and metatranscriptomic analysis Identified Macrocarpal, a potential disease suppression marker from *R. solani* AG8 suppressive soil and revealed diverse patterns of metabolomic profiles among suppressive and non-suppressive soils	Hayden et al. (2019)
2	*Fusarium oxysporum*		**Omics-based approach**	
		Flax	Isolation of phenazine and DAPG producers done on KMB agar, genetic diversity of phenazine and DAPG producers assessed by RFLP and BOX PCR analysis of *phzC* and *phlD* genes Observed increased levels of phenazines and DAPG in naturally occurring disease suppressive soil	Mazurier et al. (2009)
		Strawberry	Induction of resistance by cropping susceptible crop, pyrosequencing of disease suppressive soil, genome mining of specific isolates and their LC/MS profiling Korean soil had disease suppressive effects due to antifungal molecules produced by *Streptomyces* and some members of *Actinobacteria*. *Streptomyces* was solely responsible for inhibition of pathogen by production of novel antagonistic thiopeptide	Cha et al. (2016)
		Banana	16S rRNA and fungal ITS gene pyrosequencing Specific genera capable of mitigating infection were found to be present in disease-free soil in contrast to disease-infected soil	Zhou et al. (2019)
			Genomics analysis revealed that high-GC *Streptomyces* species possessed disease suppressive capacity and	Heinsch et al. (2017)

(continued)

Table 4.1 (continued)

S. no.	Pathogen(s)	Host plant	Remarks including type of approach	References
			mediated evolution of microbial communities	
3	*Phytophthora nicotianae*	Pepper	16S rRNA amplicon sequencing and metaproteomics Compost was found to suppress soil-borne pathogens like *Phytopthora nicotionae* in pepper plants. Analysis revealed phylogenetic and functional differences between suppressive and non-suppressive compost	Ros et al. (2018)
			Metagenomic and metabolomic analysis Specific metabolites produced by members of *Ascomycota* phylum, abundant in suppressive compost, play roles in disease suppression	Blaya et al. (2016)
4	*Thielaviopsis basicola*	Tobacco	16S rRNA-based taxonomic microarray The diversity pattern of rhizospheric *Pseudomonads* in Morens soil was different as compared to Savoie soil, which had low disease receptivity	Almario et al. (2013)
5	*Plasmodiophora brassicae*	Cabbage	Metagenomic analysis Chi18H8, the first chitinase to be identified from a metagenomic library of suppressive soil, was found to be antagonistic to various fungal pathogens	Hjort et al. (2014)
6	*Gaeumannomyces graminis var. tritici*	Wheat	16S rRNA amplicon sequencing Wheat cultivars impacted *Pseudomonas* genetic diversity in soil microbiome in the subsequent year of wheat cultivation	Mauchline et al. (2015)
7	*Rosellinia necatrix*	Avocado	Plant pathogenicity assay, 16S rRNA amplicon sequencing, GeoChip assay Composts amended with almond shells (rich in organic matter) comprised of strong biocontrol agents like	Vida et al. (2016)

(continued)

Table 4.1 (continued)

S. no.	Pathogen(s)	Host plant	Remarks including type of approach	References
			Proteobacteria which led to disease suppression A characteristic difference in the microbial community was found between compost amended soil compared to control. In compost amended soil, well-known antagonistic microbial members belonging to *Actinobacteria*, *Pseudomonas* and *Burkholderia* were present	
8	*Fusarium culmorum*	Wheat	16S rRNA amplicon sequencing Screening and metabolite analysis of field soils for suppressiveness against *Fusarium culmorum* in wheat revealed volatile emission from diverse bacteria taxa, majorly *Acidobacteria*, which could contribute to suppressiveness in various soil samples	Ossowicki et al. (2020)

4.2 Understanding General-Soil Suppressiveness by Employing Omics Approaches

To better understand the mechanism of general-disease suppressiveness, comprehensive studies are required using high-end molecular techniques in terms of identifying the key ingredients present in the suppressive soils. There are few studies attempting it using omics approaches. In this section, we will focus on some case studies related to deciphering the mechanisms of general-disease suppressiveness.

One of the significant studies in this aspect was the metagenomics analysis of trans-European soil (van Elsas et al. 2008). In this megaproject, the authors untapped the molecular mechanisms responsible for general-disease suppressiveness using soils from four European countries: France, Sweden, the UK, and the Netherlands, which have suppressive potential against *Fusarium*, *Plasmodiophora brassicae*, and *Rhizoctonia solani*. The crucial part was to analyse the unculturable fraction, along with genes related to the production of antibiotics belonging to polyketide groups and siderophores, by performing metagenomic analysis of DNA isolated from these soils. The results of this study are notable; they found 7 novel clones of polyketide synthase I, with one of the inserts exhibiting high similarity to *Acidobacterium* sp..

This is one of the few studies that assessed the active ingredients present in the general-suppressive soils across nations.

The type of soil amendment also has a role in imparting disease suppressive potential to the soil; organic amendments have specifically been reported to play a positive role. Pane et al. (2011) reported the impact of amendment of peat along with compost on disease suppressiveness against *Sclerotinia minor*, *Pythium aphanidermatum,* and *Rhizoctonia solani.* Up to 60% growth inhibition of all these phytopathogens was reported using these amendments, with maximum inhibition observed when animal manure was used as compost. Several other soil parameters have been observed related to disease suppressiveness such as activity of chitobiose, N-acetyl glucasaminidase, and antagonistic volatiles. The increased activity of selected hydrolytic enzymes and higher alkyl/O-alkyl ratios was observed in disease suppressive soils by amendment of compost.

The potential disease suppressive indicators in the soil can be the selective abundance of specific genera imparting antagonism against phytopathogens. The 16S rRNA-based amplicon sequencing of disease suppressive soils against *F. oxysporum, P. ultimum, R. solani, and A. niger* claimed that there is significantly higher population of *Lysobacter* in these soils, this genera is known to have antagonistic activity against broad host-range plant pathogens due to the production of a number of extracellular enzymes and other antifungal compounds. Based on metagenomic analysis, 18 *Lysobacter* strains isolated using specific media showed in vitro antifungal activity against the selected pathogens (Gómez Expósito et al. 2015). Thus, from this study, it was concluded that *Lysobacter*-based bioformulations can be used as a sustainable approach to control various plant diseases.

4.3 Understanding Specific-Soil Suppressiveness by Employing Omics Approaches

In this section, we will emphasize on some studies that have adopted different omics-based approaches to understand the mechanistic complexities of specific-soil suppressiveness in context of particular pathogens and their respective host plants.

4.3.1 Rhizoctonia solani

Rhizoctonia solani, a fungal root pathogen, is one of the most studied systems in the context of disease suppression in soil. In order to understand how the sugarbeet rhizospheric microbiome composition and associated activities get altered when grown in a suppressive soil proven to be effective against *R. solani* damping-off infection (Mendes et al. 2011), a metagenomics study (based on Illumina HiSeq

2000 platform) was conducted by Chapelle et al. (2016). Comparative analysis of the composition of rhizobacterial community in non-inoculated and *R. solani* inoculated sugarbeet rhizospheric sample revealed significant differences at taxonomic levels. Interestingly, an abundance of bacterial members belonging to *Oxalobacteraceae*, *Sphingomonadaceae*, *Sphingobacteriaceae,* and *Burkholderiaceae* families were witnessed in *R. solani* inoculated sugarbeet rhizospheric microbiome, in contrast to control. Many members from this family have known inhibitory effects against fungal infections (Johnsen et al. 2010). To gain further insights into the functional aspects of these abundant bacterial family members, metatranscriptomic analysis was carried out. Greater expression of transcripts related to oxidative stress response, in addition to transcripts governing HfrA/ secretion systems and guanosine-3,5-bis-pyrophosphate ((p)ppGpp) metabolism, also known as bacterial alarmone, related to stress regulation, was found in *R. solani* inoculated sugarbeet rhizospheric microbiome, vis-a-vis control. Based on evidences from their study, Chapelle et al. (2016) proposed a mechanistic model, and highlighted that during *R. solani* infection in sugarbeet grown in suppressive soil, oxalic acid, or related molecules are produced, which trigger the activation of abundantly present rhizospheric members from suppressive soil, belonging to *Oxalobacteraceae* or *Burkholderiaceae* and eventually lead to oxidative stress responses. This possibly results in the activation of several pathways in plants, which finally establish fungal disease resistance.

Chapelle et al. (2016) had reported the involvement of *Burkholderiaceae* family members in disease suppressiveness of *R. solani.* As an extension of this study, Carrión et al. (2018) employed omics approach to conduct a comprehensive investigation in order to delineate the mechanistic role of selected *Burkholderiaceae* family members in disease suppression. Based on growth on semi-selective media (*Pseudomonas cepacia* azelaic with and without acid tryptamine, PCAT and PCATm, respectively) and PCR analysis, a total of 29 isolates of *Burkholderia* and *Paraburkholderia* were found in suppressive soil analysed from sugarbeet rhizospheric region. Based on PhyloChip analysis (Mendes et al. 2011), classification of isolates was done into five species, namely *Paraburkholderia hospita*, *P. terricola*, *P. graminis*, *Burkholderia pyrrocinia,* and *P. caledonica*. Abundance of these species, majorly *P. graminis*, was found in *R. solani* suppressive soil. However, in vitro sandwich plate assay conducted to assess antifungal activity of strains mediated by volatile organic compounds (VOCs) revealed that only *P. graminis* potently inhibited *R. solani* infection by VOCs. Metabolite profiling of all identified isolates detected about 1084 putative VOCs. Comparison of VOC profiles of selected strains with *P. graminis* PHS1 (exhibiting hyphal inhibition by VOCs in vitro) found 158 significantly different VOCs produced by *P. graminis* PHS1. Representation of sulphurous VOCs was found to be high among the significantly different VOCs of *P. graminis* PHS1. Further, comparative genomics analysis of five species upon sequencing (on Pacific Biosciences {PacBio} RS sequencing platform) revealed unique genes in *P. graminis* PHS1, involved in production of sulphurous VOCs. Overall, this study revealed that selected *Burkholderiaceae* members were responsible for disease suppressiveness of soil analysed from rhizospheric region of sugarbeet, and possible mode of antifungal activity was due to production of sulphurous VOCs.

In another study, van der Voort et al. (2016) employed 16S rRNA-microarray approach to decipher the effect of heat perturbations on changes in sugarbeet rhizobacterial community compositions and in turn suppressive potential of soil against damping-off disease caused by *R. solani*. PhyloChip data analysis for 16S rRNA gene was carried out using the data generated by Mendes et al. (2011) by employing customized tool, Phyloprofiler, developed in an open web-based platform named Galaxy (Blankenberg et al. 2010; Goecks et al. 2010). The study revealed that disturbances caused due to heat treatment for short duration (50/80 °C for 1 h) resulted in enhanced alpha diversity in rhizospheric microbiome and even led to abolishment of disease suppressive trait of soil. Members of bacterial families, viz. *Cornamonadaceae*, *Bacillaceae,* and *Paenibacillaceae* were found to be dominantly responsive to heat disturbances, whereas *Actinobacteria* and *Acidobacteria* members exhibited variable responses to heat perturbations. Based on this study, a rhizobacterial community reassembly model was put forth, whereby heat perturbation was found to accentuate growth of heat-tolerant and fast growing bacterial members, in contrast to the temperature-sensitive bacterial members, which possibly play roles in soil disease suppressiveness. Thus, heat treatment was found to restructure the existing sugarbeet rhizospheric bacterial communities, and eventually disrupt disease suppressive capacity of soil.

Rhizoctonia solani anastomosis group (AG) 8, causing root rot and bare patch disease is a major menace for farmers as it adversely affects crop productivity of cereals like wheat and barley. Suppression of *R. solani* AG8 was reported in a farmer's field at Avon, South Australia. A pioneering study carried out by Hayden et al. (2019) exemplified the implementation of environmental metabolomics in understanding the differences between *R. solani* suppressive and conducive soils present in wheat and barley cropping regions of Avon. By conducting metabolic profiling at different cropping stages of both wheat and barley, by untargeted nuclear magnetic resonance (NMR) performed on Bruker Avance III 700 MHz spectrometer and high performance liquid chromatography-mass spectroscopy (HPLC-MS), conducted in Agilent series 1290 HPLC system, stark differences in *R. solani* suppressive and conducive soils could be detected. The soil extracts of *R. solani* suppressive soils were richer in lipids, sugars, and terpenes, as compared to conducive soils. 2D-NMR analysis identified glucose as the key carbohydrate molecule present in *R. solani* suppressive soils. LC-MS analysis further highlighted that levels of metacarpal metabolites, the possible biomarkers, were prominently enhanced in *R. solani* suppressive soils. Thus, this study firstly reported the approach of untargeted metabolomics which could enable distinction between conducive and suppressive soils, in context of *R. solani* AG8. They also identified macrocarpals as the naturally occurring secondary metabolites, having antimicrobial property; their elevated levels in soil can possibly be correlated with disease suppression.

4.3.2 Fusarium oxysporum

Fusarium spp. are among the most detrimental pathogens, which negatively affect a broad range of plant hosts. *F. oxysporum* f. sp. *fragariae* causes wilt disease of strawberry, which is difficult to control. Cha et al. (2016) identified in close proximity a wilt-suppressive and conducive area in strawberry fields situated in Sancheong County, Korea. Using traditional and community-based approaches, the biological basis of *Fusarium* wilt disease suppression could be deciphered. Next-generation sequencing on Roche GX-FLX plus platform followed by analysis revealed variations in rhizobacterial communities among the wilt-suppressive and conducive samples taken from strawberry rhizosphere. *Fusarium* wilt-suppressive soil had prominent representation of *Proteobacteria*, *Firmicutes*, followed by *Actinobacteria* and *Acidobacteria*. Among them, *Actinobacteria* members were slow growers and exhibited tolerance to heat treatment, besides being abundantly present in *Fusarium* wilt-suppressive soil. Isolation of *Actinobacteria* members from *Fusarium* wilt-suppressive soil was done using selective media and their antagonistic activity was assessed in vitro. Strain S4-7, identified as *Streptomyces* sp., was found to exhibit potent biocontrol property against *Fusarium* wilt of strawberry. Genome mining enabled the identification of an effective thiopeptide antibiotic, produced by *Streptomyces* sp., with possible role in *Fusarium* wilt disease suppression. Further, HPLC-HR-ESI-MS/MS analysis revealed conprimycin and ectoine as active components of thiopeptide, from *Streptomyces* sp., to be responsible for antifungal activity. The mode of action of thiopeptide was found to be against fungal cell wall synthesis, as revealed by a chemogenomics study, carried out in yeast system.

4.3.3 Phytophthora nicotianae

P. nicotianae is the causative agent of root rot, crown rot, fruit rot, leaf-and-stem infection across diverse host plants including onion, tobacco, tomato, pepper, citrus plants, etc. In a recent investigation, Ros et al. (2018) used omics approach to delineate the suppressive effects of compost samples against *Phytophthora nicotiana* in pepper. Overall, compost samples exhibited disease suppressive effects; however, their effects had significant variations depending on the composition of composts. Analysis after 16S rRNA amplicon sequencing, performed on Illumina MiSeq platform, revealed the abundant existence of *Proteobacteria*, *Bacteroidetes,* and *Actinobacteria* in the microbial population of different compost samples analysed. Sequencing of ITS2 region resulted in identification of *Ascomycota*, *Basidiomycota,* and *Mucoromycota* as dominant representatives of fungal community in compost samples. By employing metaproteomics approach, about 367 bacterial proteins were identified in compost samples after LC-MS analysis. Among *Proteobacteria*, bacterial proteins primarily belonged to orders *Xanthomonadales*, *Pseudomonadales* and *Enterobacteriales* within *Gammaproteobacteria,* followed by order *Rhizobiales* within *Alphaproteobacteria* whereas *Betaproteobacteria* had abundance of proteins produced by order *Burkholderiales*. Among *Actinobacteria*, bacterial proteins produced by order *Micromonosporales* were found to be dominantly represented. Majority of bacterial proteins were found to be involved in carbohydrate transportation and metabolism, besides determination of cell wall structure in suppressive composts. Correlating the sequencing and metaproteomic data, insights could be gained into the dynamics of compost-mediated disease suppression in pepper plants.

In accordance to the reported omics-based studies related to general- and specific-disease suppressive soils, the representative molecular and microbial markers that can be used to characterize disease suppressive soils, having antagonistic activity against bacterial and fungal plant pathogens have been summarized in Table 4.2. Using various meta-tools to determine the presence of these markers in any given soil samples, mapping of disease suppressive potential can be easily done.

4.4 Conclusions and Future Perspectives

Over the past few decades, extensive investigations have been carried out in order to understand the intricacies of soil disease suppressiveness in context of various soil-borne pathogens and their respective host plants. With rapid technological advancements in recent times, researchers have been employing omics-based approaches to further delineate the biological basis governing soil suppressiveness across the world (Gómez Expósito et al. 2017; Schlatter et al. 2017). Insights have been gained pertaining to the roles of microbial population, exhibiting biocontrol activities in disease suppressive soils. Several molecular and microbial markers related to soil disease suppressiveness have already been identified. Role of compost in disease suppression has also been recognized. Efforts have been made to systematically characterize metabolites/proteins that impart disease suppressiveness in soil and thereby interpret the biological basis of their mode of action.

It is important to devise new strategies that can ensure effective disease management practices in agricultural systems. Emerging approach in this direction is the transfer of disease suppressive soils or microbiome transfer to establish biological control of diseases across different regions of the world. Successful application of microbiome transfer has been evidenced in the case of potato common scab, black root rot disease of tobacco as well as damping-off disease in sugarbeet (Foo et al. 2017). Further, the potential of synthetic microbial communities in mediating soil disease suppressiveness against soil-borne pathogens may be assessed. Application of high throughput technologies, coupled with exhaustive metadata analysis, can be explored to further understand the complexities of microbial interactions, which govern disease suppressiveness of diverse soils. While disease suppressive capacity of soils have been assessed for major soil-borne pathogens like *Rhizoctonia, Gaeumannomyces,* and *Fusarium* spp., it is important to decipher the disease suppressive dynamics of lesser studied pathogens also, as disease incidences are

Table 4.2 Molecular and microbial markers for characterization of disease suppressive soils

A. Molecular markers associated with disease suppressive soils				
S. no.	**Marker**	**Type of compound**	**Mechanism of antagonism**	**References**
1	Zopfiellin	Cyclooctanoid	Interrupts the metabolic pathway related to oxaloacetate synthesis	Daferner et al. (2002)
2	Methyl 2-methylpentanoate, 1,3,5-trichloro-2-methoxy benzene	Volatiles	Interrupts cell signalling responsible for fungal growth	Cordovez et al. (2015)
3	Iturin-A, Surfactin	Lipopeptides	Amphipathic molecules disrupt cell integrity	Yokota and Hayakawa (2015)
4	Nunamycin and Nunapeptin	Non-ribosomal peptides	Amphipathic molecules disrupt cell integrity	Michelsen et al. (2015)
5	2,4-diacetylphloroglucinol	Phenol	Disrupts cell integrity and membrane potential	De Souza et al. (2003)
6	Chitinase	Hydrolytic enzyme	Hydrolysis of fungal cell wall	Hjort et al. (2009)
7	Glucanase	Hydrolytic enzyme	Damages cell wall structure of fungal cells	O'Kennedy et al. (2011)
8	Phenazine 1-carboxylic acid	Aromatic carboxylic acid	Disrupts cell integrity	Mavrodi et al. (2012)
9	Pyrrolnitrin	Pyrrole	Interrupts respiratory electron transport chain	Dignam et al. (2018)
10	Bacilysin	Non-ribosomal peptide	Inhibition of glucosamine synthetase	Dignam et al. (2018)
B. Microbial markers associated with disease suppressive soils				
S. no.	**Indicator microbes with biocontrol potential**		**Target(/s) pathogen(/s)**	**References**
1	*Sphingobacterium, Sphingobacteriaceae, Chaetomiaceae, Zopfiella, Lasiosphaeriaceae, Chryseobacterium soldanellicola*, different species of *Streptomyces, Burkholderia pyrrocinia, Paraburkholderia caledonica, P. graminis, P. hospita, P. terricola*		*Rhizoctonia solani*	Liu et al. (2019) Yin et al. (2013) Cordovez et al. (2015) Carrión et al. (2019)
2.	*Bacillus amyloliquefaciens, Bacillus subtilis, Pseudomonas brenneri, Pseudomonas protegens*		*Alterneria solani, F. solani, Pectobacterium carotovorum, Phytophthora infestans, Rhizoctonia solani*	Caulier et al. (2018)
3.	*Acidobacteria, Planctomycetes, Nitrospirae, Chloroflexi, Chlorobi*		*Pythium intermedium*	van Agtmaal et al. (2015)
4.	*Pseudomonas* spp., *Burkholderia* spp., and *Actinobacteria*		*Rosellinia necatrix*	Vida et al. (2016)

(continued)

Table 4.2 (continued)

5.	*Kaistobacter, Actinobacteria, Acidobacteria, Xanthomonadacea*	*Fusarium oxysporum*	Liu et al. (2016)
6.	*Streptomyces* sp. GS93–23, *Streptomyces* sp. 3211–3, *Streptomyces* sp. S3–4	Different fungal plant pathogens	Heinsch et al. (2017)
7.	*Xylaria, Bionectria, Chaetomium, Corynascus* and *Microdiplodia* spp.	*Rhizoctonia solani* AG 8, *Fusarium pseudograminearum*	Penton et al. (2014)

often pathogen- and crop-specific. By amalgamating various omics approaches, it is essential to understand the mechanisms operative in different disease suppressive soils, so that it would be possible to develop suitable disease management practices that can mitigate different biotic stresses in plants. These endeavours can eventually improve crop productivity worldwide and establish sustainable agricultural systems.

Acknowledgement SK wishes to acknowledge the fellowship received from University Grant Commission, Government of India. AB acknowledges the award of National Post-Doctoral Fellowship from the Science and Engineering Research Board, Department of Science and Technology, Government of India (PDF/2018/001905). The authors would acknowledge the grants received from IIT Delhi under IRD-MFIRP scheme (MI02024), and Department of Biotechnology, Government of India (BT/PR27680/BCE/8/1434/2018).

References

Alabouvette C (1986) *Fusarium*-wilt suppressive soils from the Châteaurenard region: review of a 10-year study. Agronomie 6:273–284

Alabouvette C (1999) *Fusarium* wilt suppressive soils: an example of disease-suppressive soils. Australas Plant Path 28:57–64

Almario J, Kyselková M, Kopecký J et al (2013) Assessment of the relationship between geologic origin of soil, rhizobacterial community composition and soil receptivity to tobacco black root rot in Savoie region (France). Plant Soil 371:397–408

Berg G, Kurze S, Buchner A, Wellington EM, Smalla K (2000) Successful strategy for the selection of new strawberry-associated rhizobacteria antagonistic to *Verticillium* wilt. Can J Microbiol 46:1128–1137

Blankenberg D, Von Kuster G, Coraor N et al (2010) Chapter 19. Galaxy: a web-based genome analysis tool for experimentalists. In: Current protocols in molecular biology, vol 10. Wiley Interscience, Hoboken, pp 11–21

Blaya J, Marhuenda FC, Pascual JA, Ros M (2016) Microbiota characterization of compost using omics approaches opens new perspectives for *Phytophthora* root rot control. PLoS One 11: e0158048

Bonanomi G, Antignani V, Pane C, Scala F (2007) Suppression of soilborne fungal diseases with organic amendments. J Plant Pathol 89:311–340

Bonanomi G, Antignani V, Capodilupo M, Scala F (2010) Identifying the characteristics of organic soil amendments that suppress soilborne plant diseases. Soil Biol Biochem 42:136–144

Bonanomi G, Lorito M, Vinale F, Woo SL (2018) Organic amendments, beneficial microbes, and soil microbiota: toward a unified framework for disease suppression. Annu Rev Phytopathol 56:1–20

Bonilla N, Gutiérrez-Barranquero J, Vicente A, Cazorla F (2012) Enhancing soil quality and plant health through suppressive organic amendments. Diversity 4:475–491

Carrión VJ, Cordovez V, Tyc O et al (2018) Involvement of *Burkholderiaceae* and sulfurous volatiles in disease-suppressive soils. ISME J 12:2307–2321

Carrión VJ, Perez-Jaramillo J, Cordovez V et al (2019) Pathogen-induced activation of disease-suppressive functions in the endophytic root microbiome. Science 366:606–612

Caulier S, Gillis A, Colau G et al (2018) Versatile antagonistic activities of soil-borne *Bacillus* spp. and *Pseudomonas* spp. against *Phytophthora infestans* and other potato pathogens. Front Microbiol 9:143

Cha JY, Han S, Hong HJ et al (2016) Microbial and biochemical basis of a *Fusarium* wilt-suppressive soil. ISME J 10:119–129

Chapelle E, Mendes R, Bakker PA, Raaijmakers JM (2016) Fungal invasion of the rhizosphere microbiome. ISME J 10:265–268

Chet I, Ordentlich A, Shapira R, Oppenheim A (1990) Mechanism of biocontrol of soil-borne plant pathogens by rhizobacteria. Plant Soil 129:85–92

Chng S, Cromey M, Dodd S, Stewart A, Butler RC, Jaspers MV (2015) Take-all decline in New Zealand wheat soils and the microorganisms associated with the potential mechanisms of disease suppression. Plant Soil 397:239–259

Cordovez V, Carrion VJ, Etalo DW et al (2015) Diversity and functions of volatile organic compounds produced by *Streptomyces* from a disease-suppressive soil. Front Microbiol 6:1081

Cretoiu MS, Korthals GW, Visser JH, van Elsas JD (2013) Chitin amendment increases soil suppressiveness toward plant pathogens and modulates the *Actinobacterial* and *Oxalobacter aceal* communities in an experimental agricultural field. Appl Environ Microbiol 79:5291–5301

Daferner M, Anke T, Sterner O (2002) Zopfiellamides A and B, antimicrobial pyrrolidinone derivatives from the marine fungus *Zopfiellalatipes*. Tetrahedron 58:7781–7784

De Corato U, Salimbeni R, De Pretis A (2018) Suppression of soil-borne pathogens in container media amended with on-farm composted agro-bioenergy wastes and residues under glasshouse condition. J Plant Dis Protect 125:213–226

de Souza JT, Weller DM, Raaijmakers JM (2003) Frequency, diversity, and activity of 2, 4-diacetylphloroglucinol-producing fluorescent *Pseudomonas* spp. in Dutch take-all decline soils. Phytopathology 93:54–63

Dignam BE, O'Callaghan M, Condron LM, Kowalchuk GA, Van Nostrand JD, Zhou J, Wakelin SA (2018) Effect of land use and soil organic matter quality on the structure and function of microbial communities in pastoral soils: implications for disease suppression. PLoS One 13: e0196581

Dukare AS, Prasanna R, Dubey SC, Nain L, Chaudhary V, Singh R, Saxena AK (2011) Evaluating novel microbe amended composts as biocontrol agents in tomato. Crop Prot 30:436–442

Foo JL, Ling H, Lee YS, Chang MW (2017) Microbiome engineering: current applications and its future. Biotechnol J 12:1600099

Goecks J, Nekrutenko A, Taylor J (2010) Galaxy: a comprehensive approach for supporting accessible, reproducible, and transparent computational research in the life sciences. Genome Biol 11:R86

Gómez Expósito R, Postma J, Raaijmakers JM, De Bruijn I (2015) Diversity and activity of *Lysobacter* species from disease suppressive soils. Front Microbiol 6:1243

Gómez Expósito R, de Bruijn I, Postma J, Raaijmakers JM (2017) Current insights into the role of rhizosphere bacteria in disease suppressive soils. Front Microbiol 8:2529

Gu YH, Mazzola M (2003) Modification of fluorescent pseudomonad community and control of apple replant disease induced in a wheat cultivar-specific manner. Appl Soil Ecol 24:57–72

Hayden HL, Rochfort SJ, Ezernieks V, Savin KW, Mele PM (2019) Metabolomics approaches for the discrimination of disease suppressive soils for *Rhizoctonia solani* AG8 in cereal crops using 1H NMR and LC-MS. Sci Total Environ 651:1627–1638

Heinsch SC, Otto-Hanson L, Hsu SY, Kinkel L, Smanski MJ (2017) Genome sequences for *Streptomyces* spp. isolated from disease-suppressive soils and long-term ecological research sites. Genome Announc 5:e00493–e00417

Hjort K, Bergström M, Adesina MF, Jansson JK, Smalla K, Sjöling S (2009) Chitinase genes revealed and compared in bacterial isolates, DNA extracts and a metagenomic library from a phytopathogen-suppressive soil. FEMS Microbiol Ecol 71:197–207

Hjort K, Presti I, Elväng A, Marinelli F, Sjöling S (2014) Bacterial chitinase with phytopathogen control capacity from suppressive soil revealed by functional metagenomics. Appl Microbiol Biotechnol 98:2819–2828

Hunjan MS, Thakur A, Singh PP (2017) Identification and characterization of *Pseudomonas fluorescens* strains effective against *Xanthomonas oryzae* pv. *Oryzae* causing bacterial blight of rice in Punjab, India. J Appl Nat Sci 9:253–261

Johnsen MG, Hansen OC, Stougaard P (2010) Isolation, characterization and heterologous expression of a novel chitosanase from *Janthinobacterium* sp. strain 4239. Microb Cell Fact 9:5

Joshi D, Hooda KS, Bhatt JC, Mina BL, Gupta HS (2009) Suppressive effects of composts on soil-borne and foliar diseases of French bean in the field in the western Indian Himalayas. Crop Prot 28:608–615

Liu X, Zhang S, Jiang Q et al (2016) Using community analysis to explore bacterial indicators for disease suppression of tobacco bacterial wilt. Sci Rep 6:1–11

Liu L, Huang X, Zhao J, Zhang J, Cai Z (2019) Characterizing the key agents in a disease-suppressed soil managed by reductive soil disinfestation. Appl Environ Microbiol 85:e02992–e02918

Mauchline TH, Chedom-Fotso D, Chandra G et al (2015) An analysis of *Pseudomonas* genomic diversity in take-all infected wheat fields reveals the lasting impact of wheat cultivars on the soil microbiota. Environ Microbiol 17:4764–4778

Mavrodi DV, Mavrodi OV, Parejko JA et al (2012) Accumulation of the antibiotic phenazine-1-carboxylic acid in the rhizosphere of dryland cereals. Appl Environ Microbiol 78:804–812

Mazurier S, Corberand T, Lemanceau P, Raaijmakers JM (2009) Phenazine antibiotics produced by fluorescent *Pseudomonads* contribute to natural soil suppressiveness to *Fusarium* wilt. ISME J 3:977–991

Mehta CM, Pudake RN, Srivastava R, Palni U, Sharma AK (2018) Development of PCR-based molecular marker for screening of disease-suppressive composts against Fusarium wilt of tomato (*Solanum lycopersicum* L.). 3 Biotech 8(7):306

Mendes R, Kruijt M, de Bruijn I, Dekkers E, van der Voort M, Schneider JHM (2011) Deciphering the rhizosphere microbiome for disease-suppressive bacteria. Science 332:1097–1100

Michelsen CF, Watrous J, Glaring MA, Kersten R, Koyama N, Dorrestein PC, Stougaard P (2015) Nonribosomal peptides, key biocontrol components for *Pseudomonas fluorescens* In5, isolated from a Greenlandic suppressive soil. mBio 6:e00079–e00015

Noble R, Roberts SJ (2004) Eradication of plant pathogens and nematodes during composting: a review. Plant Pathol 53:548–568

O'Kennedy MM, Crampton BG, Lorito M et al (2011) Expression of a β-1, 3-glucanase from a biocontrol fungus in transgenic pearl millet. S Afr J Bot 77:335–345

Ossowicki A, Tracanna V, Petrus MLC et al (2020) Microbial and volatile profiling of soils suppressive to *Fusarium culmorum* of wheat. Proc R Soc B 287:20192527

Pane C, Spaccini R, Piccolo A, Scala F, Bonanomi G (2011) Compost amendments enhance peat suppressiveness to *Pythium ultimum, Rhizoctonia solani* and *Sclerotinia minor*. Biol Control 56:115–124

Penton CR, Gupta VV, Tiedje JM et al (2014) Fungal community structure in disease suppressive soils assessed by 28S LSU gene sequencing. PLoS One 9:e93893

Pérez-Piqueres A, Edel-Hermann V, Alabouvette C, Steinberg C (2006) Response of soil microbial communities to compost amendments. Soil Biol Biochem 38:460–470

Postma J, Schilder MT, Bloem J, van Leeuwen-Haagsma WK (2008) Soil suppressiveness and functional diversity of the soil microflora in organic farming systems. Soil Biol Biochem 40:2394–2406

Raaijmakers JM, Mazzola M (2016) Soil immune responses. Science 352:1392–1393

Ros M, Blaya J, Baldrian P, Bastida F, Richnow HH, Jehmlich N, Pascual JA (2018) *In vitro* elucidation of suppression effects of composts to soil-borne pathogen Phytophthora nicotianae on pepper plants using 16S amplicon sequencing and metaproteomics. Renew Agr Food Syst, pp 1–9

Rosenzweig N, Tiedje JM, Quensen JF, Meng Q, Hao JJ (2012) Microbial communities associated with potato common scab-suppressive soil determined by pyrosequencing analyses. Plant Dis 96:718–725

Schlatter D, Kinkel L, Thomashow L, Weller D, Paulitz T (2017) Disease suppressive soils: new insights from the soil microbiome. Phytopathology 107:1284–1297

Siegel-Hertz K, Edel-Hermann V, Chapelle E, Terrat S, Raaijmakers JM, Steinberg C (2018) Comparative microbiome analysis of a *Fusarium* wilt suppressive soil and a *Fusarium* wilt conducive soil from the Chateaurenard region. Front Microbiol 9:568

Tomihama T, Nishi Y, Mori K et al (2016) Rice bran amendment suppresses potato common scab by increasing antagonistic bacterial community levels in the rhizosphere. Phytopathology 106:719–728

van Agtmaal M, van Os G, Hol G et al (2015) Legacy effects of anaerobic soil disinfestation on soil bacterial community composition and production of pathogen-suppressing volatiles. Front Microbiol 6:701

van der Voort M, Kempenaar M, van Driel M, Raaijmakers JM, Mendes R (2016) Impact of soil heat on reassembly of bacterial communities in the rhizosphere microbiome and plant disease suppression. Ecol Lett 19:375–382

van Elsas JD, Speksnijder AJ, van Overbeek LS (2008) A procedure for the metagenomics exploration of disease-suppressive soils. J Microbiol Methods 75:515–522

Vida C, Bonilla N, de Vicente A, Cazorla FM (2016) Microbial profiling of a suppressiveness-induced agricultural soil amended with composted almond shells. Front Microbiol 7:4

Weller DM, Raaijmakers JM, Gardener BB, Thomashow LS (2002) Microbial populations responsible for specific soil suppressiveness to plant pathogens. Annu Rev Phytopathol 40:309–348

Yin C, Hulbert SH, Schroeder KL et al (2013) Role of bacterial communities in the natural suppression of *Rhizoctonia solani* bare patch disease of wheat (*Triticum aestivum* L.). Appl Environ Microbiol 79:7428–7438

Yokota K, Hayakawa H (2015) Impact of antimicrobial lipopeptides from *Bacillus* sp. on suppression of *Fusarium* yellows of tatsoi. Microbes Environ 30:281–283

Zhou D, Jing T, Chen Y et al (2019) Deciphering microbial diversity associated with *Fusarium* wilt-diseased and disease-free banana rhizosphere soil. BMC Microbiol 19:161

Chapter 5
Rhizosphere Dynamics: An OMICS Perspective

Munazza Ijaz, Muhammad Iqbal, Bilal Rasool, Muhammad Zubair, Issayeva Akmaral Umirbekovna, Shazia Anwer Bukhari, and Mahmood-ur-Rahman

Abstract The interface between soil and root system of a plant is referred to as rhizosphere. A complex microbiome is present in the ecosystem of rhizosphere that produces different metabolites and proteins to enhance the growth and yield of different plant species. To understand the complexity of rhizosphere, different multi-OMICS techniques are being used in modern sciences. The success of rhizosphere science depends upon the successful implementation of multi-OMICS technique and use of robust bioinformatics software and databases which have been used to analyze the complex data. In this chapter, the recent advances, challenges, bioinformatics tools, and latest OMICS technologies to study the rhizosphere dynamics have been discussed.

Keywords Rhizosphere · OMICS · Bioinformatics · Soil microbiome · Plant growth

---

M. Ijaz · M. Zubair · Mahmood-ur-Rahman (✉)
Department of Bioinformatics and Biotechnology, GC University—Faisalabad, Faisalabad, Pakistan
e-mail: mahmoodansari@gcuf.edu.pk

M. Iqbal
Department of Environmental Science and Engineering, GC University—Faisalabad, Faisalabad, Pakistan

B. Rasool
Department of Zoology, GC University—Faisalabad, Faisalabad, Pakistan

I. A. Umirbekovna
Department of Chemistry, Biology and Agriculture, Shymkent University, Karatau, Kazakhstan

S. A. Bukhari
Department of Biochemistry, GC University—Faisalabad, Faisalabad, Pakistan

R. N. Pudake et al. (eds.), *Omics Science for Rhizosphere Biology*, Rhizosphere Biology, https://doi.org/10.1007/978-981-16-0889-6_5

5.1 Introduction

The interface between plant roots and their surrounding soil is called rhizosphere (Odelade and Babalola 2019). It is considered as a most complex terrestrial habitat for microorganisms on earth that has thousands of microbial cells and millions of functional genes per gram of soil (Mueller et al. 2019). The diverse microbial community that lives in rhizosphere is usually known as rhizosphere microbiome which survives on exudates of plant roots and comprised of numerous microbes such as viruses, bacteria, archaea, and some eukaryotes (Berendsen et al. 2012). The production and comparative abundance of different plant root exudates have positive effects on the soil environment as well as on microbiome present in the rhizosphere. Reciprocally, the plant growth and yield also get benefits from the functional and structural properties of rhizosphere microbiome (Mendes et al. 2019).

The study of complex mechanisms in rhizosphere is quite possible by using integrated multi-OMICS approach that is a new branch of System Biology (Shinano 2019). It is an interdisciplinary field based on complex interactions in the biological systems and follows a holistic approach in advanced biological research (Pinu et al. 2019). Integrated multi-OMICS approaches are very useful for analyzing both individuals and community datasets by involving multiple OMES, such as genomes [information obtained from DNA (Faure et al. 2020)], transcriptome [information obtained from mRNA (Gao et al. 2019)], proteomes [information obtained from proteins (Kumar et al. 2019)], and metabolome [information obtained from metabolites (Mhlongo et al. 2018)]. In multi-OMICS approach, data or information is collected and analyzed with the help of a variety of softwares and instruments. For example: to study the "Central Dogma of Molecular Biology," it is necessary to measure and analyze three biomolecules, e.g., DNA, RNA, and protein by using multivariate statistical approaches (Csala and Zwinderman 2019).

These techniques can also be studied at the organism and community level. The blueprint of an organism is described as genome that means the same arrangement and sequence of nucleotides is required to build an identical organism (Franzosa et al. 2018). The "meta" prefix is used when the genomic information is of "communities." So, metagenome can be defined as the complete set of genomic information of all prokaryotes, eukaryotes, and viruses present in an ecosystem like rhizosphere. The metagenome contains the complete information about genes and organisms present in a specific ecosystem (Krishnaraj and Pasha 2017). All the functional genes which are transcriptionally active at the time of sampling from an ecosystem are described by metatranscriptomics (Yergeau et al. 2018). The complete set of proteins isolated from the ecosystem at specific time is known as metaproteome. The metaproteome gives the complete information about the regulation of gene expression (Bona et al. 2019). The set of all metabolites in an ecosystem is referred to as metametabolome (Badri et al. 2013).

The non-sequence modifications of DNA and protein of whole community is known as metaepigenome and metaepiproteome, respectively (Allen et al. 2017; Hiraoka et al. 2019). The methylation pattern of DNA in a community (metaepigenome) can be measured by next-generation sequencing (NGS) (Chatterjee et al. 2017). The post-translational modification, such as protein phosphorylation in an ecosystem (metaepiproteome) can be measured by mass spectrometry (MS) (Uzasci et al. 2016). The multi-OMICS technologies are important to understand the dynamic features of rhizosphere. The OMICS study can provide an exciting insight of different cellular functions, which may determine the relationship between various processes which are occurring in rhizosphere of a plant. These techniques are robust and significantly important; therefore, a detailed explanation of different OMICS technologies is provided in this chapter.

5.2 Recent Advances in OMICS Approaches in Rhizosphere

Currently, there is a limited number of studies reported on the integrated multi-OMICS approaches in rhizosphere science, but many researchers have applied different OMICS techniques individually and elucidated functional and structural properties of rhizosphere. For example, structures of microbial community in the rhizosphere and identification of its core members have been revealed directly with the help of NGS of 16S ribosomal RNA by using polymerase chain reaction (PCR) that helps in determining the diversity of organisms in rhizospheric soil of different plants such as: coffee, soybean, maize, common annual grass, white lupins and *Arabidopsis thaliana* (Marschner et al. 2002; Bulgarelli et al. 2012; Peiffer et al. 2013; Navarrete et al. 2014; Caldwell et al. 2015; Shi et al. 2015). Additionally, the metabolic potential of rhizospheric soil has been revealed by the studies of metagenomics on rice and soybean (Knief et al. 2012; Navarrete et al. 2014).

A metatranscriptomic study has revealed that in cereal grains (soybean and maize), the application of glyphosate on bacterial cells affects the metabolism of amino acids, nutrients, and carbohydrates (Newman et al. 2016). When rhizosphere of legumes (peas) was compared with of cereal grains (oat and wheat), it was identified from metatranscriptomic study that microbiome of pea was different and stronger due to "rhizosphere effect" (Turner et al. 2013). While there are no reports available currently about metaepigenomics in neither rhizosphere nor in some other ecosystem, the focus is needed on this domain for identification and understanding the diverse microbial community and its regulation at the epigenomic level. However, epigenomics has been used to study bacterial epigenetic modifications which give a way out for the application of metaepigenomic techniques on rhizosphere (Murray et al. 2012).

There are no significant studies available on metaepiproteome in complex ecosystems like rhizosphere, but it has been successfully implemented on microbial communities living on less diverse and simple ecosystems like acid mine waste and deep vents (Li et al. 2014; Zhang et al. 2016). The function of proteins present in rhizosphere and their downstream regulation could be studied by metaepiproteome.

The black truffles, rhizospheric methanotrophs, or tissues of roots in field-grown rice and decomposition of leaf litter are some sources on which metaproteomic studies can be done (Schneider et al. 2012; Bao et al. 2014; Zampieri et al. 2016). It has been found that approximately 4600 proteins are present in the phyllosphere (above-ground plant parts) of rice plant (Knief et al. 2012). The rhizosphere metabolomics usually focus on some general root exudates. In *A. thaliana,* many root exudates have been identified from metametabolomics studies and authenticated by multivariate data analysis (Van Dam and Bouwmeester 2016). Although independent studies provide bundles of information, integrated OMICS approaches can give a better understanding of the different processes like gene expression, transcription, translation, and production of metabolites.

5.3 Challenges for Multi-OMICS Data Analysis in Rhizosphere

Most of the disciplinary practices and methods used in rhizosphere science are going through a transition phase. Nowadays, it is known as an interdisciplinary field that focuses on statistics, mathematics, software engineering, computer science, and informatics. Improvement in experimental technologies and the development of complementary software are two main reasons for the success of rhizosphere science. There are different integrated multi-OMICS techniques that are necessary to be studied for understanding the rhizospheric structure; however, in rhizosphere environment, they face four major challenges.

Firstly, the datasets obtained from rhizosphere are big and complex due to high diversity of organisms in soil. The microarray analysis of rhizosphere sample suggested that it contains approximately 30,000 OTUs (operational taxonomic units) of archaea and bacteria (Mendes et al. 2011). A study based upon 16S rRNA sequencing of rhizosphere of *A. thaliana* had suggested that more than 2000 OTUs are found per gram of rhizospheric soil that represents high biodiversity of rhizosphere (Bulgarelli et al. 2012). Secondly, the analysis of rhizosphere needs many replicates for each sample to maintain the statistical viability (Vandenbygaart and Allen 2011). So, analysis and management of rhizosphere data is a major problem. Thirdly, wet laboratory technologies are developing with a rapid pace, thus more robust, specific and sophisticated softwares are needed to analyze the data. The actual computational performance does not combat with the data production by OMICS approaches. Different OMICS technologies produce enormous amount of data that can only be analyzed by new algorithms and software (Roudier et al. 2015). Fourthly, different factors in software engineering disrupt the OMICS data analysis, such as the absence of rigor, reproducibility, and data curation. Some software used for OMICS analysis, suffer from internal systematic issues as well as some of them are prohibitively licensed, instead of being open source software (Asay 2008). While some software are very difficult to use for new-users and non-professionals, they do not reproduce the results and do not behave according to expectations (Hunt et al. 2014).

Moreover, some softwares, databases, and websites are not actively maintained or diminish from internet sources. The data obtained from new rhizospheric technologies is needed to be analyzed from the well-tested OMICS softwares which are validated and endorsed by community. They must have specific benchmarks, datasets, standards, and workflows to study the properties of rhizosphere. The analysis of data, either of single or multi-OMICS is a great challenge in the current world of rapid discoveries. The more robust, cost-effective, less time consuming, appropriate and specific softwares, websites, and databases are required for analysis of rhizosphere in this era of multi-OMICS.

5.4 Current Resources for Multi-OMICS Analysis

There are many online softwares, databases, and websites available to analyze data of metagenomics, metatranscriptomics, metaproteomics, and metametabolomics (Table 5.1). These softwares are used for general OMICS searches and very few of them are specific to rhizospheric analysis. Therefore, more softwares are required to specifically analyze the OMICS of rhizosphere.

5.5 Recent OMICS Techniques to Study Rhizosphere Dynamics

The advancement of previous approaches and development of emerging technologies, such as metabolomics, metagenomics, metatranscriptomics, and metaproteomics have been used to measure the composition and behavior of rhizosphere. These improvements answer previous questions and raise new questions about the functions of different microbial communities (Sørensen et al. 2009; Hirsch et al. 2010).

5.5.1 Metagenomics

Metagenomics is defined as the complete genetic information of all organisms and cells that are present in a specific location at specific time. The different living organisms are present in rhizosphere, so it is important to understand their ecological functions, distribution and activity in rhizosphere. Various advanced molecular biology techniques have been reported to study microbial diversity in rhizosphere. In rhizospheric soil, the bacterial diversity which is identified by different molecular

Table 5.1 The list of resources to analyze multi-OMICS data

S. no.	Multi-OMICS tools	Functions	Handling	Coding language	Web link
1.	IntegrOmics	Performs integrative analysis on OMICS variables efficiently	Difficult	R	http://math.univtoulouse.fr/biostat
2.	MapMan	Map metabolites, gene expression, and other data and represents large datasets through metabolic pathways diagram	Easy	Java	https://mapman.gabipd.org/
3.	iCluster	Identification of novel biomarkers and their annotation by using related transcriptomics and proteomics data	Difficult	R	https://www.mskcc.org/departments/epidemiologybiostatistics/biostatistics/icluster
4.	IPA, Qiagen	Mapping and integration of genomics, metabolomics, proteomics, and transcriptomics data	Easy	License, Web, Local	https://www.qiagenbioinfor matics.com/products/ingenuity-pathway-analysis/
5.	SteinerNet	Integration of transcriptome and proteome data	Medium	R	https://cran.rproject.org/src/contrib/Archive/SteinerNet/
6.	3Omics	Integration of multiple intra- or inter-metabolomics, proteomics, and transcriptomic data	Easy	Web	http://3omics.cmdm.tw/
7.	Omics Integrator	Integrate gene expression data, epigenetics data, and proteomic data by protein–protein interaction network	Easy	Python, Web	https://github.com/fraenkel-lab/OmicsIntegrator
8.	MixOmics	Linear multivariate method for dimension reduction, integration, visualization, and data analysis	Difficult	R	http://mixomics.org/
9.	Paintomics	Integrated visualization of metabolomics and transcriptomics data	Easy	Web	http://www.paintomics.org
10.	GalaxyP, GalaxyM	Proteomics analysis based upon the transcriptomic study and provides complete omics analysis	Easy	Web	https://usegalaxy.org/

11.	MONGKIE	Analyze multi-layered OMICS data such as copy number variation, somatic mutations, and data of gene expression	Easy	Java	http://yjjang.github.io/mongkie/
12.	CrossPlatformCommander	Identification of novel biomarkers and their annotation by using related transcriptomics and proteomics data	Difficult	R	http://www.ruhrunibochum.de/mpc/software/xplatcom/index.html.en
13.	IMPaLA	Joint pathway analysis of omics datasets	Easy	Web	http://impala.molgen.mpg.de
14.	OnPLS	Multi-block analysis of data by using prefiltering of unique and locally joint modifications	Difficult	Python	https://github.com/tomlof/OnPLS
15.	moCluster	Analysis of genes set depending upon the OMICS data	Difficult	R	https://www.bioconductor.org/packages/release/bioc/ html/mogsa.html
16.	Multiple dataset integration	Integration of information from different datasets	Difficult	R, C++	https://github.com/smason/mdipp
17.	Multiple co-inertia analysis	Identifies relationship between many high-dimensional datasets	Difficult	R	https://rdrr.io/bioc/omicade4/man/mcia.html
18.	Paradigm	Determines pathway activities from different multidimensional datasets	Difficult	Commercial	http://five3genomics.com/technologies/paradigm
19.	PLRS	Flexible modeling between copy number of DNA and expression of RNA	Difficult	R	http://bioconductor.org/
20.	CONEXIC	Integration of same copy number, deletion, insertion, amplification, and gene expression datasets	Medium	Java	https://www.c2b2.columbia.edu/danapeerlab/html/conexic.html
21.	CNAmet	Integration of high-throughput gene expression, copy number, and DNA methylation data	Difficult	R	http://csbi.ltdk.helsinki.fi/CNAmet/
22.	miRTarVis	Integration of miRNA–mRNA	Easy	Java	http://hcil.snu.ac.kr/~rati/miRTarVis/index.html
23.	BioMiner	Analysis of cross-OMICS and transcriptomics high-throughput datasets	Easy	Web	http://systherDB.microdiscovery.de/

(continued)

Table 5.1 (continued)

S. no.	Multi-OMICS tools	Functions	Handling	Coding language	Web link
24.	Omics Pipe	Integrate date of mRNA sequencing, miRNA sequencing, ChIP sequencing, and high-throughput gene sequencing analysis	Difficult	Python	http://sulab.scripps.edu/omicspipe
25.	Galaxy integrated OMICS	Analysis of transcriptomics and proteomics data	Easy	Galaxy	https://usegalaxy.org/
26.	CPAS	Identify disease-related biological pathways	Difficult	R, C	https://sourceforge.net/projects/%20cpasv1/files/
27.	GeneTrail2	Integrative analysis of genomic, proteomic, transcriptomic, and miRNA datasets	Easy	Web	https://genetrail2.bioinf.uni-sb.de/
28.	BioVLAB-mCpG-SNP-EXPRESS	Analysis of integrated gene expression, genetic variation, and DNA methylation	Easy	Web	http://bhi2.snu.ac.kr:3000/
29.	Multi-Assay-Experiment	Operation and storage on diverse genomic datasets	Difficult	R, Bio-conductor	https://bioconductor.org/packages/release/bioc/html/MultiAssayExperiment. html
30.	Mergeomics	Involves in genetic association, transcriptomic association, and epigenetic association	Difficult	R	http://mergeomics.research.idre.ucla.edu/
31.	OMICtools	Database that helps in the integration of omic data	Easy	Web	https://omictools.com/
32.	Global Natural	Untargeted datasets of metabolomics	Easy	Web	http://gnps.ucsd.edu
33.	NCBI Multiple Database	Transcriptomics and genomics databases	Easy	Web	https://www.ncbi.nlm.nih.gov/
34.	MetabolomicsWorkBench	Archive of metabolomics data	Easy	Web	http://www.metabolomicsworkbench.org/
35.	Proteome-Xchange-Consortium	Proteomics database	Easy	Web	http://www.proteomexchange.org/

36.	Ecomics	Multi-omics collection of E. coli with consistent meta-data evidence, accomplish tentative classification, development, proteome, and transcriptome analysis	Easy	Web	http://prokaryomics.com/
37.	WikiPathways	Collection of paths acquiescent to automatic and manual workflows for the mapping of metabolites, proteins, and genes	Easy	Web, R	http://wikipathways.org/
38.	PeptideAtlas repository	Database of proteomics	Easy	Web	http://www.peptideatlas.org/PASS/PASS00512
39.	PRIDE	Database of proteomics	Easy	Web	https://www.ebi.ac.uk/pride/archive
40.	XCMS Online	Systems biology scale	Easy	Web	https://xcmsonline.scripps.edu/

techniques is much more than the predicted OTUs. The metagenomes studied on the basis of 16S rRNA and ITS have revealed 33,346 bacterial (Mendes et al. 2011) and 3320 fungal (Schmidt et al. 2013) OTUs are present in 1 g of rhizospheric soil. The Roche 454 and Illumina sequencing platform can be used to examine the biodiversity in the rhizosphere (Unno 2015). It has been revealed from Roche 454 pyrosequencing that proteobacteria, bacteroidetes, acidobacteria, and actinobacteria are the major bacterial taxa present in the rhizosphere of oak plant (Uroz et al. 2010).

The metagenomic analysis of micro-sites present in *Lolium perenne* has suggested that proteobacteria, acidobacteria, and actinobacteria are some major bacterial taxa which were found (Lagos et al. 2014). The studies of apple rhizosphere through Illumina sequencing have identified different bacterial taxa, such as gematimonadetes, bacteroidetes, proteobacteria, actinobacteria, and acidobacteria (Sun et al. 2014). The metagenomics studies give a detailed description about various OTUs present in the rhizospheric soil. These techniques are also helpful in determining the functions and structures of different OTUs present in an ecosystem.

5.5.2 Metaproteomics

The metaproteomics is defined as complete set of proteins that are present in an ecosystem at specific time interval. It can give information about different soil microbes and their functional importance, such as biochemical processes and bioremediation (Bastida et al. 2012). The proteogenomics is an emerging technique used for studying physiology, ecology, and evolution of rhizospheric microbiome (VerBerkmoes et al. 2009). This approach gives significant results because it uses metagenomics and metaproteomics simultaneously to analyze the data. Although this is important to understand that the repositories used for identification of protein–protein interactions are incomplete, numerous metaproteomics experiments have revealed the protein diversity that are usually expressed because of plant interactions with rhizospheric microbiome. The metatranscriptomic profile of ratoon sugarcane and plant sugarcane was compared in a study which suggested that rhizosphere has more enzymatic activities, increased gene expression and great catabolic diversity due to more plant–microbe interactions. It was also discussed in this research that 24.77% proteins of rhizosphere came from bacteria and are involved in membrane transport and signal transduction (Lin et al. 2013). Another study on *Rehmannia glutinosa* has described that majority of the proteins produced by rhizospheric microbes are involved in energy and amino acid metabolism and to generate responses against different abiotic and biotic stresses (Wu et al. 2011). Moreover, a study on *Lactuca sativa* rhizospheric soil has showed that proteins present in rhizosphere are responsible for energy metabolism, responses against viruses, and different stresses (Moretti et al. 2012). The metatranscriptomic studies give a clear description of the genes and proteins present in rhizosphere.

5.5.3 Metatranscriptomics

The metatranscriptomics is referred to as complete set of mRNA present in all cells or organisms of in an ecosystem at specific time, and it provides the information about the gene expression of microbiome. The metabolic activities and functions of novel genes can be identified using metatranscriptomic analysis. Although the metatranscriptomic study has tremendous potential, it cannot be widely used for rhizosphere due to instability of mRNA as well as it is quite difficult to extract mRNA from a complex ecosystem. There are many other challenges, i.e., interference of mRNA molecules with humic acid, shorter half-life of mRNA, and it is difficult to separate mRNA from other molecules of RNA. Moreover, most of the experiments on rhizospheric soil usually focus on plant growth-promoting rhizobacteria (PGPRs). Therefore, there is a great need to study other microbes like fungi, protozoa, microalgae, and archaea present in microbiome of rhizosphere (Lagos et al. 2015).

The evolutionary studies of archaeal and bacterial genomes can be improved with the help of "Genomic Encyclopedia of Bacteria and Archaea (GEBA) project" which was launched in 2007 (Rinke et al. 2013). The strain Mg1 of *Streptomyces* species was sequenced by using a third-generation sequencing tool named as "Pacific biosciences single–molecule real–time (PacBio-RS-SMRT) sequencer" (Niedringhaus et al. 2012). This species was found responsible for the degradation of a famous bacteria known as *Bacillus subtilis* (Hoefler et al. 2013). The technique was beneficial to analyze long reads of 16S rRNA which was extracted from rhizospheric soil. It can also be used for determining the microbial taxonomy and phylogeny of bacteria present in rhizosphere. The metatranscriptomic studies can be helpful to find the gene expression of multiple genes present in microbiome in response to particular stress. There is a need to develop robust techniques to isolate RNA from rhizospheric soil so that metatranscriptomic analysis could be carried out in an efficient way.

5.5.4 Metametabolomics

The metametabolomics is the study of structures and functions of metabolites present in an ecosystem. This is not restricted to metabolic profiling only, but also considers the production and identification of metabolites present in that specific region of rhizosphere and quantify the abundance of different metabolites. Metametabolomics also determine the differences among metabolites. In rhizosphere, the metabolites come from root exudates which can be identified and analyzed by this technique. Many PGPRs live near the roots which are also responsible for the production of metabolites. This technique helps to determine the functions and production of metabolites produced from rhizosphere microbiome. Numerous compounds in rhizosphere are identified and modified through various metabolomic engineering techniques. These studies also provide a chance to critically analyze each metabolite produced from both microbiome and plant (Van Dam and Bouwmeester 2016). The metametabolomic analysis become more complicated in higher plants that produce 1–2 million metabolites approximately (Oksman-Caldentey and Inzé 2004).

It is important to understand the difference between targeted metabolite analysis and comprehensive global metabolic profiling, so that a complete understanding of primary and secondary metabolites production can be obtained. Some of organic acids, carbohydrates, lipids, vitamins, amino acids, glucosinolates, alkaloids, phenyl-propanoids, trepenoids along with different secondary metabolites should be used to develop a comprehensive metabolic profile that could be varied in the rhizosphere of different plants. Thus, it is stated that complex interaction between the microbiome and higher plants is not the only problem in metabolomics, but also the different classes of chemicals increase the complexity of rhizosphere. According to the varying microbiome and plant interaction, different metabolites are produced which can be analyzed by applying different methods. Nowadays, plant pathology can be studied by merging metabolomics, genomics, and transcriptomics approaches. Metabolomics is a post-genomic technique used to study the bacteria, archaea, and other prokaryotes in rhizosphere.

The different enzyme-based or RNA-based techniques are not enough to identify the effects of different metabolites in an ecosystem. Therefore, various technologies, such as chromatography, spectroscopy, spectrometry, and nuclear magnetic resonance are also utilized to estimate, identify, and report the various concentrations of metabolites in rhizosphere at a specific time interval. The different processes of data visualization and bioinformatics have been used to make a profile of metabolites and for identifying their functions in rhizospheric soil. With the help of these approaches a detailed report of rhizosphere metametabolomics can be generated that would have minimum errors. The metabolic profiling is referred as a complete record of various metabolites present in an organism or ecosystem. Metabolic profiling is a complex and expensive procedure, so before going to it some cheap procedures like nuclear magnetic resonance, spectroscopy, fingerprinting, and chromatography are applied to identify and screen different metabolites (Verma et al. 2018). There are some reports in which liquid chromatography and mass spectrometry have been found useful in identifying different metabolites from extract. Some bioinformatics tools are also used to develop a profile of metabolites from the data obtained through mass spectrometry and liquid chromatography (Goodacre et al. 2004). Although metametabolomics is a complex procedure, it gives useful insight into rhizosphere dynamics.

5.6 Conclusions

The rhizosphere has a complex ecosystem and many important phenomena are taking place there. So, it is required to study the complex dynamics of the rhizospheric soil. The growth and yield of different plants depend upon the healthy rhizospheric environment. Many complex agricultural issues can be resolved by applying multi-OMICS techniques to study the rhizospheric soil, but there is a great need to develop robust softwares which can be used to analyze the complex data obtained from multi-OMICS experiments. Moreover, the application of bioinformatics tools can reduce the time and cost of extensive procedures to analyze the rhizosphere.

References

Allen R, Iii W, Borkum MI et al (2017) From data to knowledge: the future of multi-omics data analysis for the rhizosphere. Rhizosphere 3:222–229. https://doi.org/10.1016/j.rhisph.2017.05.001

Asay CD (2008) The general public license version 3.0: making or breaking the foss movement. Michigan Telecommun Technol Law Rev 14:265–301

Badri DV, Zolla G, Bakker MG et al (2013) Potential impact of soil microbiomes on the leaf metabolome and on herbivore feeding behavior. New Phytol 198:264–273. https://doi.org/10.1111/nph.12124

Bao Z, Okubo T, Kubota K et al (2014) Metaproteomic identification of Diazotrophic Methanotrophs and their localization in root tissues of field-grown Rice plants. Appl Environ Microbiol 80:5043–5052. https://doi.org/10.1128/AEM.00969-14

Bastida F, Algora C, Hernández T, García C (2012) Feasibility of a cell separation-proteomic based method for soils with different edaphic properties and microbial biomass. Soil Biol Biochem 45:136–138. https://doi.org/10.1016/j.soilbio.2011.10.017

Berendsen RL, Pieterse CMJ, Bakker PAHM (2012) The rhizosphere microbiome and plant health. Trends Plant Sci 17:478–486. https://doi.org/10.1016/j.tplants.2012.04.001

Bona E, Massa N, Novello G et al (2019) Metaproteomic characterization of the *Vitis vinifera* rhizosphere. FEMS Microbiol Ecol 95:fiy204. https://doi.org/10.1093/femsec/fiy204

Bulgarelli D, Rott M, Schlaeppi K et al (2012) Revealing structure and assembly cues for Arabidopsis root-inhabiting bacterial microbiota. Nature 488:91–95. https://doi.org/10.1038/nature11336

Caldwell AC, Carneiro L, Silva F, Can C (2015) Prokaryotic diversity in the rhizosphere of organic, intensive, and transitional coffee farms in Brazil. PLoS One 10:1–17. https://doi.org/10.5281/zenodo.11120

Chatterjee A, Macaulay EC, Ludgate JL et al (2017) Comparative assessment of DNA methylation patterns between reduced representation bisulfite sequencing and Sequenom EpiTyper methylation analysis. Epigenomics 9:823–832. https://doi.org/10.2217/epi-2016-0176

Csala A, Zwinderman AH (2019) Multivariate statistical methods for high-dimensional multiset omics data analysis. In: Holger Husi DSN (ed) Computational biology. Codon Publication, Brisbane, pp 1–212

Faure C, Veyssière M, Boëlle B et al (2020) Long-read genome sequence of the sugar beet rhizosphere mycoparasite *Pythium oligandrum*. G3: Genes, Genomes, Genetics 10:431–436. https://doi.org/10.1534/g3.119.400746

Franzosa EA, Mciver LJ, Rahnavard G et al (2018) Species-level functional profiling of metagenomes and metatranscriptomes. Nat Methods 15:962–968. https://doi.org/10.1038/s41592-018-0176-y

Gao R, Hao DC, Hu W et al (2019) Transcriptome profile of polycyclic aromatic hydrocarbon-degrading fungi isolated from Taxus rhizosphere. Curr Sci 116:1218–1228. https://doi.org/10.18520/cs/v116/i7/1218-1228

Goodacre R, Vaidyanathan S, Dunn WB et al (2004) Metabolomics by numbers: acquiring and understanding global metabolite data. Trends Biotechnol 22(5):245–252. https://doi.org/10.1016/j.tibtech.2004.03.007

Hiraoka S, Okazaki Y, Anda M et al (2019) Metaepigenomic analysis reveals the unexplored diversity of DNA methylation in an environmental prokaryotic community. Nat Commun 10 (1):1–10. https://doi.org/10.1038/s41467-018-08103-y

Hirsch PR, Mauchline TH, Clark IM (2010) Culture-independent molecular techniques for soil microbial ecology. Soil Biol Biochem 42:878–887. https://doi.org/10.1016/j.soilbio.2010.02.019

Hoefler BC, Konganti K, Straight PD (2013) De novo assembly of the Streptomyces sp. strain Mg1 genome using PacBio single-molecule sequencing. Genome Announc 1(4):1–2. https://doi.org/10.1128/genomeA.00535-13

Hunt M, Newbold C, Berriman M et al (2014) A comprehensive evaluation of assembly scaffolding tools. Genome Biol 15:1–15

Knief C, Chaffron S, Stark M et al (2012) Metaproteogenomic analysis of microbial communities in the phyllosphere and rhizosphere of rice. ISME J 6:1378–1390. https://doi.org/10.1038/ismej.2011.192

Krishnaraj PU, Pasha MA (2017) Metagenome of rhizosphere and endophytic ecosystem. In: Singh RP, Kothari R, Koringa PG, Singh SP (eds) Understanding host-microbiome interactions—an omic approach. Springer Nature Singapore Pte Ltd., Singapore, pp 125–156

Kumar S, Kumar A, Mohd K et al (2019) Data in brief metaproteomic data of maize rhizosphere for deciphering functional diversity. Data Br 27:104574–104580. https://doi.org/10.1016/j.dib.2019.104574

Lagos LM, Navarrete OU, Maruyama F et al (2014) Bacterial community structures in rhizosphere microsites of ryegrass (*Lolium perenne* var. Nui) as revealed by pyrosequencing. Biol Fertil Soils 50:1253–1266. https://doi.org/10.1007/s00374-014-0939-2

Lagos ML, Maruyama F, Nannipieri P et al (2015) Current overview on the study of bacteria in the rhizosphere by modern molecular techniques: a mini-review. J Soil Sci Plant Nutr 15:504–523. https://doi.org/10.4067/S0718-95162015005000042

Li Z, Wang Y, Yao Q et al (2014) Diverse and divergent protein post-translational modifications in two growth stages of a natural microbial community. Nat Commun 5:4405. https://doi.org/10.1038/ncomms5405

Lin W, Wu L, Lin S et al (2013) Metaproteomic analysis of ratoon sugarcane rhizospheric soil. BMC Microbiol 13(1):135. https://doi.org/10.1186/1471-2180-13-135

Marschner P, Neumann G, Kania A et al (2002) Spatial and temporal dynamics of the microbial community structure in the rhizosphere of cluster roots of white lupin (*Lupinus albus* L.). Plant Soil 246:167–174. https://doi.org/10.1023/A:1020663909890

Mendes R, Kruijt M, De Bruijn I et al (2011) Deciphering the rhizosphere microbiome for disease-suppressive bacteria. Science 332(6033):1097–1100. https://doi.org/10.1126/science.1203980

Mendes LW, Chaves MGD, Fonseca MDC et al (2019) Resistance breeding of common bean shapes the physiology of the rhizosphere microbiome. Front Microbiol 10:2252–2262. https://doi.org/10.3389/fmicb.2019.02252

Mhlongo M, Piater LA, Madala NE et al (2018) The chemistry of plant—microbe interactions in the rhizosphere and the potential for metabolomics to reveal signaling related to defense priming and induced systemic resistance. Front Plant Sci 9:1–17. https://doi.org/10.3389/fpls.2018.00112

Moretti M, Minerdi D, Gehrig P et al (2012) A bacterial-fungal metaproteomic analysis enlightens an intriguing multicomponent interaction in the rhizosphere of *Lactuca sativa*. J Proteome Res 11:2061–2077. https://doi.org/10.1021/pr201204v

Mueller CW, Carminati A, Kaiser C et al (2019) Editorial: rhizosphere functioning and structural development as complex interplay between plants, microorganisms and soil minerals. Front Environ Sci 7:130–133. https://doi.org/10.3389/fenvs.2019.00130

Murray IA, Clark TA, Morgan RD et al (2012) The methylomes of six bacteria. Nucleic Acids Res 40:11450–11462. https://doi.org/10.1093/nar/gks891

Navarrete A, Van Veen JA, Mendes LW et al (2014) Taxonomical and functional microbial community selection in soybean rhizosphere. ISME J 8:1577–1587. https://doi.org/10.1038/ismej.2014.17

Newman MM, Lorenz N, Hoilett N et al (2016) Science of the total environment changes in rhizosphere bacterial gene expression following glyphosate treatment. Sci Total Environ 553:32–41. https://doi.org/10.1016/j.scitotenv.2016.02.078

Niedringhaus TP, Milanova D, Kerby MB et al (2012) Landscape of next-generation sequencing technologies. Anal Chem 83:4327–4341. https://doi.org/10.1021/ac2010857

Odelade KA, Babalola OO (2019) Bacteria, fungi and archaea domains in rhizospheric soil and their effects in enhancing agricultural productivity. Int J Environ Res Public Health 16:3873–3891. https://doi.org/10.3390/ijerph16203873

Oksman-Caldentey KM, Inzé D (2004) Plant cell factories in the post-genomic era: new ways to produce designer secondary metabolites. Trends Plant Sci 9:433–440. https://doi.org/10.1016/j.tplants.2004.07.006

Peiffer JA, Spor A, Koren O et al (2013) Diversity and heritability of the maize rhizosphere microbiome under field conditions. PNAS 110:6548–6553. https://doi.org/10.1073/pnas.1302837110/-/DCSupplemental.www.pnas.org/cgi/doi/10.1073/pnas.1302837110

Pinu FR, Beale DJ, Paten AM et al (2019) Systems biology and multi-omics integration: viewpoints from the metabolomics research community. Meta 9:76. https://doi.org/10.3390/metabo9040076

Rinke C, Schwientek P, Sczyrba A et al (2013) Insights into the phylogeny and coding potential of microbial dark matter. Nature 499:431–437. https://doi.org/10.1038/nature12352

Roudier P, Ritchie A, Hedley C, Medyckyj-scott D (2015) The rise of information science: a changing landscape for soil science. IOP Conf Ser Earth Environ Sci 25:012023. https://doi.org/10.1088/1755-1315/25/1/012023

Schmidt P, Bálint M, Greshake B et al (2013) Illumina metabarcoding of a soil fungal community. Soil Biol Biochem 65:128–132

Schneider T, Keiblinger KM, Schmid E et al (2012) Who is who in litter decomposition? Metaproteomics reveals major microbial players and their biogeochemical functions. ISME J 6:1749–1762. https://doi.org/10.1038/ismej.2012.11

Shi S, Nuccio E, Herman DJ et al (2015) Successional trajectories of rhizosphere bacterial communities over consecutive seasons. MBio 6:13–20. https://doi.org/10.1128/mBio.00746-15

Shinano T (2019) Research on ways to improve crop productivity through the regulation of rhizosphere environments. Soil Sci Plant Nutr 66:10–14. https://doi.org/10.1080/00380768.2019.1666301

Sørensen J, Nicolaisen MH, Ron E et al (2009) Molecular tools in rhizosphere microbiology-from single-cell to whole-community analysis. Plant Soil 321:483–512. https://doi.org/10.1007/s11104-009-9946-8

Sun J, Zhang Q, Zhou J et al (2014) Illumina amplicon sequencing of 16S rRNA tag reveals bacterial community development in the rhizosphere of apple nurseries at a replant disease site and a new planting site. PLoS One 9:e111744. https://doi.org/10.1371/journal.pone.0111744

Turner TR, Ramakrishnan K, Walshaw J et al (2013) Comparative metatranscriptomics reveals kingdom level changes in the rhizosphere microbiome of plants. ISME J 7:2248–2258. https://doi.org/10.1038/ismej.2013.119

Unno T (2015) Bioinformatic suggestions on MiSeq-based microbial community analysis. J Microbiol Biotechnol 25:765–770

Uroz S, Buée M, Murat C et al (2010) Pyrosequencing reveals a contrasted bacterial diversity between oak rhizosphere and surrounding soil. Environ Microbiol Rep 2:281–288. https://doi.org/10.1111/j.1758-2229.2009.00117.x

Uzasci L, Auh S, Cotter RJ et al (2016) Mass spectrometric phosphoproteome analysis of HIV-infected brain reveals novel phosphorylation sites and differential phosphorylation patterns. Proteom Clin Appl 10:126–135. https://doi.org/10.1002/prca.201400134

Van Dam NM, Bouwmeester HJ (2016) Metabolomics in the rhizosphere: tapping into below-ground chemical communication. Trends Plant Sci 21:256–265. https://doi.org/10.1016/j.tplants.2016.01.008

Vandenbygaart AJ, Allen OB (2011) Experiment design to achieve desired statistical power. Can J Soil Sci 91:309–310. https://doi.org/10.4141/CJSS2010-068

VerBerkmoes NC, Denef VJ, Hettich RL et al (2009) Systems biology: functional analysis of natural microbial consortia using community proteomics. Nat Rev Microbiol 7:196–205. https://doi.org/10.1038/nrmicro2080

Verma A, Kumar S, Kumar G, et al (2018) Rhizosphere metabolite profiling: an opportunity to understand plant-microbe interactions for crop improvement. In: Crop improvement through microbial biotechnology. Elsevier, pp 343–361

Wu L, Wang H, Zhang Z et al (2011) Comparative metaproteomic analysis on consecutively *Rehmannia glutinosa*-monocultured rhizosphere soil. Plos One 6:e20611. https://doi.org/10.1371/journal.pone.0020611

Yergeau E, Tremblay J, Joly S et al (2018) Soil contamination alters the willow root and rhizosphere metatranscriptome and the root—rhizosphere interactome. ISME J 12:869–884. https://doi.org/10.1038/s41396-017-0018-4

Zampieri E, Chiapello M, Daghino S et al (2016) Soil metaproteomics reveals an inter-kingdom stress response to the presence of black truffles. Sci Rep 6:25773. https://doi.org/10.1038/srep25773

Zhang W, Sun J, Cao H et al (2016) Post-translational modifications are enriched within protein functional groups important to bacterial adaptation within a deep-sea hydrothermal vent environment. Microbiome 4:49–58. https://doi.org/10.1186/s40168-016-0194-x

Chapter 6
Transcriptomics Analyses and the Relationship Between Plant and Plant Growth-Promoting Rhizobacteria (PGPR)

Shokufeh Moradi, Bahman Khoshru, Debasis Mitra, Bhaswatimayee Mahakur, Pradeep K. Das Mohapatra, Behnam Asgari Lajayer, and Mansour Ghorbanpour

Abstract Plants and microbes in the rhizosphere are constantly interacting in a chain, from pathogens (deleterious) to symbionts (beneficial). Relationships between the two sides are very large at various stages of plant development, and signaling molecules play a significant role on both sides. The result of this conversation is that the microbe is absorbed by the plant's roots and responds to it. Among the microbes that approach the roots of plants during this signal conversation are the plant growth-promoting rhizobacteria (PGPR). In order to emphasize the beneficial effects of PGPR on the plant, it is important for the roots of the plants to be colonized by PGPR. Colonization of plant roots by PGPR induces beneficial effects on the plant, such as increasing the overall growth of the plant, resistance to disease, and increasing tolerance to abiotic stresses. Among the mechanisms used by PGPR to increase plant growth are: eliminating the need for plant nitrogen by fixing the molecular nitrogen of the atmosphere, increasing the bioavailability of nutrients needed by plants such as phosphorus, potassium, and iron producing various growth

---

S. Moradi · B. Khoshru
Department of Soil Biology and Biotechnology, Faculty of Agriculture, University of Tabriz, Tabriz, Iran

D. Mitra · P. K. Das Mohapatra
Department of Microbiology, Raiganj University, Uttar Dinajpur, West Bengal, India

B. Mahakur
Department of Botany and Biotechnology, Ravenshaw University, Cuttack, Odisha, India

B. Asgari Lajayer (✉)
Health and Environment Research Center, Tabriz University of Medical Sciences, Tabriz, Iran
e-mail: h-asgari@tabrizu.ac.ir

M. Ghorbanpour (✉)
Department of Medicinal Plants, Faculty of Agriculture and Natural Resources, Arak University, Arak, Iran
e-mail: m-ghorbanpour@araku.ac.ir

R. N. Pudake et al. (eds.), *Omics Science for Rhizosphere Biology*, Rhizosphere Biology, https://doi.org/10.1007/978-981-16-0889-6_6

hormones and regulators such as auxins, cytokinins, and gibberellins, reducing the level of ethylene produced by plant roots under stress conditions, etc. Since PGPR may modulate plant regulatory mechanisms, an important question to be answered is how is the plant–microbe relationship, and what factors play a role in it, and whether or not the microbes can influence the expression of different plant genes at different times? To answer these questions, transcriptomics analysis is an ideal technique that has been addressed in this chapter.

Keywords Microbes · Plants · Rhizosphere · Transcriptomics · PGPR

6.1 Introduction

In the rhizosphere, the root is associated with a large number of microorganisms, collectively referred to as the rhizosphere microbiome (Berendsen et al. 2012; Bulgarelli et al. 2013; Saghafi et al. 2018, 2019a, b, 2020). To preserve the root microbiome, the plant secretes 20% of its photosynthetic material through the roots (Bais et al. 2006). Rhizospheric bacteria take action in exchange for photosynthetic substances provided by plant roots, such as improving root growth, increasing uptake of nutrient, and stimulating the plant's innate immune system (Lugtenberg and Kamilova 2009; Berendsen et al. 2012; Venturi and Keel 2016; Sarikhani et al. 2019b; Khoshmanzar et al. 2020). Popular examples of such microbes include arbuscular mycorrhiza (AM) fungi and PGPR bacteria (Zamioudis and Pieterse 2012; Sarikhani et al. 2019a). The ability of PGPR to increase plant growth and immunity is associated with negative and hostile characteristics (Huot et al. 2014). These bacteria prepare the soil for the development of biological controllers (Pieterse et al. 2016).

To date, there has been a great deal of focus on the response of plants to these beneficial microbes (Xie et al. 2015). Detailed work has been undertaken on the interactions between plants and roots-related PGPR in order to define the criteria of bacteria for adaptation and root colonization, and to provide valuable knowledge on the possible applications of PGPR strains in agriculture (Fan et al. 2012). There is irrefutable proof that interactions between bacteria and plants in the rhizosphere are mainly controlled by root exudates. For example, flavonoids play a significant part in the early signaling of the association of legumes and rhizobes, which are secreted from the roots of legumes (Badri et al. 2009). There are many ways to understand plant–microbial relationships, but to examine the momentary changes in these interactions, a technique is needed that reveals the interaction at all times. In this regard, transcriptome analysis can be the best technique.

The complete collection of RNA molecules in the organism is known as a transcriptome. It consists of mRNAs (coding messenger-RNA), rRNAs (ribosomal RNA), tRNAs (transfer RNA), and non-coding types of RNAs, such as sRNAs (small RNAs). According to biological studies, mRNAs are important for the comprehension of the functional elements of the genome and the molecular components of mRNA cells. Therefore, they are very important in terms of the relationship between genotype and phenotype. There are basically three main techniques to deal with transcriptome: microarrays, real-time RT-PCR, and next-generation RNA sequencing (RNA-seq). In addition, the aim of transcriptomics is to identify genes that are expressed under different conditions and provide a different viewpoint on the genome (Lowe et al. 2017). Strategies and techniques such as microarray analyses, cDNA-based suppression-subtractive hybridization, and promoter trapping are used to study the interaction between root exudates and microbes in the laboratory. Genes involved in this interaction include the genes of metabolism, chemotaxis and bacterial motility, transport, and antibiotic secretion (Zhang et al. 2015).

Next-generation sequencing (NGS) techniques are used to sequence the whole genome and to analyze gene expression at various periods (dynamic transcriptome) (Metzker 2010). Certain uses of the NGS system include study of RNA (including RNA-seq) in studies of small regulatory RNAs and genome annotation and calculation of transcript expression rates under various conditions for both eukaryotes and prokaryotes. This approach provides a great way to research root exudate–bacteria interactions than previous techniques such as microarray analysis and RNA-seq. Therefore, it saves time and is useful in examining uncharacterized genes (Zhang et al. 2015). These techniques have enhanced our thoughtful of plant–microbe interactions and will create new perspectives for future researches (Xie et al. 2015).

6.2 Plant Growth-Promoting Rhizobacteria

There are bacteria around the root of the plants that interact with root exudates. Such bacteria, classified as plant growth-promoting rhizobacteria (PGPR), colonize the roots of the plant, promote plant development, and eliminate disease, and benefit from the nutritional benefits of the food. Based on their interaction with plants, PGPR can be classified into two groups: co-existing bacteria and free living rhizobacteria (Khan 2005). PGPR may also be split into two classes based on its location: iPGPR and ePGPR. The iPGPR, which resides within plant cells, develops nodules, and is found inside complex structures (i.e., symbiotic bacteria). And ePGPR, which lives outside plant cells and does not develop nodules but also accelerates plant development (e.g., free-living rhizobacteria) (Gray and Smith 2005). Various PGPR have been used as stimulant inoculants for plant growth in agriculture and have benefited from their beneficial effects (Sarikhani et al. 2016, 2019a, b), PGPR genera such as *Bacillus*, *Azotobacter*, *Azospirillum*, *Pseudomonas*, *Serratia*, *Burkholderia*, *Agrobacterium*, *Arthrobacter*, *Erwinia*, *Micrococcous*, *Caulobacter*, *Chromobacterium*, *Flavobacterium*, and *Cellulomonas Flavigena* (Hassan et al. 2019).

6.3 Beneficial Characterizations of Rhizobacteria

The pathways used by PGPR to promote plant development have not yet been fully established (Dey et al. 2004). However, there are several potential pathways for growing plant development through PGPR include:

1. Increasing the bioavailability of nutrients required by the plant, i.e., atmospheric nitrogen fixation (Kennedy et al. 2004), increasing the dissolution of low-soluble mineral elements such as phosphorus and potassium (Sarikhani et al. 2016, 2019a, b; Banerjee and Yasmin 2002), production of siderophores and increased bioavailability of elements such as iron (Glick and Pasternak 2003).
2. Production of various hormones and stimulating plant growth such as IAA (indole acetic acid) cytokinin's (Dey et al. 2004) GA (gibberellic acid), and ABA (abscisic acid) (Dobbelaere et al. 2003).
3. Increasing root growth in stress conditions by producing ACC (1-aminocyclopropane-1-carboxylate) deaminase enzyme and reducing ethylene level and increasing the root length and growth (Glick et al. 1998).
4. Restricting or controlling the plant pathogens through the production of siderophores, antibiotics, chitinase, ß-1, 3-glucanase, hydrogen cyanide, fluorescent pigment, etc. (Glick and Pasternak 2003).
5. Production of water-soluble B vitamins such as niacin, pantothenic acid, thiamine, riboflavin, and biotin (Revillas et al. 2000; Heidarpour et al. 2019).
6. Improving soil structure, and bioaccumulation or microbial leaching of minerals (Sarikhani et al. 2019b).

The PGPR also have recently been used to mineralize organic pollutants, such as bioremediation of oil-contaminated soils (Zhuang et al. 2007). Given the above, it can be understood that if the plant can benefit from the help of these beneficial microbes, it will lead to increased growth and yield even at stressful conditions (Shoebitz et al. 2009).

The population of these beneficial bacteria in the soil is generally low, and one of agricultural researchers' aims is to raise the population of these beneficial bacteria in the soil. For example, the use of these beneficial microbes in the form of biofertilizers and their inoculation into plants while increasing plant yield, will lead to an increase in the population of these beneficial bacteria in the soil (Burd et al. 2000). Currently, a variety of PGPR are used around the world in various forms such as liquid and solid biofertilizers (Cocking 2003.

6.4 Interaction of Plant and Beneficial Rhizobacteria

The word "rhizosphere" was first coined by Hiltner as the field of microbial action around rhizosphere and roots (Hiltner 1904). Rhizosphere as the amount of soil that has been influenced by exudation from plant root tissues and colonized by rhizobacteria. Bacterial populations have been estimated to be 10–100 times higher in the rhizosphere than in bulk soil (Berendsen et al. 2012). Rhizobacterial interactions with roots of plants in the rhizosphere can be positive, negative, or neutral. Friendly associations can contribute to enhanced growth of plants and to suppression of plant pathogens (Morgan et al. 2005). The role of rhizosphere is crucial to plant growth promotion, nutrition, and crop quality (Berg and Smalla 2009; Hassan et al. 2019; Mitra et al. 2019a, b) due to the significance of plant–microbe interactions in carbon sequestration, nutrient cycling, and ecosystem functioning (Singh et al. 2004; Sarikhani et al. 2016, 2019a). The beneficial relationships between the plant and the microbes are complicated. Plants emit chemicals such as malic acid that draw rhizobacteria, enabling microorganisms to move to and from the roots. Rhizobacteria consume sugars and amino acids extracted from the plant during colonization. At the same time, PGPR produces substances that influence plant growth and development, such as IAA hormones, cytokinins, and gibberellins (Mitra et al. 2016; Sarikhani et al. 2019b).

PGPR-associated host plants comprise members of the families: *Fabaceae, Asteraceae, Poaceae, Crassulaceae Asteraceae, Brassicaceae,* and *Solanaceae. Fabaceae* includes essential *Glycine max*, *Pisum sativum,* and *Medicago sativa* crops. Within this family, the symbiotic relationship between endophytic nitrogen-fixing bacteria and leguminous plants has been well described (Oldroyd et al. 2011). Crop plants belonging to the Poaceae family used maize, sorghum, and barley for the phytoremediation of metal polluted soil because of their high productivity and good demand for biofuels (Vamerali et al. 2010). The PGPB related with these plants is linked to the free-living *Pseudomonas* sp. and *Burkholderia* sp. as well as the association of endophytic (*Bacillus* sp.) with hyper-accumulator plants. Microbial populations associated with plants and soil have been found to have some specificity for growing plant species that could be attributable to secondary metabolites generated by root exudates. Knowing PGPB genetic variation may expand the knowledge base for beneficial plant–microbe interactions and may be useful in formulating new inoculants and improving crop processes for the most efficient usage (Yadav et al. 2015; Mitra 2017; Khoshmanzar et al. 2020).

Endo-rhizosphere is rich in diverse nutrients associated with bulk soil due to an abundance of root exudates, including amino acids, organic acids, enzymes, carbohydrates, minerals, and vitamins (Gray and Smith 2005). Root exudates emit ions, water, and oxygen which common of all produce compounds containing carbon (Uren 2000). Some root exudates are repellent pathogenic, whereas others are appealing to helpful microbes (Panneerselvam et al. 2019; Ahemad and Kibret 2014) based on physiological status, plant types, and microorganisms (Kang et al. 2010). Various associations between rhizobacteria and plant arise in the rhizosphere. For example, signal molecule interactions between plant roots and rhizobacteria are similarly important and occur in the rhizosphere (Werner 2000), and these interactions may be root–insect, root–root, and root–microbe interactions that influence plant growth and crop yields (Shaikh et al. 2018; Bais et al. 2006).

Despite the general agreement of extracellular plant-derived signals can affect the behavior of bacteria in the rhizosphere, very little is known about the impact of these signals on the patterns of bacterial gene expression and the role of such genes with altered expression in the plant–microbe interaction (PMI) process (Goldberg 2000). It is also hypothesized that plant selection of specific microbial communities relies, at least in part, on the activation of unusual patterns of gene expression in the microbe in response to molecular signals from the host (Morrissey et al. 2004). In the period of sustainable crop growth, PMI in the rhizosphere plays a crucial part in converting, mobilizing, solubilizing, etc. nutrients from a small supply of nutrients, and eventually taking up plant-based vital nutrients to achieve their maximum genetic potential.

6.5 Genes and Plant Growth Promotion

It has already been determined that all plant–PGPR interaction mechanisms are based on genetics and signal transduction components between the two parties, and due to the low studies in this part, our understanding of transcriptional changes during colonization is not very broad (Zhuang et al. 2007). The number of PGPR-inoculated plant reports about transcriptional changes is not copious, and most of these studies are based on the Induced Systemic Resistance (ISR) of the plant (especially *Arabidopsis thaliana*) by PGPRs such as *Bradyrhizobium* strain ORS278 (Cartieaux et al. 2008), *Pseudomonas* sp. (Verhagen et al. 2004), and *B. subtilis* (Zhuang et al. 2007). In recent years, studies of *Arabidopsis* have shown that transcriptional responses in the plant are highly dependent on its PGPR partner.

The biofilm formation has been shown to have a very important effect on bacterial establishment and colonization of plant roots (Rudrappa et al. 2008). Another effect of biofilm formation is its protective effect, which protects the root from the invasion of pathogens (Compant et al. 2005) in the bacterium *B. amyloliquefaciens*. The two genes *ycmA* and *luxS* have been identified by root exudates. *ycmA* plays a role in the formation of critical biofilm (Fernández et al. 2012). It has been shown that root exudates, in addition to expressing the above genes, have increased the transcription intensity of these genes (Fan et al. 2012).

Plant growth hormones are an important factor in regulating plant development including auxin, cytokinin, and bioactive gibberellin (Han et al. 2018). Auxin is involved in lateral root growth and hypocotyls elongation, cytokinins, and gibberellins are phytohormones that regulate growth, seed germination, seed, fruit growth, etc. (Swain and Singh 2005). The changes in the rate of transcription of genes associated with plant hormones have been reported in PGPR-inoculated plants (Xie et al. 2015). It has been reported that the exact concentration of auxin and cytokinin in the plant should be adjusted between the root system and the aerial part of the plant, and the slightest change in the transcriptional rate of genes related to phytohormones can alter the growth and pattern of all plant parts, which have been reported in PGPR-induced plants (Skvortsov and Azhikina 2010).

In the PGPR, *P. putida* strain MTCC5279 has been reported to induce many genes with the help of plants. These genes are involved in maintaining the integrity of the genome, auxin production responsive to plant growth, signaling, and induction of ABA in induced systemic resistance, nutrient uptake, inducing the biosynthesis pathway of amino acids, etc. (Jatan et al. 2019). It has been reported that the presence of PGPR in the roots of *Arabidopsis thaliana* leads to RNA helicase (*PRH75, At5g62190*) overexpression, which is needed to regulate plant growth and development in young tissues (Ng and Yanofsky 2001; Sessitsch et al. 2005). It has also been reported that *At2g15890* is responsible for the emergence of petals and stamen identity, which is overexpressed in the presence of PGPR.

In plant–PGPR interaction, it is necessary to identify signals released from both sides to better understand this relationship (Cartieaux et al. 2008). Understanding these signals causes the bacteria to be attracted to the roots of the plant to settle in the roots, Chemotaxis, mobility caused by flagella, riding, and production of surfactants are among the movements of bacteria in the roots (Compant et al. 2005). The analysis of transcriptional modifications in the genes involved in chemotaxis (cheC, cheD) and motility (hag, fliP, fliD, and flgM) found that the root exudate produces compounds that enhance the expression of these genes (Sessitsch et al. 2005).

In order to investigate the regulatory functions of PGPR in rhizosphere and plant activity, in particular the mechanisms involved in stimulating plant growth due to the expression of specific genes in the plant affected by PGPR, transcriptional studies are required and may provide an in-depth understanding of plant–microbe relations.

6.6 Transcriptomics Analyses

Identification and characterization of PGPRs has been performed by different phenotypic and molecular methods. The new approaches are generally based on molecular methods to study of beneficial characterizations of rhizobacteria which may decrease the time of findings from days to a few hours. These methods such as transcriptomics are very particular, optional, and trustworthy to find PGPRs, as well as they can establish the association between function and organization of the PGPRs (Kasa et al. 2015). Transcriptomics is used as a useful methodology to analyze host–microorganism interactions. Only a few reports were carried out with PGPRs (Table 6.1) (Chauhan et al. 2019), while this method has been widely applied to study of gene expression related to pathogenic microbes infecting their host plant (Tiwari et al. 2017). Some of the genes involved in metabolism and chemotaxis from *P. aeruginosa* were recognized to react to sugar-beet root exudates (Chauhan et al. 2019).

Table 6.1 Some studies of PGPRs transcriptomics

Function	Bacteria	References
Salt tolerance	IcePGPB *Halomonas* sp.	Zhang et al. (2020)
Soybean resistance to SCN	*Bacillus simplex*	Kang et al. (2018)
Nitrogen fixation	*Paenibacillus riograndensis*	Brito et al. (2017)
Synthesis of plant hormones Biofilm formation Nonribosomal synthesis of lipopeptides and polyketides	*Bacillus amyloliquefaciens*	Zhang et al. (2015)
Responses of PGPR to root exudates	*Bacillus amyloliquefaciens*	Fan et al. (2012)
Rice seedlings	*Bacillus subtilis*	Xie et al. (2015)

6.6.1 Isolation of RNA (Properties of Bacterial RNA)

Transcriptome analysis of plant growth-promoting bacteria needs to the isolation of bacterial RNA. A typical bacterial cell contains 0.05–0.1 pg of RNA, which constitutes approximately 6% of the cell weight. The total bacterial RNA, i.e., the bacterial transcriptome, contains a large amount of non-coding RNA species, such as tRNA and rRNA; moreover, the latter one as a structural and functional component of ribosomes represents the fraction largest in its relative content. In addition, the total RNA of the bacterial cell contains regulatory RNA and mRNA. All of the mRNAs, whose amount in only a few cases exceeds 4% of the total cell RNA, are gene transcripts of this cell (Brown 2002). A distinctive feature of bacteria is a high rate of adaptation to changing environmental conditions; correspondingly, the variations in gene expression in reaction to either host defense or the action of drugs are a necessary condition for the survival and function of intracellular pathogens. However, the rate of changes in the functional state of a bacterial cell depends not only on the rate of synthesis of new gene transcripts but also on the degradation rate of old transcripts. It has been demonstrated that the half-life of bacterial RNA is relatively short, amounting on average to 7 min; moreover, this time for several mRNAs is less than 2 min (Hambraeus et al. 2003). Ribonucleases, in particular, RNase E, are the main contributors to the degradation of bacterial transcripts. Therefore, the possibility of isolating high-quality RNA appropriate for further analysis depends on how quickly the method used for RNA isolation allows the ribonucleases to be inactivated and RNA to be stabilized (Kennell 2002).

6.6.2 RNA Amplification and the Synthesis of Bacterial cDNA

The relative prokaryotic RNA content in an infected tissue specimen is low, and the isolated RNA quantity is frequently insufficient for transcriptome analysis. Therefore, researchers have to resort to various methods of RNA or cDNA amplification (Lang et al. 2009). For a high-throughput whole transcriptome analysis, the total RNA enriched with bacterial species is usually converted into a cDNA sample using reverse transcription enzymes. To construct the first cDNA strand from prokaryotic RNA, the three primer types are used, namely, specifically selected primers to certain regions of individual genes (gene-specific primers) (Stahlberg et al. 2004), oligo (dT) primers complementary to poly(A) 3′terminal regions of mRNA, and random primers. The method for cDNA synthesis using gene-specific primers is inapplicable to bacterial whole transcriptome analysis. The question regarding the degree of prokaryotic mRNA polyadenylation is still open. Although it has been shown that some subpopulations of bacterial mRNAs are polyadenylated (Adilakshmi et al. 2000), the pool of prokaryotic cDNAs produced by this method is insufficiently representative (Lakey et al. 2002). The use of random priming for synthesizing bacterial cDNA also has some disadvantages, the main one of which is the unequal probability of synthesis from a particular primer under specified conditions of the reaction medium, which also leads to the degeneration of a cDNA pool. All of this complicates the work with bacterial mRNA and interferes with the application of several popular methods for the analysis of gene expression (SAGE and its derivatives) (Skvortsov and Azhikina 2010).

6.6.3 Expressed Sequence Tags (EST)

The use of expressed sequence tags dates related to the early 1980s (Parkinson and Blaxter 2009). Expressed sequence tags (ESTs) are relatively short reads (200–800 base pairs (bp)) generated from cDNA clones from which PCR primers can be derived. Since they indicate the expressed section of a genome, ESTs are extremely useful for the identification of genes and verification of gene predictions, therefore represent an inexpensive replacement for a complete sequence of genomes. EST analysis and SAGE are unsuitable to the analysis of bacterial transcriptomes due to the specific particularity of mRNA polyadenylation (Skvortsov and Azhikina 2010).

6.6.4 Serial and Cap Analysis of Gene Expression (SAGE/ CAGE)

This method (SAGE) is a highly effective technology to get a global gene expression profile of a cell or tissue (Velculescu et al. 2000). This technique can be used to characterize a set of specific genes by comparing the profiles created for a pair of cells that are kept at various conditions and also for the explanation of quantitative gene expression template that does not depend on the previous access of transcript information (Polyak and Riggins 2001; Velculescu et al. 2000). SAGE methodology isolates short fragments of genetic data from the expressed genes. These unique sequence labels approximately 9–10 base pairs (bp) are conjoined sequentially into long DNA fragments for lump-sum sequencing. The serial analysis of many gene-specific labels allows the concurrent accumulation of information from genes expressed in the tissue of interest and creates an expression profile in cell or tissue under study (Hogenesch et al. 2001). This sequencing information is then examined to identify gene expression level and each gene expressed in the cell (Daly 2002), as well as forms a library that analyzes the variances in gene expression between cells. The abundance of each SAGE label in the cloned multimers directly reflects the transcript frequency (Velculescu et al. 2000).

Most SAGE experiments have used the 4-bp recognition site anchoring enzyme *Nla*III, predicted to occur every 256 bp and thus present on most mRNA species. However, creating a second SAGE library with a different anchoring enzyme may be useful for detecting transcripts without a *Nla*III site and also for reconfirming transcript identity in those with both anchoring restriction sites. This may significantly lessen the work associated with data analysis, but the marginal utility of such an approach remains to be demonstrated. Next, the sample is equally divided into two separate tubes and ligated to two different linkers, A or B. Both linkers contain the recognition site for *Bsm*FI, a type IIS restriction enzyme that cuts 10-bp 3 from the anchoring enzyme recognition site. *Bsm*FI generates a unique oligonucleotide known as the SAGE tag, hence called the tagging enzyme (TE). The SAGE tags released from the oligo(dT) beads are then separated, blunted, and ligated to each other to give rise to ditags. The ditags are PCR amplified, released from the linkers, gel purified, serially ligated, cloned, and sequenced using an automated sequencer (SAGE2) (Patino et al. 2002).

6.7 Methods for Investigating Gene Expression

6.7.1 Methods Based on Hybridization

6.7.1.1 Northern Blot

The Northern blot is a molecular biology technique used in a mixture of complete RNAs to check for different populations of RNAs (Moustafa and Cross 2016). The approach is focused upon a nucleotide strand's capacity to bind to its complementary strand. The name Northern blot was granted in reference to the Southern blot (Southern 1975), which, although with DNA, follows the same definition (Moustafa and Cross 2016). The approach is used to identify differences in target gene expression rates in a certain tissue, cell form, or biological samples under specified environmental conditions (e.g., stressed vs. non-stressed plants). A number of changes have been proposed since its introduction to improve Northern blot results and durability. Vacuum-blotting, RNA-transfer simulation, and RNA ultraviolet fixation are among these modifications (Kroczek and Siebert 1990).

6.7.1.2 Microarray Method

DNA microarrays are a series of small probes of oligonucleotides bound to a rigid surface (i.e., glass) unique to thousands of genes. This method is focused on nanofabrication methods and is focused on simultaneous hybridization of the detector, where it is possible to calculate the relative abundance of thousands of transcripts from two or more samples at once.

The first step in microarray protocols, as in the case of RT-PCR, is isolation of RNA and its retro-transcription into cDNA. After that phase, numerous labeling techniques are carried out by commercial channels (Valdés et al. 2013). The fluorescence produced by identified targets bound with the probes is determined after hybridization. Raw data (intensity of fluorescence) is transformed and normalized to eliminate systemic variance and enable the comparison of various samples to be rendered acceptable. Therefore, the analysis of gene expression between various samples is achieved by evaluating a ratio of fluorescence strength (Karakach et al. 2010). Microarrays usually have a strong background noise owing to cross-hybridization, rendering low-copy transcripts challenging to identify. Microarrays are closed platforms which require prior knowledge of the assessed organism to design the probes (Karakach et al. 2010).

6.7.2 Methods Based on PCR

6.7.2.1 Real-Time Quantitative PCR

Real-time quantitative polymerase chain reaction (qPCR) is actually one of the most commonly used techniques for researching identified candidate gene expression. When this process is used, complementary DNA (cDNA) is synthesized by reverse transcriptase from isolated mRNA. Gene-specific primers are then used to amplify the interest gene, and the abundance of that particular cDNA molecule is tracked in real time during amplification using fluorescent dyes (Josefsen et al. 2012). One can measure the actual sum of starting RNA transcript from the kinetics of PCR product creation and do quantitative comparisons between samples (Pfaffl 2012). It is, however, very inexpensive, and simple to pursue in-house relative to many other RNA technologies; therefore, it is accessible to most study classes. However, questions have been posed repeatedly about (1) the lack of quality assurance and standardization in a real-time qPCR workflow, and (2) the lack of adequate information published in several publications to allow for qualitative assessment of the findings (Chapman and Waldenström 2015; Derveaux et al. 2010).

6.7.2.2 Real-Time RT-PCR

Real-time RT-PCR is based on the fluorescence detection during the PCR reaction. The PCR reaction mix for this reason involves intercalating fluorophores or specified labeling probes. Because of these properties, RT-PCR in real time became the global norm for gene expression assays because it is fast, accurate, and capable of detecting tiny amounts and transcript number changes (Kubista et al. 2006). One of the key drawbacks of RT-PCR in real-time is the small number of assays per sample that can be performed concurrently for operational purposes. Transcriptomics' high-throughput criteria contribute to PCR not being considered a transcriptomic method in real time but is commonly accepted as a testing tool (Lamas et al. 2016; Nonis et al. 2014).

Several reasons have led to the development of this technique into a common testing tool: (1) as a homogeneous assay it eliminates the need for post-PCR processing; (2) a broad dynamic range (>107-fold) allows for a direct comparison between RNAs that vary greatly in their abundance; and (3) the assay understands the PCR's intrinsic quantitative ability, rendering it a quantitative.

6.7.3 Next-Generation Sequencing: RNA-Seq

A particular technique called RNA-seq rapidly extended NGS from DNA-sequencing to gene expression study. Such technologies are open systems focused on a combination of high-throughput sequencing and bioinformatics methods that collect and measure the transcripts present in an RNA extract (Lowe et al. 2017). The application of NGS allows for comparisons of genomes and transcriptomes between organisms, species, and meta-populations, thereby allowing a better underlying of ecological and evolutionary processes. For example, NGS data analyzes help researchers to investigate selection trends (Angeloni et al. 2012) and to recognize candidate genes that underpin adaptation (Stapley et al. 2010). It is significant because loci are candidates for variability in fitness under selection and could be representative of the particular selective forces involved (Nielsen 2005). NGS approaches allow demographic and genetic popularization processes to be separated from adaptive processes (Kirk and Freeland 2011; Angeloni et al. 2012).

Genome-wide coverage utilizing NGS enables processes impacting more than one locus to be observed, such as genomic regions providing signs of systematic sweeps (Schlötterer 2003; Boitard et al. 2012). Statistical approaches are possible to improve the identification of selective sweeps, such as the pooled-sample sensitive approach developed by Boitard et al. (2012), which was able to use coverage for genotyping as small as 1X per individual.

6.7.3.1 Methods Based on Sequencing

Transcriptome analysis is normally completed using methodologies based on hybridization or sequencing which involve mandatory of fluorescently tagged fragments to supplementary probe sequences (e.g., in microarray) (Okoniewski and Miller 2006) although these methods suffer from low determination, specificity, and sensibility (Mantione et al. 2014). Later, Sanger sequencing-based approaches such as SAGE, CAGE, and MPSS were developed; however, these techniques have major difficulties such as pay attention to parochial transcripts structure and incapability to individuate between isoforms (Wang et al. 2009; Anamika et al. 2016). Next-generation sequencing (NGS) is approach that allows sequencing of many gene fragments in parallel and RNA sequencing (RNA-seq) has appeared as a potent technique for assessment of the transcriptome (Okoniewski and Miller 2006).

6.7.3.2 Using RNA-Seq for the Analysis of Bacterial Transcriptomes

Some benefits of the use of RNA-seq for the study of transcriptomics includes not relying on prior information about sequences of genome or transcript and also methods of sequencing without the cross-hybridization, decreasing background levels. This method improves the range of detection because signals are not saturated as easily (Wang et al. 2009). Using RNA-seq allows the discovery of several transcriptional features such as the 5′ end of all RNAs (Wurtzel et al. 2010), and TSSs can be identified by choosing for primary RNA transcripts with a 5-′-triphosphate (Sharma et al. 2009). Most RNA-seq studies are performed on instruments, while direct sequencing of RNAs is probable (Ozsolak et al. 2009) and sequence DNAs designed for DNA-based sequencing because of the development of commercial instruments. Hence, the preparation of the cDNA library from RNA is a necessary stage for RNA-seq. Every cDNA in an RNA-seq library is created of a cDNA insert of determined size connected to adapter fragments. Methods of the cDNA library preparation differ according to the RNA species under examination, so it can vary in sequence, size, abundance, and structural properties (Hrdlickova et al. 2017). To determine the ratio between primary and processed for a transcript can be used from an RNA ligation stage that differentially labels primary and processed RNA molecules. This method should be directly appropriate to RNA-seq though it has not been used on a global scale (Filiatrault 2011).

6.7.3.3 Bioinformatics Analysis of RNA-Seq Data

Several stages are necessary for analysis of the RNA-seq data including quality assessment, information processing, transcriptome assembly, quantification, statistical analysis, and functional annotation (Fig. 6.1) which are described in the following (Anamika et al. 2016):

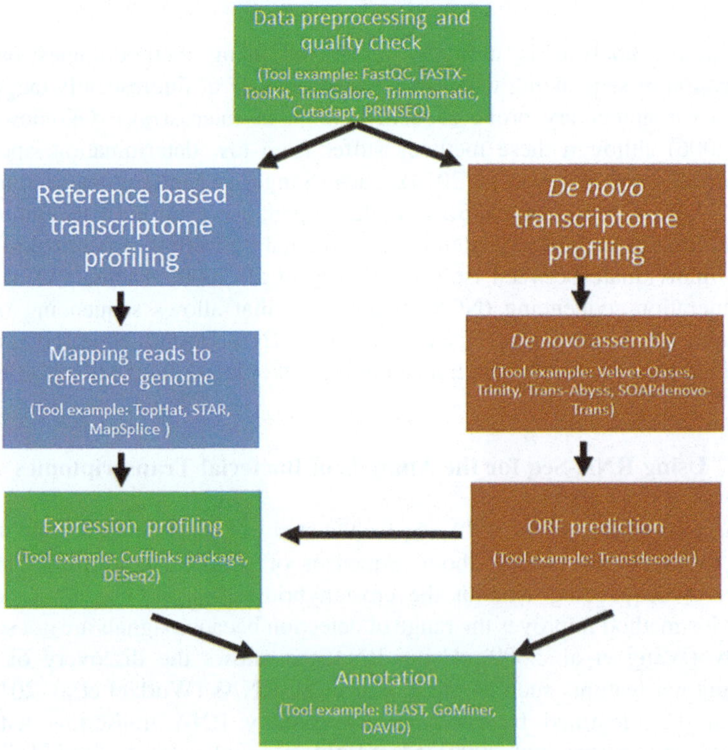

Fig. 6.1 Basic RNA-seq data analysis workflow (Anamika et al. 2016)

6.7.3.4 Quality Check and Data Preprocessing

One of the expected outcomes of transcriptomics analysis is the process of finding knowledge from a lot of data that is retrieved. Then this function requires the processing of millions of short reads (~100 nucleotides). Data preprocessing usually includes filtering, alignment, assembly, clustering, counting, and normalization for each experimental condition, and also analyses of differential expression in these conditions. In this regard, the FASTQ format has recently considered (Fig. 6.2) (Lee-Liu et al. 2012).

FastQC

FastQC can be run from both Linux (using command line) and Windows systems as an independent Java-based program. This function is an easy-to-use tool that assesses the quality of read data from the next-generation sequencers method (NGS). The inputs for FastQC can be either in the compressed or uncompressed form in Fastq and SAM format that relates primary statistics such as over-represented sequences, base content, and quality, and adapter sequence (Anamika et al. 2016).

6.7.3.5 Transcriptome Assembly

Generally, the first stage of the computational workflow in a transcriptome profiling study is aligning the quality-evaluated reads to the reference transcriptome using an appropriate read aligner. In the following, the reads are used to determination of the genomic features quantity that needs to be standardized before comparison of different experimental conditions. In order to draw statistical inference on their difference in expression between samples, the standardized counts are used. Finally, the expressed genes are processed to conclude biological insights. The success of transcriptome assembly depends on decisions that the user takes while choosing reference genome, annotation, tools, and associated parameter values at every step of the analysis (Anamika et al. 2016). Decisions that the user takes while choosing reference genome, annotation, tools, and associated parameter values at every step of the analysis brings success in transcriptome assembly (Anamika et al. 2016).

6.7.3.6 Annotation and Pathway Analysis

After computing the abundance of transcriptome and mapping the assembled contigs fragments to the assembled transcriptome or reference genome and differential expression data analysis, it can be searched coding regions within de novo assembled transcripts using ORF predictor tools such as Transdecoder (http://transdecoder.

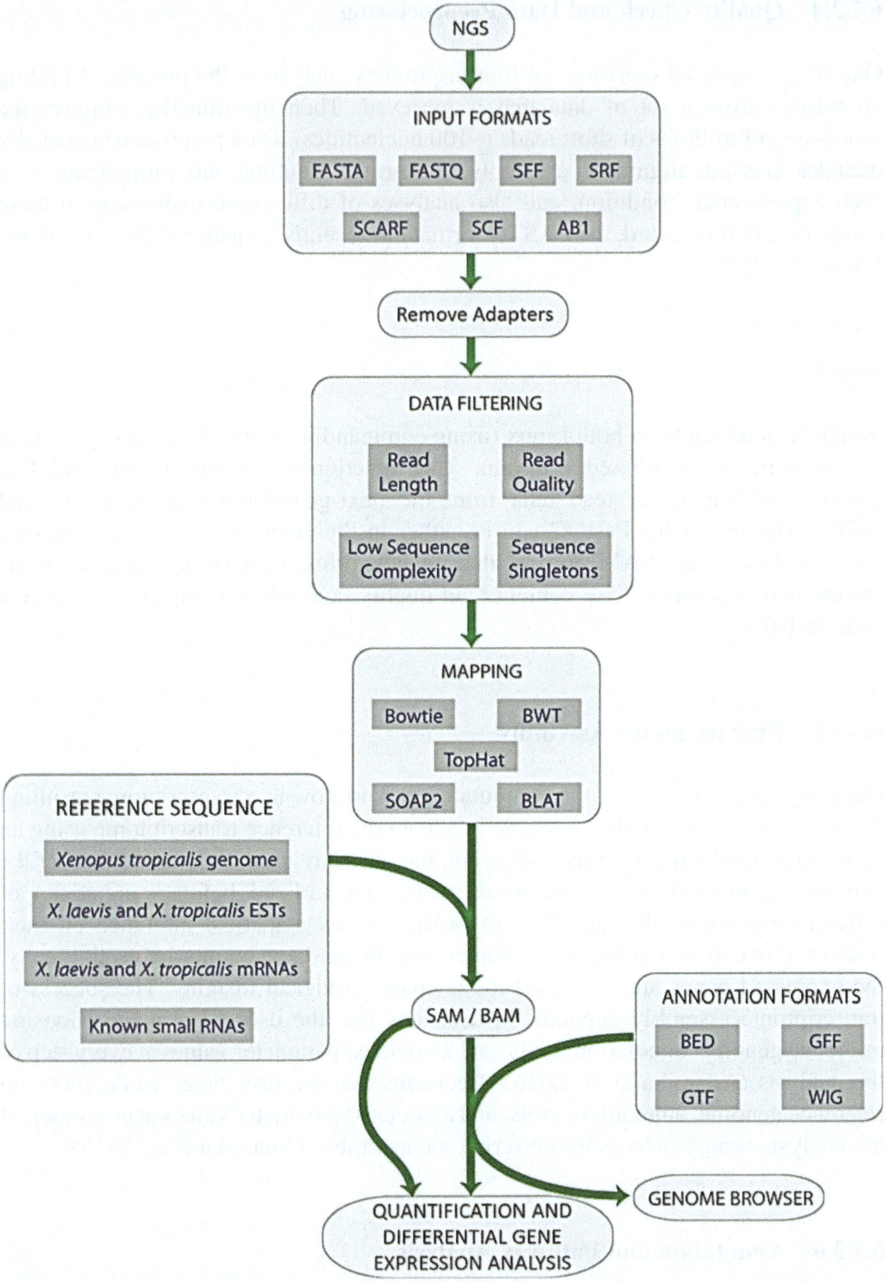

Fig. 6.2 RNA-seq data analysis flowchart. Next-generation sequencing results can be delivered in several input formats. However, FASTQ format has become a standard in the field. Adapters are first removed from raw sequences, followed by data filtering. The next stage is the mapping of the filtered sequences to *Xenopus* reference sequences. Output files (SAM/BAM) can be used for quantification and differential gene expression analyses. It is important to mention that although deep sequencing technologies and their associated software are rapidly changing, the diagram

github.io/). Two tools of homologous gene identification of assembled transcripts are BLAT and BLAST (Altschul et al. 1990).

6.8 Conclusion

The plant–microbe association in the soil is inevitable, and if this relationship is being optimal, it can guarantee the growth and yield of the plant under different conditions (stress and non-stress). In this situation that the growth of plants and production of agricultural products is facing many problems due to biotic and abiotic stresses and agricultural products are constantly decreasing, so regulating and establishing plant–microbial relationship can greatly reduce the problems of universal food production. In order to establish an ideal relationship between the plant and beneficial soil microbes, a deep insight about this relationship is needed. The more the unknowns of this relationship are resolved, the more the production of agricultural products will increase with the help of these beneficial microbes. A PGPR may affect the entire life cycle of a plant, accelerate its growth rate, and shorten its growing season, both of which are related to most crops. The whole plant–microbial relationship is under the full control of genes, and the expression or silencing of these genes is influenced by various factors. Identifying and understanding the involve genes and mechanisms in this relationship can have profound implications and is important for improving strategies for using these beneficial bacteria in agriculture.

Unlike the genome, which is stable, the transcriptome varies (highly dynamic) from moment to moment in the throughout the life of an organism. To understand these changes, analysis of content and composition of the mRNA molecules (transcriptome analysis) can be helpful. With these studies, it is possible to estimate which gene is expressed at a particular time. Also, specific RNA molecules numbers analyzing can show the intensity of transcription of a gene. Therefore, to understand the function of genes (coding or non-coding) and also to detect their expression in different stages of the organism's life, transcriptome analysis can be used. Finally, the study of transcriptome analysis results gives us a lot of practical information about the deep understanding of plant–PGPR interaction during plant lifetime, which could be a window into new science for relevant biological associations and solve the shortage of agricultural products production universal problems.

Fig. 6.2 (continued) displayed here is robust because it is general and not attached to any specific software or technology (Lee-Liu et al. 2012)

References

Adilakshmi T, Ayling PD, Ratledge C (2000) Polyadenylylation in mycobacteria: evidence for oligo (dT)-primed cDNA synthesis. Microbiology 146(3):633–638

Ahemad M, Kibret M (2014) Mechanisms and applications of plant growth promoting rhizobacteria: current perspective. J King Saud Univ Sci 26(1):1–20

Altschul SF, Gish W, Miller W, Myers EW, Lipman DJ (1990) Basic local alignment search tool. J Mol Biol 215(3):403–410

Anamika K, Verma S, Jere A, Desai A (2016) Transcriptomic profiling using next generation sequencing-advances, advantages, and challenges. In: Next generation sequencing-advances, applications and challenges, vol 9, pp 7355–7365

Angeloni F, Wagemaker N, Vergeer P, Ouborg J (2012) Genomic toolboxes for conservation biologists. Evol Appl 5(2):130–143

Badri DV, Weir TL, van der Lelie D, Vivanco JM (2009) Rhizosphere chemical dialogues: plant–microbe interactions. Curr Opin Biotechnol 20(6):642–650

Bais HP, Weir TL, Perry LG, Gilroy S, Vivanco JM (2006) The role of root exudates in rhizosphere interactions with plants and other organisms. Annu Rev Plant Biol 57:233–266

Banerjee MR, Yasmin L (2002). Sulfur oxidizing rhizobacteria: an innovative environment friendly soil biotechnological tool for better canola production. In: Proceeding of Agroenviron, Cairo, Egypt, 1–7

Berendsen RL, Pieterse CM, Bakker PA (2012) The rhizosphere microbiome and plant health. Trends Plant Sci 17(8):478–486

Berg G, Smalla K (2009) Plant species and soil type cooperatively shape the structure and function of microbial communities in the rhizosphere. FEMS Microbiol Ecol 68(1):1–13

Boitard S, Schlötterer C, Nolte V, Pandey RV, Futschik A (2012) Detecting selective sweeps from pooled next-generation sequencing samples. Mol Biol Evol 29(9):2177–2186

Brito LF, Irla M, Kalinowski J, Wendisch VF (2017) Detailed transcriptome analysis of the plant growth promoting Paenibacillus riograndensis SBR5 by using RNA-seq technology. BMC Genomics, 18(1):1–15

Brown TA (2002) Genomes, 2nd edn. Wiley-Liss, Oxford, p 2002. https://www.ncbi.nlm.nih.gov/books/NBK21128/

Bulgarelli D, Schlaeppi K, Spaepen S, Van Themaat EV, Schulze-Lefert P (2013) Structure and functions of the bacterial microbiota of plants. Annu Rev Plant Biol 64:807–838

Burd GI, Dixon DG, Glick BR (2000) Plant growth-promoting bacteria that decrease heavy metal toxicity in plants. Can J Microbiol 46(3):237–245

Cartieaux F, Contesto C, Gallou A, Desbrosses G, Kopka J, Taconnat L, Renou JP, Touraine B (2008) Simultaneous interaction of *Arabidopsis thaliana* with *Bradyrhizobium* sp. strain ORS278 and *Pseudomonas syringae* pv. tomato DC3000 leads to complex transcriptome changes. Mol Plant-Microbe Int 21(2):244–259

Chapman JR, Waldenström J (2015) With reference to reference genes: a systematic review of endogenous controls in gene expression studies. PLoS One 10(11):e0141853

Chauhan PS, Lata C, Tiwari S, Chauhan AS, Mishra SK, Agrawal L, Chakrabarty D, Nautiyal CS (2019) Transcriptional alterations reveal Bacillus amyloliquefaciens-rice cooperation under salt stress. Sci Rep 9(1):1–13

Cocking EC (2003) Endophytic colonization of plant roots by nitrogen-fixing bacteria. Plant Soil 252(1):169–175

Compant S, Duffy B, Nowak J, Clement C, Barka EA (2005) Use of plant growth-promoting bacteria for biocontrol of plant diseases: principles, mechanisms of action, and future prospects. Appl Environ Microbiol 71:4951–4959

Daly MJ (2002) Estimating the human gene count. Cell 109(3):283–284

Derveaux S, Vandesompele J, Hellemans J (2010) How to do successful gene expression analysis using real-time PCR. Methods 50(4):227–230

Dey RK, Pal KK, Bhatt DM, Chauhan SM (2004) Growth promotion and yield enhancement of peanut (Arachis hypogaea L.) by application of plant growth-promoting rhizobacteria. Microbiol Res 159(4):371–394

Dobbelaere S, Vanderleyden J, Okon Y (2003) Plant growth-promoting effects of diazotrophs in the rhizosphere. Crit Rev Plant Sci 22(2):107–149

Fan B, Carvalhais LC, Becker A, Fedoseyenko D, von Wirén N, Borriss R (2012) Transcriptomic profiling of Bacillus amyloliquefaciens FZB42 in response to maize root exudates. BMC Microbiol 12(1):116

Fernández L, Breidenstein EB, Song D, Hancock RE (2012) Role of intracellular proteases in the antibiotic resistance, motility, and biofilm formation of Pseudomonas aeruginosa. Antimicrob Agents Chemother 56(2):1128–1132

Filiatrault MJ (2011) Progress in prokaryotic transcriptomics. Curr Opin Microbiol 14(5):579–586

Glick BR, Pasternak JJ (2003) Plant growth promoting bacteria. In: Glick BR, Pasternak JJ (eds) Molecular biotechnology principles and applications of recombinant DNA, 3rd edn. ASM Press, Washington, pp 436–454

Glick BR, Penrose DM, Li J (1998) A model for lowering plant ethylene concentration by plant growth promoting rhizobacteria. J Theor Biol 190:63–68

Goldberg JB (2000) 'Pseudomonas' 99, The seventh international congress on Pseudomonas: biotechnology and pathogenesis', organized by the American Society for Microbiology, was held in Maui, HI, USA, 1–5 September 1999. Trends Microbiol 8(2):55–57

Gray EJ, Smith DL (2005) Intracellular and extracellular PGPR: commonalities and distinctions in the plant–bacterium signaling processes. Soil Biol Biochem 37(3):395–412

Hambraeus G, von Wachenfeldt C, Hederstedt L (2003) Genome-wide survey of mRNA half-lives in Bacillus subtilis identifies extremely stable mRNAs. Mol Gen Genomics 269(5):706–714

Han X, Zeng H, Bartocci P, Fantozzi F, Yan Y (2018) Phytohormones and effects on growth and metabolites of microalgae: a review. Fermentation, 4(2), p 25

Hassan MK, McInroy JA, Kloepper JW (2019) The interactions of rhizodeposits with plant growth-promoting rhizobacteria in the rhizosphere: a review. Agriculture 9(7):142

Heidarpour A, Aliasgharzad N, Khoshmanzar E, Lajayer BA (2019) Bio-removal of Zn from contaminated water by using green algae isolates. Environ Technol Innov 16:100464

Hiltner L (1904) Uber nevere Erfahrungen und Probleme auf dem Gebiet der Boden Bakteriologie und unter besonderer Beurchsichtigung der Grundungung und Broche. Arbeit Deut Landw Ges Berlin 98:59–78

Hogenesch JB, Ching KA, Batalov S, Su AI, Walker JR, Zhou Y et al (2001) A comparison of the Celera and Ensembl predicted gene sets reveals little overlap in novel genes. Cell 106 (4):413–415

Hrdlickova R, Toloue M, Tian B (2017) RNA-Seq methods for transcriptome analysis. Wiley Interdiscip Rev RNA 8(1):e1364

Huot B, Yao J, Montgomery BL, He SY (2014) Growth–defense tradeoffs in plants: a balancing act to optimize fitness. Mol Plant 7(8):1267–1287

Jatan R, Tiwari S, Asif MH, Lata C (2019) Genome-wide profiling reveals extensive alterations in Pseudomonas putida-mediated miRNAs expression during drought stress in chickpea (Cicer arietinum L.). Environ Exp Bot 157:217–227

Josefsen MH, Löfström C, Hansen T, Reynisson E, Hoorfar J (2012) Instrumentation and fluorescent chemistries used in qPCR. In: Quantitative real-time PCR in applied microbiology. Caister Academic Press, Norfolk, pp 27–52

Kang BG, Kim WT, Yun HS, Chang SC (2010) Use of plant growth-promoting rhizobacteria to control stress responses of plant roots. Plant Biotechnol Rep 4(3):179–183

Kang W, Zhu X, Wang Y, Chen L, Duan Y (2018) Transcriptomic and metabolomic analyses reveal that bacteria promote plant defense during infection of soybean cyst nematode in soybean. BMC Plant Biol 18(1):1–14

Karakach TK, Flight RM, Douglas SE, Wentzell PD (2010) An introduction to DNA microarrays for gene expression analysis. Chemom Intell Lab Syst 104(1):28–52

Kasa P, Modugapalem H, Battini K (2015) Isolation, screening, and molecular characterization of plant growth promoting rhizobacteria isolates of Azotobacter and Trichoderma and their beneficial activities. J Nat Sci Biol Med 6(2):360

Kennell D (2002) Processing endoribonucleases and mRNA degradation in bacteria. J Bacteriol 184 (17):4645–4657

Kennedy IR, Choudhury ATMA, Kecskés ML (2004) Non-symbiotic bacterial diazotrophs in crop-farming systems: can their potential for plant growth promotion be better exploited? Soil Biol Biochem 36(8):1229–1244

Khan AG (2005) Role of soil microbes in the rhizospheres of plants growing on trace metal contaminated soils in phytoremediation. J Trace Elem Med Biol 18(4):355–364

Khoshmanzar E, Aliasgharzad N, Neyshabouri MR, Khoshru B, Arzanlou M, Lajayer BA (2020) Effects of Trichoderma isolates on tomato growth and inducing its tolerance to water-deficit stress. Int J Environ Sci Technol 17(2):869–878

Kirk H, Freeland JR (2011) Applications and implications of neutral versus non-neutral markers in molecular ecology. Int J Mol Sci 12(6):3966–3988

Kroczek RA, Siebert E (1990) Optimization of northern analysis by vacuum-blotting, RNA-transfer visualization, and ultraviolet fixation. Anal Biochem 184(1):90–95

Kubista M, Andrade JM, Bengtsson M, Forootan A, Jonák J, Lind K et al (2006) The real-time polymerase chain reaction. Mol Asp Med 27(2–3):95–125

Lakey DL, Zhang Y, Talaat AM, Samten B, DesJardin LE, Eisenach KD et al (2002) Priming reverse transcription with oligo (dT) does not yield representative samples of Mycobacterium tuberculosis cDNA. Microbiology 148(8):2567–2572

Lamas A, Franco CM, Regal P, Miranda JM, Vázquez B, Cepeda A (2016) High-throughput platforms in real-time PCR and applications. In: Polymerase chain reaction for biomedical applications, vol 14, p 15

Lang JE, Magbanua MJM, Scott JH, Makrigiorgos GM, Wang G, Federman S et al (2009) A comparison of RNA amplification techniques at sub-nanogram input concentration. BMC Genomics 10(1):326

Lee-Liu D, Almonacid LI, Faunes F, Melo F, Larrain J (2012). Transcriptomics using next generation sequencing technologies. In: Xenopus protocols. Humana Press, Totowa, NJ, pp 293–317

Lowe R, Shirley N, Bleackley M, Dolan S, Shafee T (2017) Transcriptomics technologies. PLoS Comput Biol 13(5):e1005457

Lugtenberg B, Kamilova F (2009) Plant-growth-promoting rhizobacteria. Annu Rev Microbiol 63:541–556

Mantione KJ, Kream RM, Kuzelova H, Ptacek R, Raboch J, Samuel JM, Stefano GB (2014) Comparing bioinformatic gene expression profiling methods: microarray and RNA-Seq. Med Sci Monit Basic Res 20:138

Metzker ML (2010) Sequencing technologies—the next generation. Nat Rev Genet 11(1):31–46

Mitra D (2017) Isolation and characterization of potential nutrient mobilizer and plant growth promoters from some typical harsh environment of Karnataka. M.Sc. Thesis, Graphic Era University. Dehradun, Uttarakhand and ICAR-IIHR, Bengaluru

Mitra D, Sharma K, Uniyal N, Chauhan A, Sarkar P (2016) Study on plant hormone (indole-3-acetic acid) producing level and other plant growth promotion ability (pgpa) by Asparagus racemosus rhizobacteria. J Chem Pharm Res 8:995–1002

Mitra D, Jain D, Panneerselvam P (2019a) Microbial resources for sustainable agriculture. LAMBERT Academic Publishing. 1–137 pp. isbn:978-613-9-83608-6

Mitra D, Snežana A, Panneerselvam P, Manisha SA, Tanja V, Ganeshamurthy AN, Devvret V, Poonam RT, Divya J (2019b) Plant growth promoting microorganisms (PGPMs) helping in sustainable agriculture: current perspective. Int J Agric Sci Veter Med 7(2):50–74

Morgan JAW, Bending GD, White PJ (2005) Biological costs and benefits to plant–microbe interactions in the rhizosphere. J Exp Bot 56(417):1729–1739

Morrissey JP, Dow JM, Mark GL, O'Gara F (2004) Are microbes at the root of a solution to world food production? EMBO Rep 5(10):922–926

Moustafa K, Cross JM (2016) Genetic approaches to study plant responses to environmental stresses: an overview. Biology 5(2):20

Ng M, Yanofsky MF (2001) Function and evolution of the plant MADS-box gene family. Nat Rev Genet 2:186–195

Nielsen R (2005) Molecular signatures of natural selection. Annu Rev Genet 39:197–218

Nonis A, De Nardi B, Nonis A (2014) Choosing between RT-qPCR and RNA-seq: a back-of-the-envelope estimate towards the definition of the break-even-point. Anal Bioanal Chem 406 (15):3533–3536

Okoniewski MJ, Miller CJ (2006) Hybridization interactions between probesets in short oligo microarrays lead to spurious correlations. BMC Bioinformatics 7(1):276

Oldroyd GE, Murray JD, Poole PS, Downie JA (2011) The rules of engagement in the legume-rhizobial symbiosis. Annu Rev Genet 45:119–144

Ozsolak F, Platt AR, Jones DR, Reifenberger JG, Sass LE, McInerney P et al (2009) Direct RNA sequencing. Nature 461(7265):814–818

Panneerselvam P, Senapati A, Kumar U, Sharma L, Lepcha P, Prabhukarthikeyan SR, Jahan A, Parameshwaran C, Govindharaj GPP, Lenka S, Nayak PK (2019) Antagonistic and plant-growth promoting novel Bacillus species from long-term organic farming soils from Sikkim, India. 3 Biotech 9(11):416

Parkinson J, Blaxter M (2009) Expressed sequence tags: an overview. In: Expressed sequence tags (ESTs). Humana Press, Totowa, pp 1–12

Patino WD, Mian OY, Hwang PM (2002) Serial analysis of gene expression: technical considerations and applications to cardiovascular biology. Circ Res 91(7) 565–569

Pfaffl MW (2012) Quantification strategies in real-time polymerase chain reaction. Quantitative real-time PCR. Appl Microbiol 53–62

Pieterse CM, de Jonge R, Berendsen RL (2016) The soil-borne supremacy. Trends Plant Sci 21 (3):171–173

Polyak K, Riggins GJ (2001) Gene discovery using the serial analysis of gene expression technique: implications for cancer research. J Clin Oncol 19(11):2948–2958

Revillas JJ, Rodelas B, Pozo C, Martínez-Toledo MV, González-López J (2000) Production of B-group vitamins by two Azotobacter strains with phenolic compounds as sole carbon source under diazotrophic and adiazotrophic conditions. J Appl Microbiol 89(3):486–493

Rudrappa T, Biedrzycki ML, Bais HP (2008) Causes and consequences of plant-associated biofilms. FEMS Microbiol Ecol 64(2):153–166

Saghafi D, Ghorbanpour M, Asgari Lajayer B (2018) Efficiency of rhizobium strains as plant growth-promoting rhizobacteria on morphophysiological properties of *Brassica napus* L. under salinity stress. J Soil Sci Plant Nutr 18(1):253–268

Saghafi D, Delangiz N, Asgari Lajayer B, Ghorbanpour M (2019a) An overview of the improvement of crop productivity in saline soils by halotolerant and halophilic PGPRs. 3 Biotech 9:261

Saghafi D, Ghorbanpour M, Shirafkan Ajirloo H, Asgari Lajayer B (2019b) Enhancement of growth and salt tolerance in *Brassica napus* L. seedlings by halotolerant rhizobium strains containing ACC deaminase activity. Plant Physiol Rep 24:225–235

Saghafi D, Asgari Lajayer B, Ghorbanpour M (2020) Engineering bacterial ACC deaminase for improving plant productivity under stressful conditions. In: Sharma V, Salwan R, Al-Ani LKT (eds) Molecular aspects of plant beneficial microbes in agriculture. Elsevier, pp 259–272

Sarikhani MR, Khoshru B, Oustan S (2016) Efficiency of some bacterial strains in potassium release from mica and phosphate solubilization under in vitro conditions. Geomicrobiol J 33 (9):832–838

Sarikhani MR, Aliasgharzad N, Khoshru B (2019a) P solubilizing potential of some plant growth promoting bacteria used as ingredient in phosphatic biofertilizers with emphasis on growth promotion of Zea mays L. Geomicrobiol J 37(4):327–335

Sarikhani MR, Khoshru B, Greiner R (2019b) Isolation and identification of temperature tolerant phosphate solubilizing bacteria as a potential microbial fertilizer. World J Microbiol Biotechnol 35(8):126

Schlötterer C (2003) Hitchhiking mapping–functional genomics from the population genetics perspective. Trends Genet 19(1):32–38

Sessitsch A, Coenye T, Sturz AV, Vandamme P, Barka E, Wang-Pruski G, Faure D, Reiter B, Glick BR, Nowak J (2005) Burkholderia phytofirminssp. Nov., a novel plant-associated bacterium with plant beneficial properties. Int J Syst Evol Microbiol 55:1187–1192
Shaikh S, Wani S, Sayyed R (2018) Impact of interactions between rhizosphere and rhizobacteria: a review. J Bacteriol Mycol 5:1058
Sharma A, Nitharwal RG, Singh B, Dar A, Dasgupta S, Dhar SK (2009) Helicobacter pylori single-stranded DNA binding protein–functional characterization and modulation of H. pylori DnaB helicase activity. FEBS J 276(2):519–531
Shoebitz M, Ribaudo CM, Pardo MA, Cantore ML, Ciampi L, Cura JA (2009) Plant growth promoting properties of a strain of Enterobacter ludwigii isolated from Lolium perenne rhizosphere. Soil Biol Biochem 41(9):1768–1774
Singh BK, Millard P, Whiteley AS, Murrell JC (2004) Unravelling rhizosphere–microbial interactions: opportunities and limitations. Trends Microbiol 12(8):386–393
Skvortsov TA, Azhikina TL (2010) A review of the transcriptome analysis of bacterial pathogens in vivo: problems and solutions. Russian J Bioorg Chem 36(5):550–559
Southern EM (1975) Detection of specific sequences among DNA fragments separated by gel electrophoresis. J Mol Biol 98(3):503–517
Stahlberg A, Håkansson J, Xian X, Semb H, Kubista M (2004) Properties of the reverse transcription reaction in mRNA quantification. Clin Chem 50(3):509–515
Stapley J, Reger J, Feulner PG, Smadja C, Galindo J, Ekblom R et al (2010) Adaptation genomics: the next generation. Trends Ecol Evol 25(12):705–712
Swain SM, Singh DP (2005) Tall tales from sly dwarves: novel functions of gibberellins in plant development. Trends Plant Sci 10(3):123–129
Tiwari S, Prasad V, Chauhan PS, Lata C (2017) Bacillus amyloliquefaciens confers tolerance to various abiotic stresses and modulates plant response to phytohormones through osmoprotection and gene expression regulation in rice. Front Plant Sci 8:1510
Uren NC (2000) Types, amounts, and possible functions of compounds released into the rhizosphere by soil-grown plants. In: The rhizosphere. CRC Press, Boca Raton, pp 35–56
Valdés A, Ibáñez C, Simó C, García-Cañas V (2013) Recent transcriptomics advances and emerging applications in food science. TrAC Trends Anal Chem 52:142–154
Vamerali T, Bandiera M, Mosca G (2010) Field crops for phytoremediation of metal-contaminated land. A review. Environ Chem Lett 8(1):1–17
Velculescu VE, Vogelstein B, Kinzler KW (2000) Analysing uncharted transcriptomes with SAGE. Trends Genet 16(10):423–425
Venturi V, Keel C (2016) Signaling in the rhizosphere. Trends Plant Sci 21(3):187–198
Verhagen BWM, Glazebrook J, Zhu T, Chang HS, van Loon LC, Pieterse CMJ (2004) The transcriptome of rhizobacteria-induced systemic resistance in Arabidopsis. Mol Plant-Microbe Interact 17:895–908
Wang Z, Gerstein M, Snyder M (2009) RNA-Seq: a revolutionary tool for transcriptomics. Nat Rev Genet 10(1):57–63
Werner D (2000) Organic signals between plants and microorganisms. In: The rhizosphere. CRC Press, Boca Raton, pp 213–238
Wurtzel O, Sapra R, Chen F, Zhu Y, Simmons BA, Sorek R (2010) A single-base resolution map of an archaeal transcriptome. Genome Res 20(1):133–141
Xie S, Wu H, Chen L, Zang H, Xie Y, Gao X (2015) Transcriptome profiling of Bacillus subtilis OKB105 in response to rice seedlings. BMC Microbiol 15(1):21
Yadav AN, Sachan SG, Verma P, Saxena AK (2015) Prospecting cold deserts of north western Himalayas for microbial diversity and plant growth promoting attributes. J Biosci Bioeng 119 (6):683–693
Zamioudis C, Pieterse CM (2012) Modulation of host immunity by beneficial microbes. Mol Plant-Microbe Interact 25(2):139–150
Zhang N, Yang D, Wang D, Miao Y, Shao J, Zhou X, Shen Q (2015) Whole transcriptomic analysis of the plant-beneficial rhizobacterium Bacillus amyloliquefaciens SQR9 during enhanced biofilm formation regulated by maize root exudates. BMC Genomics 16(1):685

Zhang J, Wang P, Tian H, Tao Z, Guo T (2020) Transcriptome analysis of ice plant growth-promoting endophytic bacterium Halomonas sp. Strain MC1 to identify the genes involved in salt tolerance. Microorganisms 8(1):88

Zhuang X, Chen J, Shim H, Bai Z (2007) New advances in plant growth-promoting rhizobacteria for bioremediation. Environ Int 33(3):406–413

Chapter 7
Proteomics for Understanding the Interaction Between Plant and Rhizospheric Microflora

Ramesh Namdeo Pudake, Pallavi, and Mrinalini Singh Pundir

Abstract Rhizosphere is a complex system of biological activities of plants and microflora. Interaction between plants and microbes residing in its rhizosphere has been point of interest among the scientific communities for a long time. In-depth knowledge of these interactions is crucial to the current world scenario in context of food availability. Metagenomics and metatranscriptomic studies are being done with the objective elucidate the diversity of culturable and nonculturable microbiome. But this information is incomplete without understanding their functional role in plant–microbiome interaction. Complete proteome represents the ongoing metabolic processes happening in soil at particular time and needs to be studied for knowing the key players in functionality of microbiome. Metaproteomics is emerging tool that sketch the information about entire proteins present in a specific environmental situation at a particular time. It correlates the diversity and functionality of soil microorganisms in both dominant species and at community level. With the help of traditional tools, the development of high-throughput proteomics tools like mass spectrometry, the better understanding of functional aspects of soil complex system has become feasible. However, the progress is little bit slow due to the presence of some bottle neck like presence of various interfering molecules present in the soil samples, scarcity of soil proteome databases, etc. This chapter discusses proteomics tools that are available and review recent studies where the proteomics tools have been applied to decode the underlying processes responsible for differential functioning of soil microbiome in diverse environments.

R. N. Pudake (✉)
Amity Institute of Nanotechnology, Amity University Uttar Pradesh, Noida, Uttar Pradesh, India
e-mail: rnpudake@amity.edu

Pallavi
J.N.L. College Khagaul, Patliputra University, Patna, Bihar, India

M. S. Pundir
Department of Biological Sciences, College of Basic Science and Humanities, G.B. Pant University of Agriculture and Technology, Pantnagar, U.S Nagar, UK, India

R. N. Pudake et al. (eds.), *Omics Science for Rhizosphere Biology*, Rhizosphere Biology, https://doi.org/10.1007/978-981-16-0889-6_7

Keywords Proteomics · Metaproteomics · Rhizospheric microflora · Plant–microbe interaction

7.1 Introduction

To quench the hunger of ever-increasing world's population, which is estimated to reach the number of 10 billion by 2050, we need to substantially increase food production. This is one of major challenge of the humankind in twenty-first century (www.unfpa.org). This requires an increase in agricultural productivity of staple crops. However, in the agriculture practices that are being used result in serious environmental issues through the use of chemicals, salinization, and the lowered water table. Additionally, crop yield in many African, Asian, and South American developing countries cannot be further increased without an increase in forest area and that may destruct the biodiversity, which is already under danger due to other anthropogenic activities; thus, the greatest challenge for today's science to improve crop production in an environmentally viable manner.

Most of the scientific efforts are being focused on plant biotechnology, in developing new crop varieties with enhanced biotic and abiotic stress tolerance and better nutritional value using either breeding or genetic manipulation for introduction of desirable traits. Considering the crucial role played by microbial communities in maintaining plant health, productivity, and biodiversity, their exploitation in developing new environmentally sustainable crop productivity enhancing strategies is minimal. While the current advancement in genomics tools and onsite studies is helping in better understanding of plant–microbe interactions mechanism, but still its full utilization in field conditions yet to be achieved.

In the past, majority of the diversity research was focused on plant and above-ground visible organisms, but still many studies have clearly demonstrated the correlation between the "above-ground" and "below-ground" invisible diversity (Van Der Heijden et al. 2008). One basic and visible advantage for the plant is an increase in access to soil nutrients. Age old symbiont that fixes nitrogen from air, i.e., rhizobia is extensively studied, and recent data shed more light on the mechanism of nutrient exchange in between the plant and bacteria (Lodwig et al. 2003). The plant released metabolites act as a signal to nitrogen-fixing bacteria, and they colonize the roots to get carbohydrates from the plant in exchange of nitrogen. Similarly, some fungal interactions with plants like arbuscular mycorrhizal fungi (AMF) also provide nutrients and water to the host plant in exchange of carbohydrates (Bonfante and Genre 2010). Along with these directly related microbes, there are many other microbes like bacteria of the *Azospirillum* genus which promote the plant root growth and better nutrient absorption (Arzanesh et al. 2011).

Along with the better nutrient availability to plant, the soil microbes also protect their hosts from various disease-causing pathogens. For example, the microbes from various genera like *Pseudomonas*, *Bacillus*, and *Trichoderma* protect the crop plants through releasing a range of metabolites against other phytopathogens (Ab Rahman et al. 2018). Some non-pathogenic microbes on interaction with plants also confer a kind of immunity to the host plant against pathogens which is termed as induced systemic resistance (ISR). Some biocontrol agents are already available in the market, not fully utilized because of persisting problems associated with their efficacy and consistency. But through future research and improvement such microbes can act as alternatives to chemical control that may help in achieving sustainability in current agriculture.

Recent data on studies on this plant–rhizosphere microbes interaction have indicated that there is the presence of high levels of host specificity. Even the different cultivars of the same plant show different microbial dynamics associated with them. This soil microbial population diversity associated with that particular plant is partly dependent on inducing specific genes of soil microbes by the biochemical signals released by the plant. A cross talk between the rhizobium and legumes can be a base model of plant-induce gene expression in associated microbes. Also, the phenolics compounds from plant wounds induces the expression of virulence genes from *Agrobacterium* spp. is another example (Loh et al. 2002). But in general, limited knowledge is available on the signaling mechanism involved behind the beneficial microbes and plants interaction. This deeper understanding of the processes related to the response and specificity of microbial population with the plant signals will help in obtaining the benefits of plant–microbe interactions in full extent.

During the last decade, the face of genomics has been completely overhauled; lots of new technologies have been developed with improved precision and speed. Today, it is possible to understand the molecular foundation of plant–microbe interaction. By describing the root exudates, induced comprehensive molecular response by soil microbes, will give us the complete picture of the functional mechanism behind the specific diversity. It will describe the differential responses of soil microbes to different plant cultivars and species. This information will also be helpful in understanding the basic differences between associative, mutual, and pathogenic interactions of microbes with the host plants.

Although there is availability of complete genome sequences owing to the development of advanced sequencing technologies, this information is not enough to understand the underlying mechanisms that synchronize all cellular functions involved in supporting the complexity of microbial association with plants. Many proteins arbitrate their defined function through the establishment of both stable or transient protein complexes and networks. Many times the same protein may interact with different partner proteins in different situations of biological activity, making their expression pattern a spatially and temporally regulated phenomenon (Hsieh et al. 2003). Besides, protein–protein interactions are regulated by post-translational modifications and can be tissue specific (Zhang et al. 2019). To assist the basic understanding in a plant–microbe interaction, the proteome and metabolome of the plant can be potential tools to provide additional information on levels of regulation and regulatory processes. This chapter will discuss the use of proteomics research that has increased our understanding of plant–soil microbe interaction.

7.2 Proteomics

Proteomics is defined as the technique when the entire protein component expressed by a genome or by a cell of an organism can be used for assessing the expression and localization of proteins, as well as for analysis of post-translational modifications (Wilkins et al. 1996). Classical proteomics comprises protein identification, while functional proteomics involves the detailed characterization of protein structure and cellular function as well as protein–protein interactions (Yarmush and Jayaraman 2002). The functional proteomics is concentrated on the elucidation of interactions between proteins in the cellular networks and it requires high-throughput instruments to elucidate it(LaBaer and Ramachandran 2005; Yarmush and Jayaraman 2002). It intends to characterize the function of each protein in a given organism in a particular cellular condition (LaBaer and Ramachandran 2005). The techniques that are being used in proteomic studies are discussed below in detail.

7.2.1 Gel-Based Protein Separation Techniques

7.2.1.1 Two-Dimensional Gel Electrophoresis (2-DE)

The first and the oldest technique used for large-scale protein separation along with detection of difference in abundance as well as pattern of protein distribution is 2-DE. Through the year, many technical advancements have been made to improve its accuracy there by widening its scope of application. In this technique, proteins are first separated on the basis of their charge (iso-electric focusing), followed by separation according to their molecular weight (SDS-PAGE) (Görg et al. 2004; Wittmann-Liebold et al. 2006). After separation, the next step is visualization of protein spots, which is done by staining. Coomassie Blue (CBB-R, Colloidal), SYPRO (ruby, red, orange), or silver stain are the commonly used dyes (Miller et al. 2006; Westermeier and Marouga 2005). Among these, silver stain is the most sensitive option with the detection limit of 0.1 ng/spot but it interferes with downstream processes, i.e., Edman degradation, mass spectrometry (MS) analysis and accurate determination of spot volumes, making Coomassie Blue preferred dye for the visualization of proteins (Wittmann-Liebold et al. 2006). Software such as Melanie (Geneva Bioinformatics, Switzerland), PDQuest (Bio-Rad Life Science, USA), Phoretix (http://www.perkinelmer.com/proteomics), Progenesis (Nonlinear Dynamics, USA), Z3 or Z4000 are used for analysis of scanned digital images of 2-DE gels (Righetti et al. 2004). The protein spots of interest from the gels are removed either manually or robotically. The differentially expressed protein samples are then subjected to digestion with trypsin and analyzed by MS to ascertain their identities (Rose et al. 2007).

Although the effectiveness of 2-DE is definite, there are several limitations to this technique that include poor separation of protein with very high or low MW, extreme pI values, low-abundance, and extremely hydrophobic proteins (Molloy and Witzmann 2002). Still the robustness and reproducibility of protein separation by immobilized pH gradient (IPG) strips make this technique attractive to many researchers.

7.2.1.2 Fluorescence 2-D Difference Gel Electrophoresis (DIGE)

The major limitation associated with conventional 2-DE is its requirement of images from at least two different gels for comparison to generate results, and it is very hard to get reproducible gels when running 2-DE leading to discrepancies. Many times, minute changes in protein level go undetected due to this (Marouga et al. 2005). The newer DIGE techniques employ different sets of fluorophores (Cy2, Cy3, and Cy5) which are able to covalently label each sample and separating them on the single 2-DE gel (Ünlü et al. 1997). The development and progress in DIGE technique has made it more sensitive and linear by eliminating the post-electrophoretic processing like fixing and destaining of protein gels. This similar electrophoresis conditions increased the reproducibility by directly comparing samples (Van den Bergh and Arckens 2005; Zhou et al. 2002) . The subsequent images are then electronically assessed by using software such as De-Cyder (GE Healthcare, USA) that are exclusively designed for 2-D DIGE analysis (Marouga et al. 2005). However, this technique has its own drawbacks, like proteins lacking lysine cannot be labeled and need special equipment for visualization. In addition, high cost of fluorophores leads to significantly higher experimental cost (Lery et al. 2011; Van den Bergh and Arckens 2005). Other multiplexing methods like labeling of two samples with different radio isotopes have also been reported (Spandidos and Rabbitts 2002); however, protein labeling in in vivo *condition with* radioactive isotope is not feasible in all biological experiments; therefore, DIGE technique has more promise in terms of applicability in gel-based protein analysis.

7.2.2 Gel-Free Proteomics

With the progress in the proteomics technologies and instrumentations, more precise gel-free quantitative proteomics approaches with higher sensitivity are in different phases of development and could be utilized for protein identification, dynamic regulation, and analyzing the post-translational modifications (Picotti et al. 2009). Gel-free-based technique includes Isotope Coded Affinity Tag (ICAT), Isobaric Technique for Relative and Absolute Quantification (ITRAQ), Stable Isotope Labeling by Amino Acid in Cell Culture (SILAC), and Multidimensional Protein Identification Technique (Mud PIT). These techniques also follow the same steps as the gel-based techniques, first separation is done, followed by identification or quantification by mass spectrometry or tandem mass spectrometry (Quirino et al. 2010).

Finally, database comparison is done to identify protein and its function. They are briefly discussed below.

7.2.2.1 Isotope-Coded Affinity Tag (ICAT)

In this method, proteins of two samples are chemically tagged with separate isotopes, and this differential tagging can help to measure quantitative changes in protein levels (Gygi et al. 1999, 2002). ICAT reagents consist of three regions: one protein-reactive group that specifically recognizes cysteine residues in proteins, a second linker region, and third as an affinity biotin tag for separation. Two separate linker regions containing a light or heavy isotope that vary by eight mass units are used for differential labeling of the two separate protein samples. These protein samples are then subjected to subsequent analysis by liquid chromatography-electrospray-tandem mass spectrometry (LC-ESI-MS/MS) (Gygi et al. 1999). The differential changes in particular protein levels induced by a treatment to the organism are assessed by comparing the changes in intensities of protein peaks in the samples that were labeled with the light and heavy isotopes. This technique doesn't require 2-DE, but has selective biasness toward the proteins with high cysteine content and the detection of acidic proteins is difficult (Gygi et al. 1999; Zhou et al. 2002). Although the automation is easily possible in case of ICAT as compared to 2-DE, it is still not widely used in proteomics due to above said limitations.

7.2.2.2 Isobaric Tags for Relative and Absolute Quantitation (iTRAQ)

The iTRAQ technology is a recent variation of ICAT (Ross et al. 2004), and both the techniques are based on a similar concept of tagging. While ICAT depends on tagging cysteine residues, the iTRAQ method uses tagging on primary amines (Schneider et al. 2012). This technology offers many advantages, which include the possibility of multiplexing several samples, quantification, simplified analysis, and enhanced precision and accuracy in analysis (Aggarwal et al. 2006; Wiese et al. 2007). The iTRAQ labeling approach uses four isobaric amine-specific tags that eliminates dependency of ICAT on nonabundant cysteine, and potentially allows for the extended coverage of the proteome (Ross et al. 2004). Another advantage of this technique is that we can tag four samples simultaneously that reduces the time required for mass spectrometry analysis.

7.2.2.3 Multidimensional Protein Identification Technology (MudPIT)

Another robust and widely acceptable alternative to 2-DE in proteomics is MudPIT that can analyze the complex protein mixtures (Issaq et al. 2005; Kislinger et al. 2005). In this method, the protein samples are subjected to sequence-specific enzymatic digestion, generally with trypsin and endo-proteinase lysC, and the resultant peptide mixtures are separated by strong cation exchange (SCX) and reversed phase (RP) high performance liquid chromatography (HPLC)(Chen et al. 2006). Peptides from the RP column go to the mass spectrometer and then MS data is used to search the protein in databases by SEQUEST algorithm. This technique identifies a comprehensive and unbiased list of proteins present in a proteome of any organism, within a short time. It is highly sensitive and good reproducibility in analysis; however, it is not able to provide quantitative information (Kislinger et al. 2005; Washburn et al. 2001). As discussed above, the blend of techniques like HPLC, liquid phase isoelectric focusing and capillary electrophoresis has provided the high-throughput alternative for protein separation from the complex mixtures (Kislinger and Emili 2005). The techniques like MudPIT, ICAT, and iTRAQ have the potential to replace 2-DE, at least theoretically, but did not achieve that in the real world. That is due the confidence by researchers in tried and tested 2-DE technique. In future, due to development of automation and availability of computing tools, it is expected that these gel-free techniques will gain more importance. Even though the label-based, or label-free modern gel-free quantitative proteomic methods can provide more information on the changes in protein expression quantity between the samples when compared to 2-DE, they have their own limitations. These methods are designed for less hydrophobic, more aqueous buffer-soluble sub-proteomes making it difficult to get comprehensive proteome coverage. But that is not the case for gel-based protein separation techniques where the more efficient solubilization of hydrophobic protein occurs in buffers and detergents used. The development and use of combination of label-free quantitative analysis techniques and 2-DE gels can be a solution to solve one of the main disadvantages of gel-based method for separation and quantitative analysis of proteins, i.e., related to co-concentrating/co-localization of several proteins and their altered forms in one spot or band on the gel.

7.2.3 Mass Spectrometry (MS) for Protein Identification

Irrespective of gel separation techniques, a mass spectrometer is always the primary tool for protein identification. Prior to discovery of MS, the protein sequences were decided by Edman degradation (Laursen 1971). However, recently a significant development has been achieved in the application of MS for the identification of protein sequences. Basic units of mass spectrometers are an ion source, the mass analyzer, and an ion detection system. Three steps are involved in analysis of proteins by MS and that are—(a) protein ionization and generation of gas-phase ions, (b) separation of ions according to their mass to charge ratio and (c) detection of ions (Yates et al. 2009). Proteins from gel-free methods like ICAT and MudPIT can be directly used for analysis, while in gel-based separation (2-DE and 2-D DIGE), the proteins are first recovered from gel and then digested with enzyme. The peptides resulting after trypsin digestion are then separated by LC or directly analyzed by

MS. After the analysis, the peptide masses are searched against the peptide fingerprints of known proteins in the online databases using computer tools like Mascot and Sequest (Helsens et al. 2007; Tabb 2015). Currently, the two major sources for ionization, i.e., matrix-assisted laser desorption/ionization (MALDI) and electrospray ionization (ESI) are being used in MS. Regarding the mass analyzers, time-of-flight (TOF), ion trap, quadrupole, and Fourier transform ion cyclotron (FTIC) are being commonly used for protein identification and characterization (Aebersold and Mann 2003). A proteome analysis of different mass analyzers is significantly improved when they are used in combination like use of quadrupole-TOF and quadrupole-ion trap together. The blend of ionization sources with different types of mass analyzers are used depending on the specific application, and this provides a variety of dedicated mass spectrometers for protein analysis (Domon and Aebersold 2006).

Simple MS such as MALDI-TOF can give us data only on mass of protein whereas tandem MS are used for knowing the amino acid sequence (Domon and Aebersold 2006). In MALDI, a laser ion source causes excitation of the matrix and the sample of interest that is crystallized with it on a metal surface, which are then released into the gas phase. It quantifies the mass of peptides produced from a trypsin digested protein and produces a list of experimental peptide masses, commonly referred as "mass fingerprints" (Fenselau 1997). In electrospray ionization (ESI), the analyte is ionized from a solution and moved into the gas phase by creating a fine spray from a high voltage needle that yields in multiple charging of the analyte and generation of multiple consecutive ions (Fenn et al. 1989). Two MS separation principles are used for performing tandem mass spectrometry or MS/MS. In this method, separate trypsin-digested peptides are fragmented after a liquid phase separation. Tandem MS instruments such as triple quadrupole, quadrupole ion trap, Fourier transform ion-cyclotron resonance or quadrupole time-of-flight are used in LC-MS/MS or nanospray experiments with electrospray ionization (ESI) to generate ion spectra of peptide fragment (Corthals et al. 2000). This tandem spectrum then used to search databases to know the protein identity. Recent developments made MS very accurate, high-throughput, and robust techniques that make characterization of entire proteomes a feasible task.

7.3 Tools for Analyzing Protein–Protein Interactions and Protein Function

Protein–protein interactions (PPIs) are very critical in a wide range of biological processes and essential in cell-to-cell interactions, metabolism, and development of organisms. Therefore, it is very important to study these interactions. Traditionally, methods like yeast two-hybrid and immunoprecipitation (IP) are being popular methods for identifying and characterizing protein–protein interactions. The development of protein microarrays made high-throughput analysis possible for thousands of proteins simultaneously (Ramachandran et al. 2005). Protein microarrays can provide important and systemic information in functional protein analysis that may be not possible in other techniques. Antibodies or purified proteins are immobilized in protein microarray chips are used to check the cellular response to the eternal stimulus like pathogen or abiotic stress, and also to study protein–protein interactions. To study protein function in high throughput, along with antibodies and proteins, cDNAs are also being used for making nucleic acid programmable protein array (NAPPA) (Ramachandran et al. 2008). The integrated use of surface plasmon resonance (SPR) and MS has created a unique opportunity for protein interactions, where SPR measures the interaction of proteins with surface-immobilized ligands, and MS is used to analyze the structural features of the bound proteins (Nedelkov and Nelson 2003). From the past experience on these breakthrough technologies, it is sure that they have potential to be a powerful tool utilized throughout rhizosphere sciences.

7.4 Metaproteomics and Its Use in Rhizosphere

Metaproteomics is a powerful tool for obtaining data on all proteins isolated directly from environmental and soil samples at a particular time. Microbial proteins mediated many functions in soil, thus measuring or understanding of them though "omics" science can be the true indicator of potential activity of the microbial community. Metaproteomics can provide solution to elucidate the functional roles of soil microorganism, such as biogeochemical processes, degradation, or bioremediation processes (White et al. 2017). Similar to DNA and RNA isolation method from soil and its uses in metagenomics and metatranscriptomics analysis, there has been a steady progress in protein extraction and analytical methodology. However, extraction of protein is problematic as they form strong interaction with other organic and inorganic component of soil, thus a standard soil protocol is not available and every method incorporates certain soil-based modifications (Chourey et al. 2010; Taylor and Williams 2010). Along with this limited databases are available for soil protein identification. And recent metaproteomics studies are filling those gaps and found that diversity of proteins playing important role in plant and soil microbial communities interactions (Abiraami et al. 2020). Some of the studies with comparative metaproteomics analysis have been summarized in Table 7.1.

7.5 Metaproteogenomics Studies in Rhizosphere

In this method, the combined information of genomics and proteomics for a particular sample is used for better understanding of genetic and functional mechanisms involved in any biological phenomenon. When we have to study the soil microbial community, this global metagenome- and proteome-based analyses identify its

Table 7.1 Summary of proteomics studies to understand plant–soil microbe interaction

Sr. no	Plant	Aim of Study	Method/Technique used	Results	References
1	Arabidopsis	To find out root and shoot responses to *Paenibacillus polymyxa* E681	2DE MALDI-TOF/TOF in conjunction with metagenomics	PGPR promote plant growth-by induced metabolism and activation of defense-related proteins	Kwon et al. (2016)
2	Cucumber	Identification of proteins responsible for in situ root colonization and biofilm formation	SDS-PAGE, iTRAQ labeling with 2D LC/MS	A novel regulator ResE from *Bacillus amyloliquefaciens* SQR9 was found to be involved in root colonization and biofilm formation	Qiu et al. (2014)
3	Grape	To characterize the bacterial community associated with grape roots	LC/MS	Bacteria belonging to *Streptomyces, Bacillus*, *Bradyrhizobium, Burkholderia,* and *Pseudomonas* genera were the most active in protein expression and were mainly involved in phosphorus and nitrogen rhizosphere metabolism	Bona et al. (2019)
4	Maize	Study the effect of metal-resistant bacteria on maize biomass	2-DE, MALDI-TOF	Upregulation of proteins related to plant development and stress response in presence of *Pseudomonas* sp. TLC 6-6.5-4	Li et al. (2014)
5	Maize	To study the effects of biostimulants on the rhizosphere	LC/MS	Enzymatic activity related to phosphorus and glucose processes was increased in rhizosphere of biostimulant-treated samples	Mattarozzi et al. (2020)
6	Maize	Understanding the mechanism of *Herbaspirillum seropedicae* strain SmR1 interaction with maize	MALDI-TOF	*H. seropedicae* dinitrogenase and type III secretion ATP synthase was differentially expressed	Ferrari et al. (2014)
7	Maize	Identification of differentially expressed proteins during the	2 DE with MS	Three unidentified proteins were found to be upregulated during the interaction and are supposed to play	Faleiro et al. (2015)

		interaction of plant roots and PGPR-*Azospirillum brasilense*		some role in establishing the connection between PGPR and host plant	
8	*Miscanthus sinensis—energy grass*	Understanding the mechanism of *Herbaspirillum frisingense* GSF30 (T)-induced growth response	2-DE	*H. frisingense* improves plant growth by modulating plant hormone signaling pathways	Straub et al. (2013)
9	*Rehmannia glutinosa, a medicinal plant*	To identify the reason behind plant growth reduction due to continuous monoculture	2-DE gel MALDI-TOF/TOF-MS	The results suggested that accumulation of root exudates affected the protein responsible in nutrient cycle and that resulted in plant growth reduction	Wu et al. (2011)
10	Rice	Identification of physiology and adaptations microbial communities in the phyllosphere and rhizosphere of rice	1D-SDS gel electrophoresis, HPLC-ESI-MS/MS, and DNA sequencing	Rhizosphere bacteria involved in regulating transport processes and stress responses of plant	Knief et al. (2012)
11	Rice	Identification of microorganisms responsible for methane oxidation and nitrogen fixation in rice fields	1-DE, nano-liquid chromatography (LC)–electrospray ionization–tandem mass spectrometry (MS/MS)	Study revealed that type II methanotrophs of methylocystaceae family that were inhabiting the vascular bundles and epidermal cells of rice roots were playing important role in CH_4 oxidation and N_2 fixation	Bao et al. (2014)
12	*Rice*	Mechanisms of interaction of PGPR-*Herbaspirillum. seropedicae* SmR1 with rice	2-D electrophoresis. MALDITOF/TOF and MASCOT program	The dinitrogenase reductase NifH and glutamine synthetase GlnA, which participate in nitrogen fixation and ammonium assimilation, respectively, were the most abundance in *H. seropedicae.*	Alberton et al. (2013)
13	Soybean	Understanding N_2 fixation efficiency by *Bradyrhizobium elkanii*	TMT labeling and quantitative mass spectrometry	The symbiotic nitrogen fixation efficiency is associated with increased amounts of bacterial enzymes that coordinate *Nod* factor production, porphyrin biosynthesis, NH_3	Cooper et al. (2018)

(continued)

Table 7.1 (continued)

Sr. no	Plant	Aim of Study	Method/Technique used	Results	References
				assimilation, ATP production, amino acid metabolism, and purine, metabolism as well as being associated with increased amounts of proteins needed to exchange molecules with soybean cells	
14	Sugarcane	To uncover the cause behind yield loss due to continuous ratooning	2-D gel electrophoresis (2-DE MALDI TOF/TOF mass spectrometer	Ratooning practice induced significant changes in the soil enzyme activities and proteins related to carbohydrate and amino acid metabolism and stress response were upregulated	Lin et al. (2013)
15	Sugarcane	Insights into early signaling of the *Gluconacetobacter diazotrophicus* -sugarcane interaction	1D-GE and ESI-Q-TOF	Bacterial protein involved in chemotaxis, cell differentiation, and osmotic and oxidative stress regulation are responsive to host signals	Lery et al. (2011)
16	Tomato	To find out role of microbial protein participating in the development of disease as well as resistance of tomato plant to *Fusarium*	2-DE MALDI-TOF	Reaffirm the role of nonculturable microbes in plant resistance to soil-borne pathogen	Manikandan et al. (2017)
17	Tomato	To understand role of plant growth-promoting bacteria (PGPB) in yield promotion	Nano LC MS/MS	The ethylene response pathway plays an important role in plant and PGPB interaction	Ibort et al. (2018)

important members, the physiological potential, and the metabolic pathways in the rhizosphere under given conditions.

In one study, the rice rhizosphere and phyllosphere were analyzed with metaproteogenomics, and its finding suggested that protein expression of dinitrogenase reductase was specific to rhizosphere despite its genomic presence in phyllosphere (Knief et al. 2012). Also, it was found that proteins involved in methanogenesis and methanotrophy were abundant in the rhizosphere, highlighting the importance of microbes in one-carbon compound cycling. In a consolidative analyses on the proteome, transcriptome, and metabolome of Arabidopsis and well-known PGPR strain *Paenibacillus polymyxa* E681; it has been reported that induced metabolism and activation of defense-related proteins in Arabidopsis resulted in growth promotion (Kwon et al. 2016). The potential of this technique was discussed in relation to the studies of plant endosphere for discovering more about endophyte functionality (Kaul et al. 2016).

7.6 Conclusion

The soil is a very complex system and studies planned in this system always need a robust and high-throughput technique to understand it to deeper extend. As discussed, metaproteomics can be a tool that gives an idea about the functionality of soil microbiome. But due to the existing high analytical cost might be discouraging the researchers to do the more replicative studies to confirm their findings. This may lead to biased analysis favoring the spatial and temporal variations present in proteome. But the recent development and reduction in cost of MS analysis and cheaper protein isolation kits in near future will remove this bottle neck, and more replicated and comprehensive studies will be planned. The bioinformatic tools and databases are also improving and they will be helpful in clear results in protein identification. Also, in combination with other omics tools like metagenomics and metatranscriptomics, soil metaproteomics can provide a comprehensive picture of microbiome functioning. This would make metaproteomics a very useful and invaluable tool in rhizospheric microbe studies to unravel its interactions with plants and use that knowledge to improve crop yield through sustainable ways.

References

Ab Rahman SFS, Singh E, Pieterse CM, Schenk PM (2018) Emerging microbial biocontrol strategies for plant pathogens. Plant Sci 267:102–111

Abiraami T, Singh S, Nain L (2020) Soil metaproteomics as a tool for monitoring functional microbial communities: promises and challenges. Rev Environ Sci Biotechnol:1–30

Aebersold R, Mann M (2003) Mass spectrometry-based proteomics. Nature 422:198–207

Aggarwal K, Choe LH, Lee KH (2006) Shotgun proteomics using the iTRAQ isobaric tags. Brief Funct Genomics 5:112–120

Alberton D, Müller-Santos M, Brusamarello-Santos LCC, Valdameri G, Cordeiro FA, Yates MG, de Oliveira Pedrosa F, de Souza EM (2013) Comparative proteomics analysis of the rice roots colonized by *Herbaspirillum seropedicae* strain SmR1 reveals induction of the methionine recycling in the plant host. J Proteome Res 12:4757–4768

Arzanesh MH, Alikhani H, Khavazi K, Rahimian H, Miransari M (2011) Wheat (*Triticum aestivum* L.) growth enhancement by Azospirillum sp. under drought stress. World J Microbiol Biotechnol 27:197–205

Bao Z, Okubo T, Kubota K, Kasahara Y, Tsurumaru H, Anda M, Ikeda S, Minamisawa K (2014) Metaproteomic identification of diazotrophic methanotrophs and their localization in root tissues of field-grown rice plants. Appl Environ Microbiol 80:5043–5052

Bona E, Massa N, Novello G, Boatti L, Cesaro P, Todeschini V, Magnelli V, Manfredi M, Marengo E, Mignone F (2019) Metaproteomic characterization of the *Vitis vinifera* rhizosphere. FEMS Microbiol Ecol 95:fiy204

Bonfante P, Genre A (2010) Mechanisms underlying beneficial plant–fungus interactions in mycorrhizal symbiosis. Nat Commun 1:1–11

Chen EI, Hewel J, Felding-Habermann B, Yates JR (2006) Large scale protein profiling by combination of protein fractionation and multidimensional protein identification technology (MudPIT). Mol Cell Proteomics 5:53–56

Chourey K, Jansson J, VerBerkmoes N, Shah M, Chavarria KL, Tom LM, Brodie EL, Hettich RL (2010) Direct cellular lysis/protein extraction protocol for soil metaproteomics. J Proteome Res 9:6615–6622

Cooper B, Campbell KB, Beard HS, Garrett WM, Mowery J, Bauchan GR, Elia P (2018) A proteomic network for symbiotic nitrogen fixation efficiency in *Bradyrhizobium elkanii*. Mol Plant-Microbe Interact 31:334–343

Corthals, G., Gygi, S., Aebersold, R., and Patterson, S. (2000). Identification of proteins by mass spectrometry. In: Proteome research: two-dimensional gel electrophoresis and identification methods. Springer, pp 197–231

Domon B, Aebersold R (2006) Mass spectrometry and protein analysis. Science 312:212–217

Faleiro AC, Neto PAV, de Souza TV, Santos M, Arisi ACM (2015) Microscopic and proteomic analysis of Zea mays roots (P30F53 variety) inoculated with *Azospirillum brasilense* strain FP2. J Crop Sci Biotechnol 18:63–71

Fenn JB, Mann M, Meng CK, Wong SF, Whitehouse CM (1989) Electrospray ionization for mass spectrometry of large biomolecules. Science 246:64–71

Fenselau C (1997) Peer reviewed: MALDI MS and strategies for protein analysis. Anal Chem 69:661A–665A

Ferrari CS, Amaral FP, Bueno JCF, Scariot MC, Valentim-Neto PA, Arisi ACM (2014) Expressed proteins of *Herbaspirillum seropedicae* in maize (DKB240) roots-bacteria interaction revealed using proteomics. Appl Biochem Biotechnol 174:2267–2277

Görg A, Weiss W, Dunn MJ (2004) Current two-dimensional electrophoresis technology for proteomics. Proteomics 4:3665–3685

Gygi SP, Rist B, Gerber SA, Turecek F, Gelb MH, Aebersold R (1999) Quantitative analysis of complex protein mixtures using isotope-coded affinity tags. Nat Biotechnol 17:994–999

Gygi SP, Rist B, Griffin TJ, Eng J, Aebersold R (2002) Proteome analysis of low-abundance proteins using multidimensional chromatography and isotope-coded affinity tags. J Proteome Res 1:47–54

Helsens K, Martens L, Vandekerckhove J, Gevaert K (2007) MascotDatfile: an open-source library to fully parse and analyse MASCOT MS/MS search results. Proteomics 7:364–366

Hsieh JJ-D, Ernst P, Erdjument-Bromage H, Tempst P, Korsmeyer SJ (2003) Proteolytic cleavage of MLL generates a complex of N-and C-terminal fragments that confers protein stability and subnuclear localization. Mol Cell Biol 23:186–194

Ibort P, Imai H, Uemura M, Aroca R (2018) Proteomic analysis reveals that tomato interaction with plant growth promoting bacteria is highly determined by ethylene perception. J Plant Physiol 220:43–59

Issaq HJ, Chan KC, Janini GM, Conrads TP, Veenstra TD (2005) Multidimensional separation of peptides for effective proteomic analysis. J Chromatogr B 817:35–47

Kaul S, Sharma T, Dhar MK (2016) "Omics" tools for better understanding the plant–endophyte interactions. Front Plant Sci 7:955

Kislinger T, Emili A (2005) Multidimensional protein identification technology: current status and future prospects. Expert Rev Proteomics 2:27–39

Kislinger T, Gramolini AO, MacLennan DH, Emili A (2005) Multidimensional protein identification technology (MudPIT): technical overview of a profiling method optimized for the comprehensive proteomic investigation of normal and diseased heart tissue. J Am Soc Mass Spectrom 16:1207–1220

Knief C, Delmotte N, Chaffron S, Stark M, Innerebner G, Wassmann R, Von Mering C, Vorholt JA (2012) Metaproteogenomic analysis of microbial communities in the phyllosphere and rhizosphere of rice. ISME J 6:1378

Kwon YS, Lee DY, Rakwal R, Baek SB, Lee JH, Kwak YS, Seo JS, Chung WS, Bae DW, Kim SG (2016) Proteomic analyses of the interaction between the plant-growth promoting rhizobacterium *Paenibacillus polymyxa* E681 and *Arabidopsis thaliana*. Proteomics 16:122–135

LaBaer J, Ramachandran N (2005) Protein microarrays as tools for functional proteomics. Curr Opin Chem Biol 9:14–19

Laursen RA (1971) Solid-phase Edman degradation: an automatic peptide sequencer. Eur J Biochem 20:89–102

Lery LM, Hemerly AS, Nogueira EM, von Krüger WM, Bisch PM (2011) Quantitative proteomic analysis of the interaction between the endophytic plant-growth-promoting bacterium Gluconacetobacter diazotrophicus and sugarcane. Mol Plant-Microbe Interact 24:562–576

Li K, Pidatala VR, Shaik R, Datta R, Ramakrishna W (2014) Integrated metabolomic and proteomic approaches dissect the effect of metal-resistant bacteria on maize biomass and copper uptake. Environ Sci Technol 48:1184–1193

Lin W, Wu L, Lin S, Zhang A, Zhou M, Lin R, Wang H, Chen J, Zhang Z, Lin R (2013) Metaproteomic analysis of ratoon sugarcane rhizospheric soil. BMC Microbiol 13:135

Lodwig EM, Hosie AH, Bourdes A, Findlay K, Allaway D, Karunakaran R, Downie J, Poole PS (2003) Amino-acid cycling drives nitrogen fixation in the legume–rhizobium symbiosis. Nature 422:722–726

Loh J, Pierson EA, Pierson LS III, Stacey G, Chatterjee A (2002) Quorum sensing in plant-associated bacteria. Curr Opin Plant Biol 5:285–290

Manikandan R, Karthikeyan G, Raguchander T (2017) Soil proteomics for exploitation of microbial diversity in Fusarium wilt infected and healthy rhizosphere soils of tomato. Physiol Mol Plant Pathol 100:185–193

Marouga R, David S, Hawkins E (2005) The development of the DIGE system: 2D fluorescence difference gel analysis technology. Anal Bioanal Chem 382:669–678

Mattarozzi M, Di Zinno J, Montanini B, Manfredi M, Marengo E, Fornasier F, Ferrarini A, Careri M, Visioli G (2020) Biostimulants applied to maize seeds modulate the enzymatic activity and metaproteome of the rhizosphere. Appl Soil Ecol 148:103480

Miller I, Crawford J, Gianazza E (2006) Protein stains for proteomic applications: which, when, why? Proteomics 6:5385–5408

Molloy MP, Witzmann FA (2002) Proteomics: technologies and applications. Brief Funct Genomics 1:23–39

Nedelkov D, Nelson RW (2003) Surface plasmon resonance mass spectrometry: recent progress and outlooks. Trends Biotechnol 21:301–305

Picotti P, Bodenmiller B, Mueller LN, Domon B, Aebersold R (2009) Full dynamic range proteome analysis of S. cerevisiae by targeted proteomics. Cell 138:795–806

Qiu M, Xu Z, Li X, Li Q, Zhang N, Shen Q, Zhang R (2014) Comparative proteomics analysis of Bacillus amyloliquefaciens SQR9 revealed the key proteins involved in in situ root colonization. J Proteome Res 13:5581–5591

Quirino B, Candido E, Campos P, Franco O, Krüger R (2010) Proteomic approaches to study plant–pathogen interactions. Phytochemistry 71:351–362
Ramachandran N, Larson DN, Stark PR, Hainsworth E, LaBaer J (2005) Emerging tools for real-time label-free detection of interactions on functional protein microarrays. FEBS J 272:5412–5425
Ramachandran N, Raphael JV, Hainsworth E, Demirkan G, Fuentes MG, Rolfs A, Hu Y, LaBaer J (2008) Next-generation high-density self-assembling functional protein arrays. Nat Methods 5:535–538
Righetti PG, Campostrini N, Pascali J, Hamdan M, Astner H (2004) Quantitative proteomics: a review of different methodologies. Eur J Mass Spectrometry 10:335–348
Rose, A., Stahlberg, E.A., and Meier, I. (2007). Genome-wide identification and comparative analysis of coiled-coil proteins. Scalable computing: practice and experience 8
Ross PL, Huang YN, Marchese JN, Williamson B, Parker K, Hattan S, Khainovski N, Pillai S, Dey S, Daniels S (2004) Multiplexed protein quantitation in Saccharomyces cerevisiae using amine-reactive isobaric tagging reagents. Mol Cell Proteomics 3:1154–1169
Schneider T, Keiblinger KM, Schmid E, Sterflinger-Gleixner K, Ellersdorfer G, Roschitzki B, Richter A, Eberl L, Zechmeister-Boltenstern S, Riedel K (2012) Who is who in litter decomposition? Metaproteomics reveals major microbial players and their biogeochemical functions. ISME J 6:1749
Spandidos A, Rabbitts TH (2002) Sub-proteome differential display: single gel comparison by 2D electrophoresis and mass spectrometry. J Mol Biol 318:21–31
Straub D, Yang H, Tsap T, Liu Y, Ludewig U (2013) Root ethylene signalling is involved in Miscanthus sinensis growth promotion by the bacterial endophyte Herbaspirillum frisingense GSF30T. J Exp Bot 64:4603–4615
Tabb DL (2015) The SEQUEST family tree. J Am Soc Mass Spectrom 26:1814–1819
Taylor EB, Williams MA (2010) Microbial protein in soil: influence of extraction method and C amendment on extraction and recovery. Microb Ecol 59:390–399
Ünlü M, Morgan ME, Minden JS (1997) Difference gel electrophoresis. A single gel method for detecting changes in protein extracts. Electrophoresis 18:2071–2077
Van den Bergh G, Arckens L (2005) Recent advances in 2D electrophoresis: an array of possibilities. Expert Rev Proteomics 2:243–252
Van Der Heijden MG, Bardgett RD, Van Straalen NM (2008) The unseen majority: soil microbes as drivers of plant diversity and productivity in terrestrial ecosystems. Ecol Lett 11:296–310
Washburn MP, Wolters D, Yates JR (2001) Large-scale analysis of the yeast proteome by multidimensional protein identification technology. Nat Biotechnol 19:242–247
Westermeier R, Marouga R (2005) Protein detection methods in proteomics research. Biosci Rep 25:19–32
White RA, Borkum MI, Rivas-Ubach A, Bilbao A, Wendler JP, Colby SM, Köberl M, Jansson C (2017) From data to knowledge: the future of multi-omics data analysis for the rhizosphere. Rhizosphere 3:222–229
Wiese S, Reidegeld KA, Meyer HE, Warscheid B (2007) Protein labeling by iTRAQ: a new tool for quantitative mass spectrometry in proteome research. Proteomics 7:340–350
Wilkins MR, Pasquali C, Appel RD, Ou K, Golaz O, Sanchez J-C, Yan JX, Gooley AA, Hughes G, Humphery-Smith I (1996) From proteins to proteomes: large scale protein identification by two-dimensional electrophoresis and amino acid analysis. Biotechnology 14:61–65
Wittmann-Liebold B, Graack HR, Pohl T (2006) Two-dimensional gel electrophoresis as tool for proteomics studies in combination with protein identification by mass spectrometry. Proteomics 6:4688–4703
Wu L, Wang H, Zhang Z, Lin R, Zhang Z, Lin W (2011) Comparative metaproteomic analysis on consecutively Rehmannia glutinosa-monocultured rhizosphere soil. PLoS One 6:e20611
Yarmush ML, Jayaraman A (2002) Advances in proteomic technologies. Annu Rev Biomed Eng 4:349–373

Yates JR, Ruse CI, Nakorchevsky A (2009) Proteomics by mass spectrometry: approaches, advances, and applications. Annu Rev Biomed Eng 11:49–79

Zhang Y, Natale R, Domingues AP, Toleco MR, Siemiatkowska B, Fàbregas N, Fernie AR (2019) Rapid identification of protein-protein interactions in plants. Curr Protocol Plant Biol 4:e20099

Zhou G, Li H, DeCamp D, Chen S, Shu H, Gong Y, Flaig M, Gillespie JW, Hu N, Taylor PR (2002) 2D differential in-gel electrophoresis for the identification of esophageal scans cell cancer-specific protein markers. Mol Cell Proteomics 1:117–123

Chapter 8
A Proteomics Perspective for Understanding Rhizosphere Biology at Higher Altitudes

Shiv Shanker Gautam, Deep Chandra Suyal, Ravindra Soni, and Reeta Goel

Abstract Earth is enriched with diverse climate, weather, and natural resources responsible for variable flora and fauna. The temperature change and variation in physical and chemical environmental factors give rise to a diverse microbial community. Soil microorganisms play an essential role in plant growth by several means including nitrogen fixation, element solubilization, nutrient mobilization and uptake, and suppression of disease, etc. However, higher altitudes face the issues of lower crop productivity due to less availability of soil nitrogen. Studies of rhizosphere communities may explore the potential microbial candidates to enhance and improve crop yield. The earlier development in molecular biology and proteomic approaches has been energized to explore such microbial communities. This chapter aimed to provide the current scenario of proteomic approaches to study the rhizosphere biology of higher altitudes.

Keywords Higher altitudes · PGPRs · Proteomics · Rhizosphere community

S. S. Gautam (✉)
Department of Microbiology, College of Basic Sciences and Humanities, GB Pant University of Agriculture and Technology (GBPUA&T), Pantnagar, Uttarakhand, India

D. C. Suyal
Department of Microbiology, Eternal University, Sirmour, Himachal Pradesh, India

R. Soni
Department of Agricultural Microbiology, College of Agriculture, Indira Gandhi Krishi Vishwavidyalaya, Raipur, Chhattisgarh, India

R. Goel
Institute of Applied Sciences & Humanities, GLA University, Mathura, Uttar Pradesh, India

R. N. Pudake et al. (eds.), *Omics Science for Rhizosphere Biology*, Rhizosphere Biology, https://doi.org/10.1007/978-981-16-0889-6_8

8.1 Introduction

Rhizosphere describes the interface between inhabiting a specific population of microorganisms and plant roots influencing each other (McNear Jr. 2013; Joshi et al. 2019). The term *rhizosphere* (Greek word "rhiza," meaning root; and "sphaira" around) first coined by the German agronomist and plant physiologist Lorenz Hiltner in 1904 (Hartmann et al. 2008). Plant–microbes relationship can be friendly or hostile (Giri et al. 2015; Dash et al. 2019) and beneficial rhizobia and fungi present the rhizosphere can provide mineral nutrients, nitrogen, and other essential elements via solubilization in exchange of carbon to host plant (Kumar et al. 2014; Suyal et al. 2014a, b; Puschel et al. 2017). However, in hostile relationship plants are exposed to a various range of fungi, bacteria, and viruses and cause economic loss worldwide (Makovitzki et al. 2007; Petriacq et al. 2018). At higher altitudes, the crops are directly affected by low temperature and faced various challenges for survival including low photosynthesis rate, water transport, and low availability of inorganic nitrogen in soil (Volder et al. 2000; Goel et al. 2017, 2018; Jeyakumar et al. 2020). The development of polyphasic agricultural strategies, i.e., reshaping the rhizosphere microbial community may improve growth and productivity of such crops and provide abiotic and biotic stresses tolerance in this region (Suyal et al. 2014a; Kumar et al. 2018, 2019).

Moreover, the microbial community proteomics have been revolutionized the research in rhizospheric microbial diversity, ecological functions, metabolic potential, and microbe–environment relationships (Wasinger and Corthals 2002; Wang et al. 2016; Suyal et al. 2015b, 2017). Proteomics comprises multiple technologies, viz. protein–protein interaction, algorithms for databases for rapid and accurate protein determination, post-translational modifications, upregulated and downregulated proteins, and their functionality, etc. (Wasinger and Corthals 2002; Soni et al. 2015; Suyal et al. 2018). Thus, proteomics can boost our understanding of rhizosphere biology at higher altitudes and improvement of crop productivity and growth in these regions.

8.2 Ecology of Higher Altitudes

Altitude can directly influence the species richness and biomass production (Bhandari and Zhang 2019). The higher altitude has been reported as low biomass production and decreased rhizosphere microbial diversity owing to suboptimal or freezing soil temperatures (Suyal et al. 2014a). Similarly, Margesin and Miteva (2011) had suggested the negative relationship among the diversity and ecology of psychrophilic microorganisms and bacterial abundance with the altitudes. In this connection, the Himalaya has been seen as the reservoir of diversity and active gene pool for cold adapted microorganisms (Suyal et al. 2014b; Joshi et al. 2017). The microorganisms of higher altitudes have the potential to tolerate extreme environmental conditions for their survivability (Barauna et al. 2017). The survival mechanisms may include the maintenance of membrane fluidity, change in metabolic rate, cellular processes, and synthesis of essential proteins, expression of housekeeping genes, and cold-shock proteins. The cold environment gives rise to a very diverse habitat for diazotrophs including the cold desert of high mountains, polar ice and snowfields, salt brine of Arctic and Atlantic, and deep oceans. The major dominant species included mosses, lichens, algae, fungi, and bacteria of such environment (D'Amico et al. 2006; Yadav et al. 2017; Joshi et al. 2017; Hamdan 2018). In high altitude ecosystems, the diazotrophs play a key role in N_2 fixation in nutrient-limited soil for plant growth and soil microorganisms (Duc et al. 2009). They are also responsible for the influence of primary productivity of the soil.

8.2.1 Cold-Adapted Microorganisms

Cold inhabiting microorganisms have been divided into two groups, psychrophiles and psychrotrophs. Psychrophiles are the microorganisms which have optimal growth temperature less than 15 °C and do not grow above 20 °C. However, psychrotrophs are those which can survive even at sub-zero temperature but grow optimally at 20–25 °C. However, this division scheme is not accurate for all microorganisms living in cold environment as many ambiguities are observed in the microorganisms with respect to the growth temperature. Psychrophilic organisms are present in the extremely cold ecosystem of glaciers, high altitude mountains, polar regions, alpine, and deep oceans. Besides the cold stress, these ecosystems are characterized by the additional stress including nutrient stress, increased, or decreased atmospheric pressure, and variable metal ion concentrations (Suyal et al. 2019b). Psychrophiles have evolved different adaptation to survive in cold stress. Adaptation at the cell membrane, RNA metabolism, transcription, translation, and protein degradation/stability are important to carry out the cellular metabolism at very low temperature.

Cold stress affects the stability of the secondary structure of nucleic acid hence affecting the vital process of transcription, translation, and RNA degradation (Suyal et al. 2017, 2019b). During this cold stress, bacteria modify its cellular physiology and biochemistry. Low temperature triggers the cold-shock response in which bacteria adapt to the cold stress. During the initial cold response, growth arrest is observed for 3–6 h. This phase of growth arrest is termed as acclimation phase, where only the cold inducible proteins (CIPs) are expressed, and rest of the protein expression is downregulated. After the acclimation phase, cell down regulates the expression of CIPs, resume expression of the other proteins and start dividing normally. Understanding the molecular mechanisms of cold adaptation in microorganisms could enhance the current knowledge of their increased survival under cold stress. Maintaining the viability of microorganisms at low temperature is a major task for the refrigerated probiotic food-based industries and microbial culture collections. Moreover, this understanding will further facilitate the development of strategies to reduce the growth of the pathogenic microorganisms under refrigerated temperature which can cause food spoilage and diseases.

8.3 Synergistic Effect of Rhizosphere with Cold-Adapted Microbial Life

The rhizosphere is the hotspot area for microbial growth, abundance, and diversity due to the influence of rhizodeposits, plant litter, nutrients, and exudates secreted by roots of the host plant (Preece and Penuelas 2016; Rajwar et al. 2018; Alawiye and Babalola 2019; Rawat et al. 2019). In rhizodeposits, microorganisms are abundant due to high nutrient availability (Preece and Penuelas 2016) that responsible for stimulating plant growth, disease suppression by the fungal and viral attack, via the release of cyanide, siderophores, ammonia, and other volatile components (Shukla et al. 2015; Suyal et al. 2015a, c). At higher altitudes, soil microorganisms cope with the harsh, oligotrophic, and nutrient-limited conditions. In such a context, rhizospheric microorganisms may be helpful in abiotic stress and improve crop yield by acting as biofertilizers (Tomer et al. 2016, 2017). There are several factors that can affect the growth rate and survival of microorganisms. The major environmental factors of survival of microbial communities are carbon and nitrogen contents, pH, temperature, moisture contents, and precipitation rate. The rhizospheric bacterial communities belong to various genera including *Arthrobacter*, *Acetobacter*, *Achromobacter*, *Anabaena*, *Azospirillum*, *Azotobacter*, *Bacillus*, *Burkholderia*, *Clostridium*, *Enterobacter*, *Flavobacterium*, *Frankia*, *Hydrogenophaga*, *Kluyvera*, *Microcoleus*, *Phyllobacterium*, *Pseudomonas*, *Serratia*, *Streptomyces*, *Rhizobium*, etc. (Soni et al. 2016, 2017; Alawiye and Babalola 2019; Suyal et al. 2019a, b; Jeyakumar et al. 2020).

Furthermore, Praeg et al. (2019) studied the rhizosphere soil of *Ranunculus glacialis* from high alpine altitudes and figured out the major prokaryotes belong to order Rhizobiales, Gamma proteobacteria, Pseudomonadales, Actinomycetales, Sphingobacteriales, Sphingomonadales, Flavobacteriales, Burkholderiales, Planctomycetales, Rhodospirillales, Xanthomonadales, respectively. While fungal communities belonged to order Hypocreales, Incertae, Pleosporales, Sporidiobolales, and Lecideales. Therefore, it can be stated that higher altitudes not only affect the vegetation and their productivity but also the microbial life of rhizospheric soil.

8.4 Microbial Community and Proteomics

Proteomics deals with the protein complements expressed in a cell under certain physiological conditions at a certain time is referred to as the proteome (Nilson and Graveley 2010). Currently, proteomics has become a very valuable tool for exploring the environmental microbial community (Wang et al. 2016; Eldakak et al. 2013). The microbial community proteomics, also known as metaproteomics is an emerging area of microbiology and helpful for studying the whole protein content directly isolated from the complex ecological community at a particular time (Wang et al. 2016). The metaproteomics can expand our knowledge of taxonomic composition and microbial functions including nutrient cycles, mutualistic relationship, metal utilization, eutrophication, nutrient mobilization and uptake, and suppression of diseases, etc. (Schneider and Riedel 2010; Baldrian and Lopez-Mondejar 2014; Glass et al. 2014; Heyer et al. 2019).

8.4.1 Strategies for Rhizosphere Proteomic Studies

For rhizosphere community studies, various strategies and steps have been developed including sample collection, protein extraction and separation, mass spectrometry analysis, database search, and data interpretation (Fig. 8.1). The protein separation and identification include one-dimensional or two-dimensional polyacrylamide gel electrophoresis (2D-PAGE) followed by enzymatic digestion, i.e., trypsin. The digested peptides are subjected to analysis by complementary mass spectrometry, i.e., MALDI-TOF-MS, LC-ESI-MS/MS, MS/MS analysis, protein identification, and interpretation by database search and bioinformatic analysis (Suyal et al. 2014b; Wang et al. 2016).

In mass spectrometry method, the high-throughput MALDI-TOF-MS analysis in combination with LC-ESI-MS/MS can enhance the identification of unknown proteins by a database search of the unknown bacterial genome (Encarnacion et al. 2005). The proteomics tools and techniques are briefly discussed here (Table 8.1).

8.4.1.1 2-DE

Two-dimensional gel electrophoresis (2-DE) is a gel-based technique extensively used for biological proteins and peptides compositional samples. This technique is helpful to separate two different proteins as spot via the biophysical separation processes including isoelectric focusing and polyacrylamide gel electrophoresis. 2-DE is broadly used for proteomic study, especially for environmental and pathologically important bacteria (Curreem et al. 2012; Eldakak et al. 2013).

8.4.1.2 2D-DIGE

Two-dimensional difference gel electrophoresis (2D-DIGE) is similar to 2-DE that help to compare one to two or three proteins concurrently on the similar gel (Minden 2012; Mozejko-Ciesielska and Mostek 2019). The proficiency of this technique is related to detect as small as 0.2 fmol of protein, and protein variability less to $\pm 15\%$, approximately 10,000-fold protein concentration array (Minden 2012).

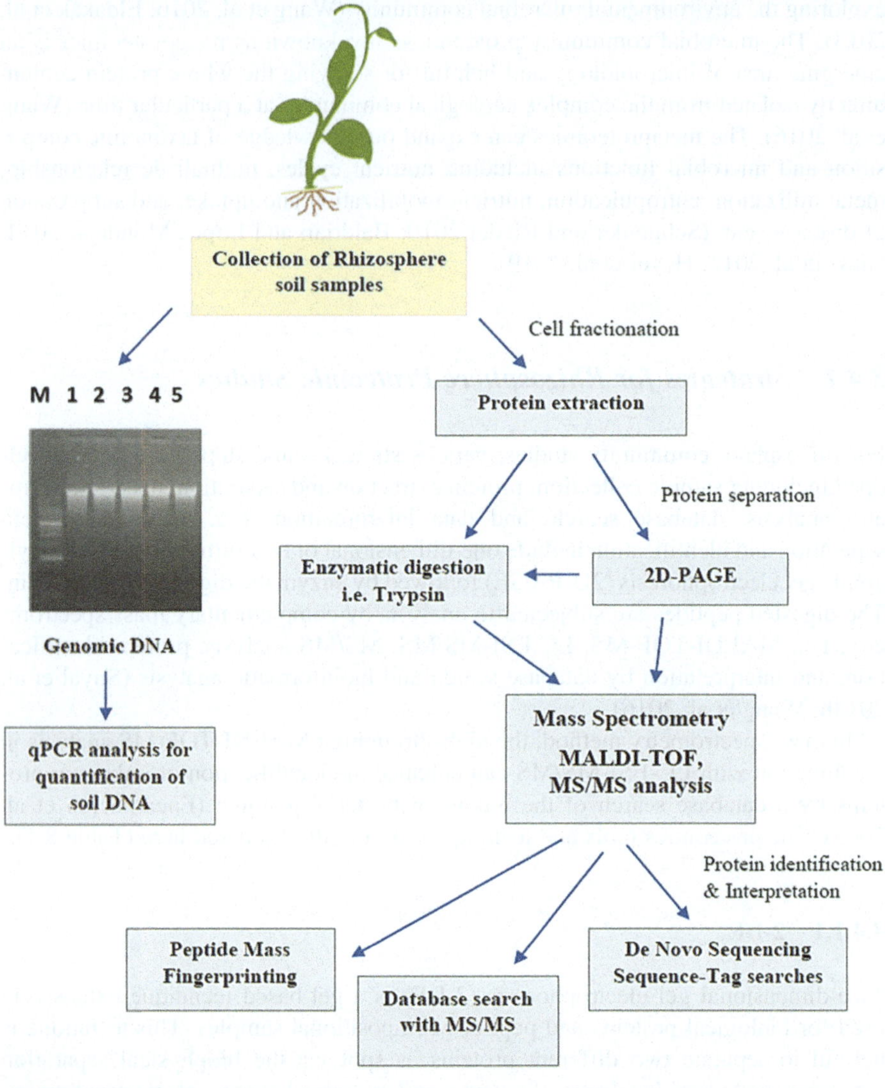

Fig. 8.1 Steps of proteomic analysis for microbial diversity from rhizospheric soil

8.4.1.3 LC-MS

Liquid chromatography-mass spectrometry (LC-MS) enables the appraisal of microorganisms by multiple discriminative peptides which allow characterization of species or at strain level (Lasch et al. 2019). This technique is slightly time-consuming when tested with sequence databases of different microbiological taxa. From digested bacterial cell extracts, LC-MS showed significant results for identification of tryptic peptides and sequence to bacterium assignments (Dworzanski

Table 8.1 Overview of proteomics techniques used for rhizosphere microbial study

Proteomic techniques/ tools	Characteristics	Ref.
2-DE	Analysis of labeled protein	Eldakak et al. (2013)
2D-DIGE	Protein separation and visualization	Mozejko-Ciesielska and Mostek (2019)
LC-MS	Identification of microbial species by multiple discriminative peptides	Lasch et al. (2019)
Shotgun proteomics	Identification of microbial proteins between sets of samples based on difference in abundance under stress or other conditions	Gouveia et al. (2020)
Protein-SIP	Identification by ^{13}C or ^{15}N labeled substrate into nucleic acid for finding functional relationship inside a microbial community	Jehmlich et al. (2008)
MALDI-TOF-MS	Identification of microbial predominated proteins, i.e., ribosomal proteins, cytosolic proteins Detection of antibiotic resistance; Epidemiological studies	Wang et al. (2016), Suyal et al. (2014b), Dingle and Butler-Wu (2013)
LC-ESI-MS/MS	Peptide and cellular protein-based identification	Encarnacion et al. (2005)
MS/MS	Identification and ^{13}C content as indicator for function and metabolic activity of microorganisms	Wang et al. (2016), Suyal et al. (2014b)
SELDI-TOF	Modified technique of MALDI-TOF and combination of MS and high-throughput nature of protein arrays	Al-Tarawneh and Bencharit (2009)
Proteins and antibody microarray	Proteins of post-translational modifications, study of host–microbe interactions, profiling antibody specificity, identification of biomarkers in autoimmune diseases	Sutandy et al. (2013)
ICAT	Quantitative proteomics, MS-based protein identification method	Hsu and Chen (2016)
MetaProteomeAnalyzer	Open-source tool for metaproteomics data analysis	Muth et al. (2018)
Unipept 4.0	Web-based application of metaproteome data analysis grounded on tryptic-peptide-based biodiversity study of MS/MS samples	Singh et al. (2019)

et al. 2006). Another similar study has been conducted to find out the relatedness among stains of *B. cereus* sensu stricto, *B. thuringiensis* and *B. anthracis* by studying peptide fractions derived from prototype database (Dworzanski et al. 2010).

8.4.1.4 Shotgun Proteomics

Shotgun proteomics is a method for identifying proteins, i.e., microbial proteins between groups of samples that depend on difference in abundance under stress or other conditions. This technique depends on a set of combination including LC and MS, covering a broader range of proteins than 2D-GE, specially the hydrophobic and low-copy proteins (Lee and Lee 2004; Hendrickson et al. 2008). In microbial community study, this technique can explain the clearly altered abundances of key protein players (Gouveia et al. 2020) and the quantification of isolated microbial proteins can be done by tandem mass spectrometry.

8.4.1.5 Protein-SIP

Protein-based stable isotope probing (Protein-SIP) is based on isotope labeled substrate (^{13}C or ^{15}N) into nucleic acid for finding functional relationship within a microbial community (Jehmlich et al. 2008). This technique showed link between microbe-specific metabolic function to their phylogeny. Moreover, proteomic-SIP can help in proteomic investigations to enumerate enzymatic pathways and such values may be combined with flux balance models of biogeochemical cycles, i.e., carbon, sulfur, and nitrogen to understand the dynamics and rates of reactions at both the grouping and ecosystem scales (Marlow et al. 2016).

8.4.1.6 MALDI-TOF-MS

Matrix-assisted laser desorption ionization time-of-flight mass spectrometry (MALDI-TOF-MS) is a powerful tool for microbial identification, i.e., bacteria, fungi, viruses, and diagnosis based on predominated proteins, i.e., ribosomal proteins, cytosolic proteins, and detection of antibiotic resistance. This technique is also used for the screening of biological warfare agents, water- and food-borne pathogens, and epidemiological studies (Dingle and Butler-Wu 2013; Suyal et al. 2014b; Wang et al. 2016).

8.4.1.7 LC-ESI-MS/MS

The liquid chromatography-electrospray ionization-tandem mass spectrometry (LC-ESI-MS/MS) is a peptide and cellular protein-based identification method (Encarnacion et al. 2005). This technique is helpful to analyze tryptic digestion of bacterial samples and provides useful information of peptide sequences, allowing identification of large number of proteins. This technique has attracted much attention for the identification of bacteria, fungi, and viruses of different ecological and diseased samples.

8.4.1.8 MS/MS

The mass spectrometry (MS/MS) provides the way of structural elucidation of a wide range of peptides (Graham et al. 2007). The MS/MS peptides are ionized by using the ESI or MALDI in the source region. Further, such peptides are separated, based on m/z ratio. The MS/MS provides the identification via ^{13}C content as indicator for functional and metabolic study of microorganisms (Wang et al. 2016; Suyal et al. 2014b).

8.4.1.9 Surface-Enhanced Laser Desorption/Ionization Time-of-Flight (SELDI-TOF)

Surface-Enhanced Laser Desorption/Ionization Time-Of-Flight (SELDI-TOF) is a modification of MALDI-TOF and combination of MS and high-throughput nature of protein arrays, known as protein chips (Al-Tarawneh and Bencharit 2009). SELDI-TOF has three major components including protein chip arrays, mass analyzer, and data analysis software. SELDI-TOF can analyze proteins from various crude samples with minimal sample size. This technique is very efficient and can detect rapidly the native undigested samples.

8.4.1.10 Proteins and Antibody Microarray

Protein microarray is an emerging technology that offers a stand for the screening of hundreds of thousands of proteins in a high-throughout manner. This technique consists of two major classes, the analytical and functional. The tissue or cell lysates can be fractionated and placed on a slide to form a reverse phase protein microarray. This technique is helpful for analyzing the proteins of post-translational modifications, host–microbe interactions, profiling of antibody specificity, identification of biomarkers in autoimmune diseases (Sutandy et al. 2013). The antibody microarray is the precise form of protein microarray. In this, the sets of antibodies have been spotted and fixed on a solid surface and reactive groups of the surface have been blocked. The sample of soluble proteins of interest is incubated on the array and targeted protein samples are captured by the antibodies. In result, the bounded samples are reported directly by fluorescent labeling or by the addition of secondary detection reagent (Chen et al. 2018). The benefits of antibody microarray can be seen in diverse biological processes including protein–protein interactions (Paul et al. 2016), analysis of signal pathways (El-Haibi et al. 2012), post-translational modifications (Pelech and Yue 2018), study of toxins (Lian et al. 2010), etc.

8.4.1.11 Isotope-Coded Affinity Tags (ICAT)

Isotope-coded affinity tags (ICAT) depend on the principle that only free cysteine thiols are disposed to labeling by the iodoacetamide-based ICAT, and mass spectrometry (MS) is used to quantitate the relative labeling of free thiols. Gygi et al. (1999) first used ICAT for differential protein expression study of *Saccharomyces cerevisiae*. This technique is helpful for quantitative proteomics, and MS-based protein identification method (Hsu and Chen 2016). The reagent named as amine reactive isobaric tag for relative and absolute quantitation (iTRAQ) has been suggested to use for protein labeling to analyze the protein expression from multiple samples and treatments at a time (Graham et al. 2007).

8.4.1.12 Unipept 4.0

Unipept (https://unipept.ugent.be) is a web-based application of metaproteome data analysis which initially focuses on tryptic-peptide-based biodiversity analysis of MS/MS samples (Singh et al. 2019). This tool is fast and efficient in depiction of taxon-specific catalytic functions and helpful to correlate BLAST-based functional study of similar data.

8.4.2 Microbial Potential Proteins for Plant Survival and Total Productivity

In cold and nitrogen depleting environment, cold-adapted bacteria can fix atmospheric N_2 while other microorganisms assimilate nitrogen or ammonia to accomplish their needs. Under cold stress, cellular normal protein products are decreased. While some special protein products are increased until adaptation in certain organisms (Table 8.2). Such cold-induced homologous class of proteins is known as cold-shock proteins (Csps) (Latha et al. 2009). Csps are multifunctional RNA/DNA binding proteins with the presence of one or more cold-shock domains (Lindquist and Mertens 2018). Csps have been reported in a wide variety of Gram-positive and Gram-negative bacteria including *E. coli* (Newkirk et al. 1994), *Bacillus* spp. (Mueller et al. 2000), *Thermotoga* spp. (Kremer et al. 2001), *Arthrobacter* spp., *Streptococcus* spp., *Listeria* sp., and *Pseudomonas* spp. (Latha et al. 2009). *E. coli* contains nine Csps (CspA to CspI) in which CspA, CspB, and CspG are cold-inducible and CspD stationary-phase inducible (Wang et al. 1999). Expression of 7.28 kDa CspD protein of psychrotolerant Antarctic *Janthinobacterium* sp. Ant-5 of class Betaproteobacteria has been observed at various temperatures ranged from 37 to −1 °C (Mojib et al. 2011). Metagenomics of Csp library from temperate and glacier soils of central Himalaya has been figured out low temperature adaptation and homology of Csp genes of *P. fluorescens*, *Psychrobacter cryohalolentis* K5, and

Table 8.2 Cold adaptation-associated bacterial proteins reported from high altitude rhizospheric soils

S. no.	Proteins	Genes	Functions	Ref.
1.	Cold-shock protein csp1–6	*Csp*1–6	Cold-shock regulation	Barauna et al. (2017)
2.	desK, desR	*des*K, *des*R	Cold induction	Barauna et al. (2017)
3.	accA - accD	*acc*A–*acc*D	Synthesis of fatty acids	Barauna et al. (2017)
4.	Cold-shock protein CspB – CspI	*csp*B–*csp*I	Cold stress regulation	Latha et al. (2009)
5.	YacG	*yac*G	DNA gyrase inhibitor	Soni et al. (2015)
6.	Clp protease	*CLPP*	Cleaves peptides and various proteins in an ATP-dependent process	Soni et al. (2015)
7.	Superoxide dismutase	*SOD1*	Breakdown harmful oxygen molecules	Soni et al. (2015)
8.	tRNA (cmo5U34)-methyltransferase	*cmo*A	Catalyzes the conversion of S-adenosyl-L-methionine (SAM) to carboxy-S-adenosyl-L-methionine (Cx-SAM)	Soni et al. (2015)
9.	Ketol-acid reductoisomerase	*AHRI*	Ketol-acid reductoisomerase activity	Soni et al. (2015)
10.	Octaprenyl-diphosphate synthase	*isp*B	Supplies octaprenyl diphosphate	Soni et al. (2015)
11.	F420-dependent glucose-6-phosphate dehydrogenase	*fgd*1	Catalyzes the coenzyme F420-dependent oxidation of glucose 6-phosphate to 6-phosphogluconolactone	Soni et al. (2015)
12.	Nucleoside diphosphate kinase	*NME1*	Synthesis of nucleoside triphosphates other than ATP	Soni et al. (2015)
13.	MsrB	*msr*B	Catalytic activity	Soni et al. (2015)
14.	MsrA	*msr*A	Catalyzes the reversible oxidation-reduction of methionine sulfoxide in proteins to methionine	Soni et al. (2015)
15.	Quinolinate synthase A	*nad*A	Catalyzes the condensation of iminoaspartate with dihydroxyacetone phosphate to form quinolinate	Soni et al. (2015)
16.	nifHD region glnB-like protein 2	*gln*B	Regulation of nitrogen fixation	Soni et al. (2015)
17.	N(2)-fixation sustaining protein CowN	*cow*N	Sustain N_2-dependent growth in the presence of low level of carbon monooxide	Soni et al. (2015), Suyal et al. (2019b)
18.	Nitrogenase-stabilizing/protective protein NifW	*nif*W	Protect the nitrogenase Fe-Mo protein from oxidative damage	Soni et al. (2015)

(continued)

Table 8.2 (continued)

S. no.	Proteins	Genes	Functions	Ref.
19.	Glutamine synthetase 2	*Gs2*	Metabolism of nitrogen by catalyzing the condensation of glutamate and ammonia to form glutamine	Soni et al. (2015)
20.	Chaperone protein DnaK	*dna*K	Involve in chromosomal replication and refolding of misfolded proteins	Suyal et al. (2017, 2018)
21.	Chaperone protein HscA homolog	*hsc*A	Essential in stress regulation	Suyal et al. (2017)
22.	60 kDa chaperonin	*gro*L	Avoid misfolding and encourage the refolding and correct assembly of unfolded polypeptides generated under stress environment	Suyal et al. (2017)
23.	Chaperone protein TorD	*tor*D	Biogenesis of TorA; TorA bears a bifunctional Tat signal peptide, which directs protein export and serves as a binding site for the TorD biosynthetic chaperone	Suyal et al. (2017), Buchanan et al. (2008)
24.	Iron-sulfur cluster repair protein YtfE	*ytf*E	Di-iron-containing protein involve in the repair of iron-sulfur clusters which damaged by oxidative and nitrosative stress conditions	Suyal et al. (2017)
25.	Ferredoxin-like protein in nif region	*fdx*N	Metal binding property with iron-sulfur involve in nitrogen fixation	Suyal et al. (2017)
26.	Protein mrp homolog	*mrp*	ATP binding	Suyal et al. (2017)
27.	Aspartate carbamoyltransferase	*pyr*B	Catalyze the condensation of L-aspartate and carbamoyl phosphate (CP) to produce N-carbamoyl-L-aspartate	Suyal et al. (2017)
28.	Phenylalanyl-tRNA synthetase alpha chain	*syf*A	ATP binding, phenylalanine tRNA ligase activity	Suyal et al. (2017)
29.	Glycine cleavage H-protein	*GCSH*	Catalyzes the degradation of glycine	Suyal et al. (2017)
30.	Dephospho-CoA kinase	*coa*E	Catalyze the phosphorylation of the 3′-hydroxyl group of dephosphocoenzyme A to form coenzyme A	Suyal et al. (2017)

Shewanella spp. MR-4 (Latha et al. 2009). Under environmental stresses, the nonsporulating bacteria can activate the viable but nonculturable state. Thus, environmental stress and low N_2 in the soil are also responsible for stimulating N_2 fixation (Suyal et al. 2014a). Earlier studies for metabolic responses of Himalayan cold-adapted diazotrophs *P. palleroniana* N26 has been investigated for nitrogen deficiency in the cold niche. Under the cold condition, expression of *nif*A, *nif*L, *nif*H, *nif*B, *nif*D, *nif*K, and *cow*N of the nitrogenase system has been observed (Suyal et al. 2018). Thus, the regulation/expression of such regulatory proteins ensures the microbial survival capability and enhancement of agronomic practices under low temperature.

CIPs prevent the secondary structure formation and facilitate the degradation of structural RNA at lower temperatures (Latha et al. 2009). Among the CIPs, some proteins from the small acidic protein family of 7.4 kDa are most strongly induced under cold stress and are termed as cold-shock proteins (CSPs). CSPs bind to the only single-stranded RNA or DNA and not to the double-stranded conformation. All CSPs have nucleic acid-binding domains termed as cold-shock domain (CSD), which facilitates their binding to the nucleic acid. Binding of CSPs to RNA maintains the single-stranded conformations thus protecting the cell from cold-induced secondary structure formation. In mesophilic organisms, CSPs are transiently induced through cold-shock and soon after the acclimatization their expression is downregulated. However, in psychrophiles they are constitutively expressed and act as cold adaptive proteins (CAPs). Besides the major function in the cold-shock response, CSPs also have major role in the other cellular processes.

8.5 Proteomics Strategies for Crop Improvement at Higher Altitude

The proteomics are helpful in gaining information about microbial community activity and their interactions between roots and soil (Bona et al. 2019). At low temperatures, rhizospheric microorganisms still survive and provide essential nutrients via mutualistic relationship, metal utilization, eutrophication, nutrient mobilization, and suppression of diseases. The metaproteomic study of such a harsh environment can be helpful to improve other cold environment crops or vegetation by rhizospheric replacement (Suyal et al. 2014a, b). The cold-regulated microbial proteins also ensure the plant survival capability and enhancement of agronomic practices under low temperature condition. The earlier advancement in proteomic approaches have been accelerated crop improvement programs globally. The microbial proteomics data available on different databases have been significantly played its role in crop improvement including the PRIDE (Proteomics IDEntification database) (https://www.ebi.ac.uk/pride/) database part of EMBL-EBI, blastp suite of NCBI, UniProt (www.uniport.org), HAMAP (High-quality Automated and Manual Annotation of microbial Proteomes) (http://www.expasy.org/sprot/hamap), PLGS (ProteinLynx Global SERVER), etc. These proteomic databases provide a platform for researchers in identifying, analyzing, and finding similarity/dissimilarity between proteins and peptides with previous data. The 2D-DIGE, MALDI-TOF-MS, and protein microarray are some powerful tools also used for characterization of metaproteomes of different crop rhizosphere soils. Furthermore, metaproteomics has been provided the scenario for soil functional composition and taxonomic structure of microbial communities within their environment. The soil enzymes, i.e., protease and peptidases, can be directly identified for participating in nitrogen fixation and nitrification (Wang et al. 2016). Additionally, the molecular mechanisms and interactions between plants and their pathogens can be easily understood (Rampitsch and Bykova 2012).

8.6 Conclusion and Future Perspectives

Proteomics is an emerging field that deals with proteins expressed in a cell under certain physiological conditions at a particular time. Currently, it has become a very valuable tool for exploring environmental microbial community, popularly known as metaproteomics. Metaproteomic tools have allowed the quantification of per species biomass to determine community structure, in situ carbon sources, and uptake of labeled substrates by community individuals. Proteomic platforms can speed up crop improvement programs worldwide. Nowadays, sustainable crop production is a major challenge for cold-climate agriculture system. The cold-adaptive microorganisms have significantly provided a rational scenario for the improvement of crop health and productivity. Thus, proteomics can be helpful in such stress conditions and holds great promise for the betterment of the agriculture system.

References

Alawiye TT, Babalola OO (2019) Bacterial diversity and community structure in typical plant rhizosphere. Diversity 11(10):179. https://doi.org/10.3390/d11100179Al-

Al-Tarawneh S, Bencharit S (2009) Application of surface-enhance laser desorption/ionization time-of-flight (SELDI-TOF) mass spectrometry in defining salivary proteomic profiles. Open Dent J 3:74–79

Baldrian P, Lopez-Mondejar R (2014) Microbial genomics, transcriptomics and proteomics: new discoveries in decomposition research using complementary methods. Appl Microbiol Biotechnol 98:1531–1537

Barauna RA, Freitas DY, Pinheiro JC, Folador ARC, Silva A (2017) A proteomic prospective on the bacterial adaptation to cold: integrating OMICs data of the psychrophic bacterium *Exiguobacterium antarcticum* B7. Proteomes 5(1):9. https://doi.org/10.3390/proteomes5010009

Bhandari J, Zhang Y (2019) Effect of altitude and soil properties on biomass and plant richness in the grasslands of Tibat, China and Manang District, Nepal. Ecosphere 10(11):e02915. https://doi.org/10.1002/ecs2.2915

Bona E, Massa N, Novello G, Boatti L, Cesaro P, Todeschini V, et al. (2019) Metaproteomic characterization of the Vitis vinifera rhizosphere. FEMS Microbiology Ecology 95(1):fiy204. https://doi.org/10.1093/femsec/fiy204

Buchanan G, Maillard J, Nabuurs SB, Richardson DJ, Palmer T, Sargent F (2008) Features of a twin-arginine signal peptide required for recognition by a Tat proofreading chaperone. FEBS Lett 582(29):3979–3984

Chen Z, Dodig-Crnkovic T, Schwenk JM, Tao S (2018) Current applications of antibody microarrays. Clin Proteomics 15:7. https://doi.org/10.1186/s12014-018-9184-2

Curreem SO, Watt RM, Lau SK, Woo PC (2012) Two-dimensional gel electrophoresis in bacterial proteomics. Protein Cell 3(5):346–363. https://doi.org/10.1007/s13238-012-2034-5

D'Amico S, Collins T, Marx JC, Feller G, Gerday C (2006) Psychrophilic microorganisms: challenge for life. EMBO Rep 7(4):385–389

Dash B, Soni R, Kumar V, Suyal DC, Dash D, Goel R (2019) Mycorrhizosphere: microbial interactions for sustainable agricultural production. In: Varma A, Choudhary D (eds) Mycorrhizosphere and Pedogenesis. Springer, Singapore, pp 321–338

Dingle TC, Butler-Wu SM (2013) MALDI-TOF mass spectrometry for microorganism identification. Clin Lab Med 33(3):589–609. https://doi.org/10.1016/j.cll.2013.03.001

Duc L, Noll M, Meier BE, Burgmann H, Zeyer J (2009) High diversity of diazotrophs in the forefield of a receding alpine glacier. Microb Ecol 57:179–190

Dworzanski JP, Deshpande SV, Chen R, Jabbour RE, Snyder AP, Wick CH, Li L (2006) Mass spectrometry-based proteomics combined with bioinformatic tools for bacterial classification. J Proteome Res 5:76–87

Dworzanski JP, Dickinson DN, Deshpande SV, Snyder AP, Eckenrode BA (2010) Discrimination and phylogenomic classification of *Bacillus anthracis-cereus-thuringiensis* strains based on LC-MS/MS analysis of whole cell protein digests. Anal Chem 82:145–155

Eldakak M, Milad SIM, Nawar AI, Rohila JS (2013) Proteomics: a biotechnology tool for crop improvement. Front Plant Sci 4:35. https://doi.org/10.3389/fpis.2013.00035

El-Haibi CP, Singh R, Gupta P, Sharma PK, Greenleaf KN, Singh S et al (2012) Antibody microarray analysis of signaling network regulated by Cxcl13 and Cxcr5 in prostate cancer. J Proteomics Bioinform 5(8):177–184. https://doi.org/10.4172/jpb.1000232

Encarnacion S, Hernandez M, Martinez-Batallar G, Contreras S, Vargas MC, Mora J (2005) Comparative proteomics using 2-D gel electrophoresis and mass spectrometry as tools to dissect stimulons and regulons in bacteria with sequenced or partially sequenced genomes. Biol Proced Online 7:117–135. https://doi.org/10.1251/bpo110

Giri K, Paliwal R, Suyal DC, Mishra G, Pandey S, Rai JPN, Verma PK (2015) Potential application of plant-microbe interaction for restoration of degraded ecosystems. In: Singh S, Srivastava K (eds) Handbook of research on uncovering new methods for ecosystem management through bioremediation. IGI Global, Hershey, pp 255–285

Glass JB, Yu H, Steele JA et al (2014) Geochemical, metagenomic and metaproteomic insights into trace metal utilization by methane-oxidizing microbial consortia in sulphidic marine sediments. Environ Microbiol 16:1592–1611

Goel R, Kumar V, Suyal DC, Dash B, Kumar P, Soni R (2017) Root-associated bacteria: rhizoplane and endosphere. In: Singh D, Singh H, Prabha R (eds) Plant-microbe interactions in agro-ecological perspectives. Springer, Singapore, pp 161–176

Goel R, Suyal DC, Kumar V, Jain L, Soni R (2018) Stress-tolerant beneficial microbes for sustainable agricultural production. In: Microorganisms for green revolution, microorganisms for sustainability, pp 141–159. https://doi.org/10.1007/978-981-10-7146-1_8

Gouveia D, Grenga L, Pible O, Armengaud J (2020) Quick microbial molecular phenotyping by differential shotgun proteomics. Environ Microbiol 22(8):2996–3004. https://doi.org/10.1111/1462-2920.14975

Graham RL, Graham C, McMullan G (2007) Microbial proteomics: a mass spectrometry primer for biologists. Microb Cell Fact 6:26. https://doi.org/10.1186/1475-2859-6-26

Gygi SP, Rist B, Gerber SA, Turecek F, Geib MH, Aebersold R (1999) Quantitative analysis of complex protein mixtures using isotope-coded affinity tags. Nat Biotechnol 17(10):994–999

Hamdan A (2018) Psychrophiles: ecological significance and potential industrial application. S Afr J Sci 114(5/6). https://doi.org/10.17159/sajs.2018/20170254

Hartmann A, Rothballer M, Schmid M (2008) Lorenz Hiltner, a pioneer in rhizosphere microbial ecology and soil bacteriology research. Plant Soil 312:7–14

Hendrickson EL, Lamont RJ, Hackett M (2008) Tools for interpreting large-scale protein profiling in microbiology. J Dent Res 87(11):1004–1015

Heyer R, Schallert K, Budel A et al (2019) A robust and universal metaproteomics workflow for research studies and routine diagnostics within 24 h using phenol extraction, FASP digest, and the metaproteomeanalyzer. Front Microbiol 10:1883. https://doi.org/10.3389/fmicb.2019.01883

Hsu JL, Chen SH (2016) Stable isotope dimethyl labelling for quantitative proteomics and beyond. Phil Trans R Soc A 374:20150364

Jehmlich N, Schmidt F, von Bergen M, Richnow HH, Vogt C (2008) Protein-based stable isotope probing (protein-SIP) reveals active species within anoxic mixed cultures. ISME J 2:1122–1133

Jeyakumar SP, Dash B, Singh AK, Suyal DC, Soni R (2020) Nutrient cycling at higher altitudes. In: Goel R, Soni R, Suyal DC (eds) Microbiological advancements for higher altitude agro-ecosystems & sustainability. Springer Nature Singapore Pvt Ltd, Singapore, pp 293–305

Joshi D, Kumar S, Suyal DC, Goel R (2017) The microbiome of the Himalayan ecosystem. In: Kala VC et al (eds) Mining of microbial wealth and metagenomics. Springer Nature Singapore Pvt Ltd, Singapore. https://doi.org/10.1007/978-981-10-5708-3_6

Joshi D, Chandra R, Suyal DC, Kumar S, Goel R (2019) Impact of bioinoculants *Pseudomonas jesenii* MP1 and *Rhodococcus qingshengii* S10107 on *Cicer arietinum* yield and soil nitrogen status. Pedosphere 29(3):388–399

Kremer W, Schuler B, Harrieder S, Geyer M, Gronwald W, Welker C, Jaenicke R, Kalbitzer HR (2001) Solution NMR structure of the cold-shock protein from the hyperthermophilic bacterium *Thermotoga maritima*. Eur J Biochem 268:2527–2539

Kumar S, Suyal DC, Dhauni N, Bhoriyal M, Goel R (2014) Relative plant growth promoting potential of Himalayan psychrotolerant *Pseudomonas jesenii* strain MP1 against Native *Cicer arietinum* L., *Vigna mungo* (L.) Hepper; *Vigna radiata* (L.) Wilczek., *Cajanus cajan* (L.) Millsp. and *Eleusine coracana* (L.) Gaertn. Afr J Microbiol 8(50):3931–3943

Kumar S, Suyal DC, Bhoriyal M, Goel R (2018) Plant growth promoting potential of psychrotolerant *Dyadobacter* sp. for pulses and finger millet and impact of inoculation on soil chemical properties and diazotrophic abundance. J Plant Nutr 41(8):1035–1046

Kumar S, Suyal DC, Yadav A, Shouche Y, Goel R (2019) Microbial diversity and soil physiochemical characteristic of higher altitude. PLoS One 14(3):e0213844

Lasch P, Schneider A, Blumenscheit C, Doellinger J (2019) Identification of microorganisms by liquid chromatography-mass spectrometry (LC-MS1) and in silico peptide mass data. bioRxiv 870089. https://doi.org/10.1101/870089

Latha PK, Soni R, Khan M, Marla SS, Goel R (2009) Exploration of Csp genes from temperate and glacier soils of the Indian Himalayas and in silico analysis of encoding proteins. Curr Microbiol 58(4):343–348

Lee WC, Lee KH (2004) Applications of affinity chromatography in proteomics. Anal Biochem 324(1):1–10

Lian W, Wu D, Lim DV, Jin S (2010) Sensitive detection of multiplex toxins using antibody microarray. Anal Biochem 401(2):271–279

Lindquist JA, Mertens PR (2018) Cold shock proteins: from cellular mechanisms to pathophysiology and disease. Cell Commun Signal 16:63. https://doi.org/10.1186/s12964-018-0274-6

Makovitzki A, Viterbo A, Brotman Y, Chet I, Shai Y (2007) Inhibition of fungi and bacterial plant pathogens in vitro and in planta with ultrashort cationic lipopeptides. Appl Environ Microbiol 73(20):6629–66636. https://doi.org/10.1128/AEM.01334-07

Margesin R, Miteva V (2011) Diversity and ecology of psychrophilic microorganisms. Res Microbiol 162:346–361

Marlow JJ, Skennerton CT, Li Z, Chourey K, Hettich RL, Pan C, Orphan VJ (2016) Proteomic stable isotope probing reveals biosynthesis dynamics of slow growing methane based microbial communities. Front Microbiol 7:563. https://doi.org/10.3389/fmicb.2016.00563

McNear DH Jr (2013) The rhizosphere—roots, soil and everything in between. Nature Education Knowledge 4(3):1

Minden JS (2012) Two-dimensional difference gel electrophoresis (2D DIGE). In: Laboratory methods in cell biology, pp 111–141. https://doi.org/10.1016/b978-0-12-405914-6.00006-8

Mojib N, Andersen DT, Bej AK (2011) Structure and function of a cold shock domain fold protein, CspD, in *Janthinobacterium* sp. Ant5-2 from East Antarctica. FEMS Microbiol Lett 319 (2):106–114

Mozejko-Ciesielska J, Mostek A (2019) A 2D-DIGE-based proteomic analysis brings new insights into cellular responses of *Pseudomonas putida* KT2440 during polyhydroxyalkanoates synthesis. Microb Cell Factories 18:93. https://doi.org/10.1186/s12934-019-1146-5

Mueller U, Perl D, Schmid FX, Heinemann U (2000) Thermal stability and atomic-resolution crystal structure of the *Bacillus caldolyticus* cold shock protein. J Mol Biol 297:975–988

Muth T, Kohrs F, Heyer R, Benndorf D, Rapp E, Reichl U, Martens L, Renard BY (2018) MPA portable: a stand-alone software package for analyzing metaproteome samples on the go. Anal Chem 90:685–689. https://doi.org/10.1021/acs.analchem.7b03544

Newkirk K, Feng W, Jiang W, Tejero R, Emerson SD, Inouye M, Montelione GT (1994) Solution NMR structure of the major cold shock protein (CspA) from *Escherichia coli*: identification of a binding epitope for DNA. Proc Natl Acad Sci U S A 91:5114–5118

Nilson TW, Graveley BR (2010) Expansion of the eukaryotic proteome by alternative splicing. Nature 463:457–463. https://doi.org/10.1038/nature08909

Paul C, Rho H, Neiswinger J, Zhu H (2016) Characterization of protein-protein interactions using protein microarrays. Cold Spring Harb Protoc 2016(10). https://doi.org/10.1101/pdb.prot087965

Pelech S, Yue L (2018) Tracking protein expression, post-translational modifications and interactions with high content antibody microarrays. FASEB J 32(1_supplement):802.6

Petriacq P, Lopez A, Luna E (2018) Fruit decay to diseases: can induced resistance and priming help? Plan Theory 7(4):77. https://doi.org/10.3390/plants7040077

Praeg N, Pauli H, Illmer P (2019) Microbial diversity in bulk and rhizosphere soil of *Ranunculus glacialis* along a high-alpine altitudinal gradient. Front Microbiol 10:1429. https://doi.org/10.3389/fmicb.2019.01429

Preece C, Penuelas J (2016) Rhizodeposition under drought and consequences for soil communities and ecosystem resilience. Plant Soil 409:1–17

Puschel D, Janouskova M, Voriskova A, Gryndlerova H, Vosatka M, Jansa J (2017) Arbuscular mycorrhiza stimulates biological nitrogen fixation in two *Medicago* spp. through improved phosphorus acquisition. Front Plant Sci 8:390. https://doi.org/10.3389/fpls.2017.00390

Rajwar J, Chandra R, Suyal DC, Tomer S, Kumar S, Goel R (2018) Comparative phosphate solubilizing efficiency of psychrotolerant *Pseudomonas jesenii* MP1 and *Acinetobacter* sp. ST02 against chickpea for sustainable hill agriculture. Biologia 73(8):793–802

Rampitsch C, Bykova NV (2012) Proteomics and plant disease: advances in combating a major threat to the global food supply. Proteomics 12(4–5):673–690. https://doi.org/10.1002/pmic.201100359

Rawat N, Sharma M, Suyal DC, Singh DK, Joshi D, Singh P, Goel R (2019) Psyhcrotolerant bio-inoculants and their co-inoculation to improve *Cicer arietinum* growth and soil nutrient status for sustainable mountain agriculture. J Soil Sci Plant Nutr 19(3):639–647

Schneider T, Riedel K (2010) Environmental proteomics: analysis of structure and function of microbial communities. Proteomics 10:785–798

Shukla A, Dhauni N, Suyal DC, Kumar S, Goel R (2015) Comparative plant growth promoting potential of psychrotolerant diazotrophs *Pseudomonas* sp. JJS2 and *Enterobacter* sp. AAB8 against native *Cajanus cajan* (L.) and *Eleusine coracana* (L.). Afr J Microbiol 9(20):1371–1375

Singh RG, Tanca A, Palomba A et al (2019) Unipept 4.0: functional analysis of metaproteome data. J Proteome Res 18(2):606–615

Soni R, Suyal DC, Agrawal K, Yadav A, Souche Y, Goel R (2015) Differential proteomic analysis of Himalayan psychrotolerant diazotroph *Pseudomonas palleroniana* N26 strain under low temperature diazotrophic conditions. CryoLetters 36(2):74–82

Soni R, Suyal DC, Sai S, Goel R (2016) Exploration of *nif*H gene through soil metagenomes of the western Indian Himalayas. 3 Biotech 6(1):1–4

Soni R, Kumar V, Suyal DC, Jain L, Goel R (2017) Metagenomics of plant rhizosphere microbiome. In: Singh R, Kothari R, Koringa P, Singh S (eds) Understanding host-microbiome interactions—an omics approach. Springer, Singapore, pp 193–205

Sutandy FXR, Qian J, Chen CS, Zhu H (2013) Overview of protein microarrays. Curr Protoc Protein Sci 72(1):27–21. https://doi.org/10.1002/0471140864.ps2701s72

Suyal DC, Shukla A, Goel R (2014a) Growth promotory potential of the psychrophilic diazotroph *Pseudmona smigulae* S10724 against Native *Vigna radiata* (L.) Wilczek. 3 Biotech 4 (6):665–668

Suyal DC, Yadav A, Shouche Y, Goel R (2014b) Differential proteomics in response to low temperature diazotrophy of Himalayan psychrophilic nitrogen fixing *Pseudomonas migulae* S10724 strain. Curr Microbiol 68(4):543–550. https://doi.org/10.1007/s00284-013-0508-1

Suyal DC, Soni R, Sai S, Goel R (2015a) Microbial inoculants as biofertilizer. In: Singh DP et al (eds) Microbial inoculants in sustainable agricultural productivity. Springer, Berlin, pp 311–318

Suyal DC, Yadav A, Shouche Y, Goel R (2015b) Diversified diazotrophs associated with the rhizosphere of Western Indian Himalayan native red kidney beans (*Phaseolus vulgaris* L.). 3 Biotech 5(4):433–441

Suyal DC, Yadav A, Shouche Y, Goel R (2015c) Bacterial diversity and community structure of Western Indian Himalayan red kidney bean (*Phaseolus vulgaris* L.) rhizosphere as revealed by 16S rRNA gene sequences. Biologia 70(3):305–313

Suyal DC, Kumar S, Yadav A, Shouche Y, Goel R (2017) Cold stress and nitrogen deficiency affected protein expression of psychrotrophic *Dyadobacter psychrophilus* B2 and *Pseudomonas jessenii* MP1. Front Microbiol 8(430):1–6

Suyal DC, Kumar S, Joshi D, Soni R, Goel R (2018) Quantitative proteomics of psychotrophic diazotroph in response to nitrogen deficiency and cold stress. J Proteome 187:235–242

Suyal DC, Joshi D, Debbarma P, Soni R, Dash B, Goel R (2019a) Soil metagenomics: unculturable microbial diversity and its function. In: Varma A, Choudhary D (eds) Mycorrhizosphere and pedogenesis. Springer, Singapore, pp 355–362

Suyal DC, Joshi D, Kumar S, Soni R, Goel R (2019b) Differential protein profiling of soil diazotroph *Rhodococcus qingshengii* S10107 towards low-temperature and nitrogen deficiency. Sci Rep 9(20378):1–9

Tomer S, Suyal DC, Goel R (2016) Biofertilizers: a timely approach for sustainable agriculture. In: Choudhary DK, Varma A, Tuteja N (eds) Plant-microbe interaction: an approach to sustainable agriculture. Springer Nature Singapore Pte Ltd, Singapore, pp 375–395

Tomer S, Suyal DC, Rajwar J, Yadav A, Shouche Y, Goel R (2017) Isolation and characterization of phosphate solubilizing bacteria from Western Indian Himalayan soils. 3 Biotech 7(2):1–8

Volder A, Bliss LC, Lambers H (2000) The influence of temperature and nitrogen source on growth and nitrogen uptake of two polar-desert species, *Saxifraga caespitose* and *Cerastium alpinum*. Plant Soil 227:139–148

Wang N, Yamanaka K, Inouye M (1999) CspI, the ninth member of the CspA family of *Escherichia coli*, is induced upon cold shock. J Bacteriol 181(5):1603–1609

Wang DZ, Kong LF, Li YY, Xie ZX (2016) Environmental microbial community proteomics: status, challenges and perspectives. Int J Mol Sci 17(8):1275. https://doi.org/10.3390/ijms17081275

Wasinger VC, Corthals GL (2002) Proteomic tool for biomedicine. J Chromatogr B Analyt Technol Biomed Life Sci 771(1–2):33–48

Yadav AN, Verma P, Kumar V, Sachan SG, Saxena AK (2017) Extreme cold environments: a suitable niche for selection of novel psychrotrophic microbes for biotechnological applications. Adv Biotechnol Microbiol 2(2):1–4. https://doi.org/10.19080/AIBM.2017.02.555584

Chapter 9
Structural and Functional Rhizospheric Microbial Diversity Analysis by Cutting-Edge Biotechnological Tools

Tanya Gupta, Debatri Chakraborty, and Angana Sarkar

Abstract The rhizosphere is a composite ecosystem which supports multiple bacterial populations that nourishes the terrestrial biosphere and plays a crucial role in the continuous recycling of minerals, nutrients, and organic matter through the soil. Diverse varieties of molecular tools based on immediate isolation and analysis of various compounds from environmental samples such as lipids, nucleic acids, and peptides have been discovered which have provided structural and functional data about microbial communities present in rhizospheric soil. With the advent of next-generation sequencing technologies (NGS), it has become possible to delve deeper into the rhizosphere microbiome to understand the unknown aspects of it. This has resulted in a shift from traditional approaches to the modern omics-based approach based on NGS sequencing technologies for discovering and distinguishing the vast microbial diversity to understand their interactions with different environmental factors. The major objective of this chapter is to provide insights on structural and functional rhizospheric microbial diversity analysis by the application of cutting-edge biotechnological tools. We have first glanced through the basic concepts of rhizosphere and its importance in plant system, the common rhizospheric microbial population, and looked at the plant–microbe interactions which are of prime importance in the rhizosphere ecosystem. Next, we come to the molecular tools used for rhizospheric microscopic diversity analysis—a detailed view into a few of the traditional approaches used for diversity approaches before proceeding to the rapidly emerging and more popular omics-based approaches used for rhizosphere microbial diversity analysis. We have also identified the merits and demerits, future opportunities of omics-based approaches in rhizosphere microbiology.

Keywords Microbial diversity · Omics · Rhizosphere · Plant · NGS

T. Gupta · D. Chakraborty · A. Sarkar (✉)
Department of Biotechnology and Medical Engineering, National Institute of Technology, Rourkela, Odisha, India
e-mail: sarkara@nitrkl.ac.in

R. N. Pudake et al. (eds.), *Omics Science for Rhizosphere Biology*, Rhizosphere Biology, https://doi.org/10.1007/978-981-16-0889-6_9

9.1 Introduction

Plant-associated microbes have been known to function as nitrogen fixers, zinc solubilizers, phosphorous solubilizers, potassium solubilizers, absorbers, accumulators, conversion of essential nutrients into a form that the plant can uptake, source of antibiotic and antifungal agents, source of commercially viable enzymes, inducer for production of plant hormones, the key to developing a tolerance for biotic and abiotic stress such as salinity, acidity, alkalinity, excess soil moisture, drought conditions, and extremely high or extremely low temperatures (Ali et al. 2018; Tanim et al. 2019). These microbes influence plant growth, and the growth of the plant and environmental stressors influence the structural and functional dynamics of such microbial niches. Majority of the plant-associated microbe population is formed by the microbiota colonizing the rhizosphere of the plant. The microbes show enforced as an ecological consequence of the influence of root exudates on which they thrive. This microbiota is extremely influential and diverse has immense applications in the field of biotechnology with significant agricultural and commercial importance.

The multiform microbiome carries out and regulates a variety of processes occurring in the rhizosphere of the plant that are relevant to plant proliferation. *Betaproteobacteria* act as nitrogen fixers by denitrification and nitrate reduction. *Acinetobacter* is known for the bioremediation of xenobiotic compounds. *Gammaproteobacteria* adapt for suitable growth in a rhizospheric environment requiring resistance to metal toxicity and metal reduction to get rid of the contaminant for bioremediation and enriching the soil to make it fertile. *Pseudomonas* and *Bacillus* species are known to act as heavy metal detoxifiers and participate in phosphorous solubilization. *Clostridia* exhibits the ability to reduce heavy metals through hydrogen metabolism and fermentation (Ghosh et al. 2019).

The increase in industrial activities over the past few decades and the intensive agricultural activities undertaken to meet heightened demand for crop production has resulted in pertinent inorganic and organic pollution. Metal pollution has led to excessive degradation of soil quality and adversely affected plant health as well as microbial population composition (Benidire et al. 2020). As a consequence of this, the use of biofertilizers involving the application of the rhizosphere-associated biome for assisting its growth by improving soil conditions to benefit the environment has been championed as an alternative for chemical fertilizers. Certain bacterial colonies in the plant roots act as bio-inoculants increasing the fertility of the soil via their metabolism and can be isolated for bioremediation purposes. The application of biofertilizers is also encouraged owing to the increasing popularity of organic farming. Some plant species are hyperaccumulators, with the rhizospheric communities regulating the uptake of metal ions contaminating the soil. Microbial diversity analysis is essential for determining the contaminant degradation ability of the various species in a microbial population. Microbial diversity can also help understand the biological impacts of certain farming practices on crop production (Wang et al. 2019), verify the presence of antibiotic and antifungal agent producing strains

(Ali et al. 2018), or to isolate any other strain from the concerning rhizosphere responsible for the production of specific enzyme or metabolite.

Many molecular tools have been developed such as the traditionally employed techniques Amplified Ribosomal DNA Restriction Analysis (ARDRA) and Denaturing Gradient Gel Electrophoresis (DGGE) for microbial diversity analysis to facilitate understanding of the community dynamics and the microbial metabolic processes that occur in the rhizosphere. Both cultivation-based approach and DNA-based approach have been employed for this purpose. With the recent advancements in technological high-throughput sequencing tools with a reduced expense, time consumption, and hassle have been developed. These sequencing tools are next-generation sequencers that are utilized in the study of the entire genetic material present in a microbial community referred to as metagenomics. The sequencing tools are also employed in metatranscriptomics which is the study of gene expression and interaction, metaproteomics which refers to the characterization analysis of the protein expressed in the microbial, and metabolomics which refers to the study of the significant metabolites and exudates present in an ecological system. Metagenomics, metatranscriptomics, metaproteomics, and metabolomics studies can be and have been applied in numerous plant-related researches to understand and define the rhizospheric microbiota of the given plant and the specificity of the plant–microbe interactions. The significance of the rhizospheric biota for plant proliferation has been well acknowledged and analyzing their microbial diversity is important for characterizing their relevant functional traits (Singh et al. 2020).

9.2 What is Rhizosphere?

The rhizosphere is the region of the soil that can be defined as the zone that is immediately surrounding the root of a plant that houses a diverse range of microbial colonies (Ali et al. 2018). Different plant species have complex and unique microflora associated with them. This microflora may compose of bacteria, fungi, viruses, and archaea. Along with the host plant, these microbes form a delicately balanced ecosystem, and they may or may not benefit from the presence of these microbes that are found on their leaves, flowers, stems, and roots. The rhizosphere is inhabited by the root-associated microbes which maintain either a symbiotic or non-symbiotic relationship with the plant. These microbes thrive on the root exudates of the host plant and therefore the rhizosphere forms an agreeable niche for the microbial population. These microbes, being present at the soil–root interface of the plant exerts a significant influence over the rhizospheric processes that take place (de los Reyes et al. 2020). These microbial interactions are very distinctive based upon the microbial species. The evolution of the soil microflora of an individual plant has occurred simultaneously with the evolution of the plant itself over time (Ghosh et al. 2019). Therefore, the microbiome of the rhizosphere is unique to the specific host plant.

9.3 Rhizospheric Diversity

Any soil sample when taken and analyzed for microbial diversity shows the soil biota representing a huge number of microbial species. The diversity which is shown by a microbial community heavily depends upon the physicochemical properties of the soil which they inhabit. At the root–soil interface, the soil habitat shows an increase in favorable conditions to promote microbial colonization, and a higher density and diversity of microbes is thus seen in rhizospheric soil when compared to the bulk soil. These microbes vary in their function from site to site and species to species (Ghosh et al. 2019). The rhizosphere of a plant houses a plethora of culture-dependent and culture-independent microbial species. Microbial colonies found in the rhizospheric niches of the plant vary from species to species and influence crop production, productivity, and plant sustainability (Fig. 9.1). An abundance is witnessed in the number and highly diverse variety of microbial species occupying the rhizospheric niche.

9.3.1 Common Rhizospheric Population

Azobacter, *Azospirillium*, *Azolla*, *Rhizobium,* and *Cyanobacteria* species are common microbes generally that are known for their nitrogen-fixing abilities. These symbiotic microbes are Gram-negative, aerobic bacteria. These nitrogen-fixing bacteria act as natural fertilizers impacting the plant metabolism, production of antibiotics, and plant growth hormones, root development and allowing for enhanced nutrient uptake. *Pseudomonas*, *Bacillus*, *Actinomyces*, *Agrobacterium*, and *Acetobacter* are known to be phosphorous-solubilizing bacteria. Apart from partaking in photosynthesis, energy transfer, and other plant processes, phosphorus is an important element because it is known to limit the fixation of nitrogen. Presence of the aforementioned bacterial species in the rhizosphere is not only limited to solubilizing phosphorus for plant uptake to regulate the metabolic processes of the plant but also enhance the nitrogen fixation process. *Pseudomonas* and *Bacillus* species also commonly facilitate the solubilization of potassium. Various *Pseudomonas*, *Bacillus*, *Rhizobium,* and *Azospirillum* species have been classified as plant growth-promoting rhizobacteria (PGPR). They are important for plant proliferation with their ability to serve as biocontrol. These species have also been reported as being capable of solubilizing zinc (Reddy et al. 2020).

9.3.2 Plant-Microbe Interaction

The evolution of the microbial strains is majorly affected by environmental conditions and plant species. The microbes evolve as to facilitate growth-promoting

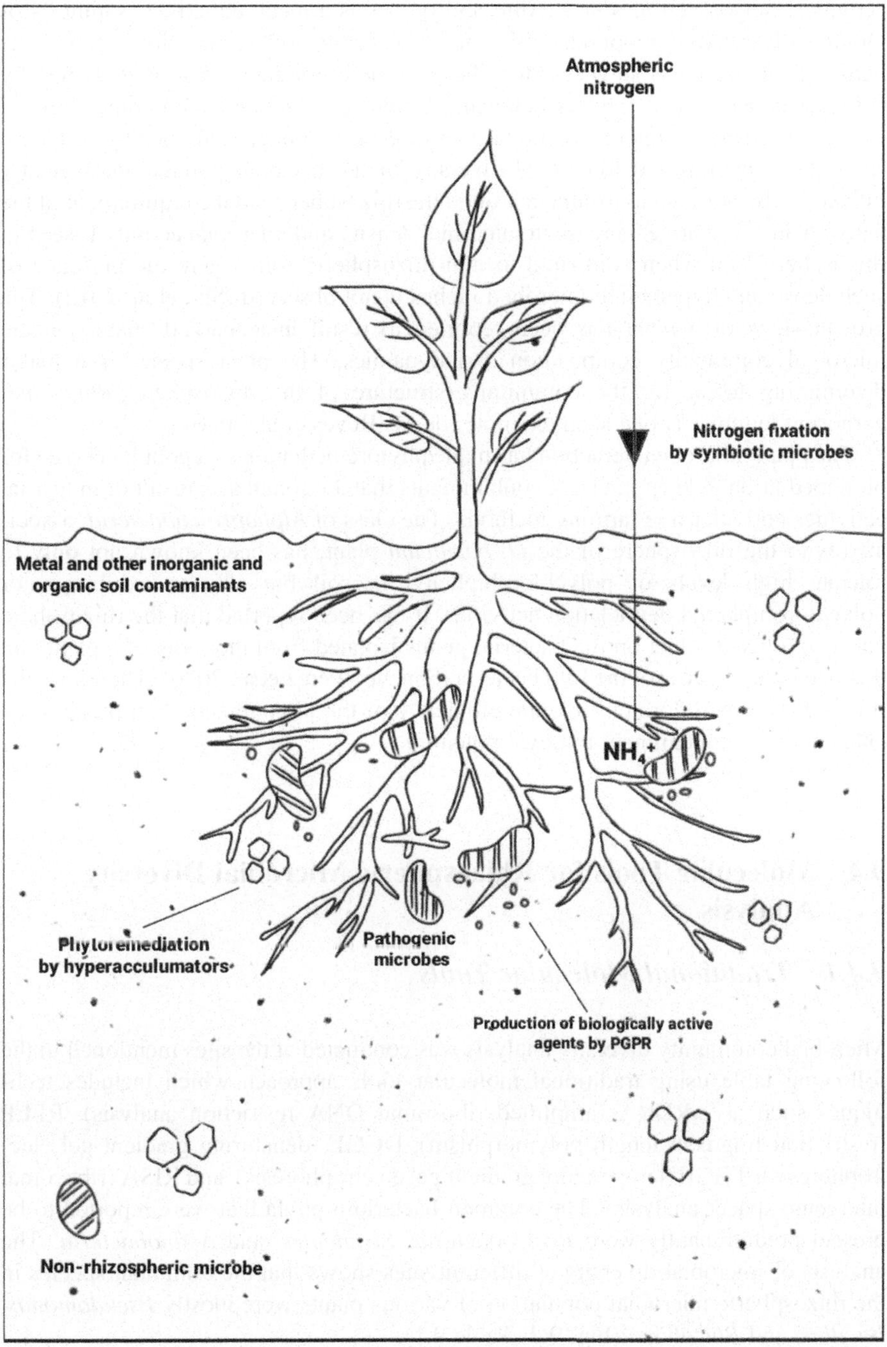

Fig. 9.1 Microbial interaction in the plant rhizosphere for nitrogen fixation, production of biologically active compounds and phytoremediation

activities, heightened nitrogen-fixing ability, act as a metal detoxifier, synthesis of biologically active compounds, and combat other the abiotic and biotic stresses its native plant species is subjected to. The plant–microbe interaction is governed by complex physiological and biochemical activities that facilitate communication.

The rhizospheric microbial interactions are very defined and specific in nature since the structural and functional diversity of the bacterial population is greatly influenced by the soil environment around the rhizosphere and the requirement of the native plant. A relative increase in microbial density and microbial activity is seen in rhizospheric soil when compared to non-rhizospheric soil, where the presence of such dense and hyperactive microbial niches is not observed (Shu et al. 2012). The growth stage of a plant has been reported to result in a marked change in the microbial community composition and dynamics. The plant species is a major determining factor for the community structure of the rhizosphere. *Rhizobium* harboring legumes is one such example (de los Reyes et al. 2020).

The plant–microbe interaction has made phytoremediation as a popular choice for bioremediation to keep in check contaminants that are found as a result of industrial activities and intensive farming methods. The class of *Alphaproteobacteria*, associated with the rhizosphere of the *O. basilicum* plant, has been known not only to tolerate high levels of polychlorobiphenyl in soil but also to exhibit major polychlorobiphenyl degradation activities. It has been reported that the rhizosphere has a significant effect on the bacterial genus isolated from the roots of a plant and the metabolic process of the said bacteria (Sánchez-Pérez et al. 2020). Therefore, the rhizospheric microflora ensures the efficiency of the phytoremediation process, by enhancing the contaminant removal potential.

9.4 Molecular Tools for Rhizospheric Microbial Diversity Analysis

9.4.1 Traditional Molecular Tools

Microbial community diversity analysis was conducted at the sites mentioned in the following table using traditional molecular tools approach which includes techniques such as ARDRA (amplified ribosomal DNA restriction analysis), RFLP (restriction fragment length polymorphism), DGGE (denaturing gradient gel electrophoresis), TGGE (temperature gradient gel electrophoresis), and RISA (ribosomal intergenic spacer analysis). The common bacterium phyla that were reported to be present predominantly were *Proteobacteria*, *Firmicutes,* and *Actinobacteria.* The analysis of microbial diversity at different sites shows that the dominant species in the rhizospheric microbial population of various plants were mostly *Pseudomonas, Bacillus,* and *Rhizobium* (Fig. 9.2; Table 9.1).

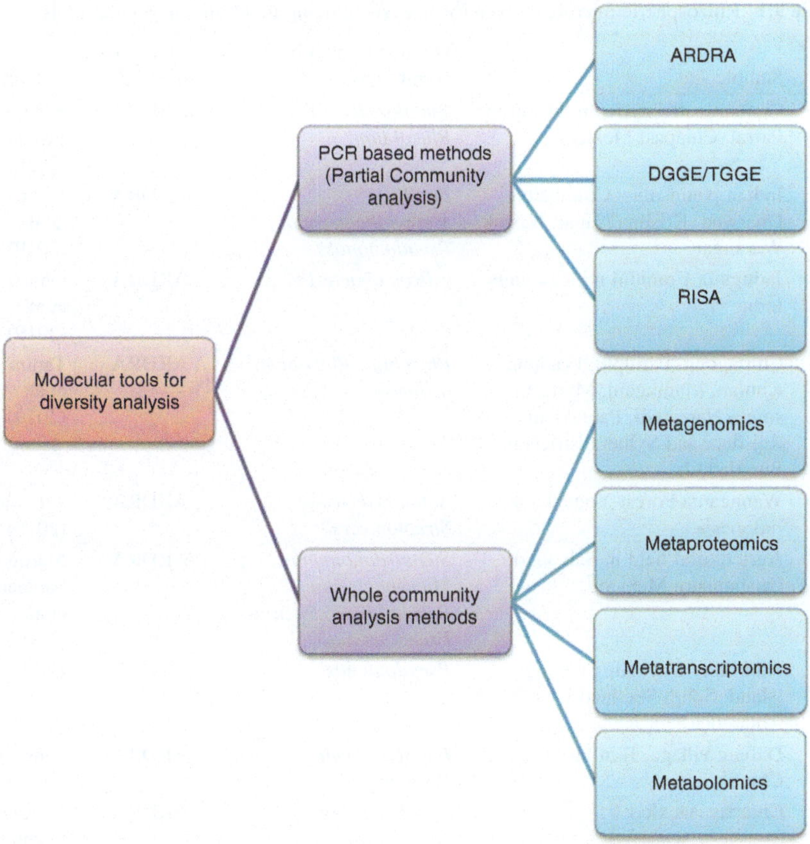

Fig. 9.2 Molecular tools for characterizing rhizospheric microbial diversity

9.4.1.1 ARDRA

Amplified ribosomal DNA restriction analysis (ARDRA) is a molecular technique that mirrors the RFLP technique except that it applies to the 16s ribosomal subunit of culture-independent bacteria. The steps involved are amplification, digestion, and gel electrophoresis. The PCR amplified fragment with restriction endonuclease is resolved with electrophoresis gel and a pattern is obtained. The methods used to analyze RAPD patterns are also used to analyze ARDRA patterns. NT-SYS and PAST are used for information about whether bands are present or not (1's and 0's for presence and absence, respectively). These patterns can be used to create phylograms or phylogenetic trees which describe the restriction pattern and give a relationship between organisms. The most common softwares used are GelCompar II and BioNumerics. Clones are amplified with primers and digested by restriction endonucleases and the resultant fragments are separated by acrylamide gels. The resultant profiles can be used for community clustering in genotyping or strain

Table 9.1 Rhizospheric microbial diversity analysis utilizing traditional molecular tools

Sl. no.	Sample site	Dominant microbial community	Molecular methods	Reference
1.	Contaminated agricultural soil in Parral, Chiapas, Mexico	*Bacillus, Lysinibacillus, Rhizobium*	ARDRA	Sánchez-Pérez et al. (2020)
2.	Indian peninsula—Coimbatore, Dharwad, Krishna Nagar, Nashik, Warangal	*Bacillus, Methylobacterium, Pseudomonas*	ARDRA	Verma et al. (2019)
3.	Jaduguda Uranium mine tailings, India	*γ-Proteobacteria*	ARDRA	Ghosh et al. (2019)
4.	Chittagong, Faridpur, Gazipur, Khulna, Manikganj, Mymen-singh, Narsingdi, Patuakhali, Rajshahi and Sylhet districts in Bangladesh	*Rhizobia—Rhizobium azibense*	ARDRA	Tanim et al. (2019)
5.	Wanagama Forest-Yogyakarta, Indonesia.	*Actinomycetes—Streptomyces.*	ARDRA	Ali et al. (2018)
6.	Agricultural field in Salvatierra, Guanajuato, Mexico	*Stenotrophomonas, Microbacterium, Burkholderia, Bacillus, Pseudomonas*	ARDRA	Marquez-Santacruz et al. (2010)
7.	Admiralty Bay, King George Island, South Shetland Islands, Antarctica	*Pseudomonas*	ARDRA	Da Silva et al. (2017)
8.	Dafang village, Jilin Province, China	*Bacillus, Acidobacteria, Proteobacteria*	ARDRA	Ying et al. (2012)
9.	Chiapas, Mexico	*Agrobacterium*	ARDRA	Lopez-Fuentes et al. (2012)
10.	Ni-rich serpentine soil	*Actinobacteria, Proteobacteria*	ARDRA	Abou-Shanab et al. (2010)
11.	Tranca, Bay, Laguna, Philippines	*Bacillus, Arthrobacter*	DGGE	de los Reyes et al. (2020)
12.	Kettara mine, Marrakech, Morocco	*Pseudomonas, Bacillus, Streptomyces, Tetrathiobacter*	DGGE	Benidire et al. (2020)
13.	Vineyards in Puglia region, Italy	*Bacillus, Actinomycetes, Pseudomonas*	DGGE	Saccà et al. (2019)
14.	Fujian Agriculture and Forest University (FAFU), Xitao town, Wuzhi County, Henan Province, China	*Pseudomonas*	DGGE	Wang et al. (2019)

(continued)

Table 9.1 (continued)

Sl. no.	Sample site	Dominant microbial community	Molecular methods	Reference
15.	Northeast Institute of Geography and Agroecology, Jilin Province, China	*Bacillus.*, *Flavobacterium*, *Rhizobium*	DGGE	Tian et al. (2019)
16	Phetchabun, Nakhon Nayok, Rayong and Chiang Mai provinces of Thailand	*Actinobacteria*	DGGE	Nimnoi et al. (2011)
17.	U.S. Salinity Laboratory, Riverside, CA, USA	Uncultured bacteria, *Proteobacteria*, *Bacteroides*	DGGE	Ibekwe et al. (2010)
18.	YingFengWuDou organic farm, ChongMing Island, Shanghai, China	*Rhizobiales*, *Pseudomonadales*	DGGE	Shu et al. (2012)
19.	Varanasi, Chandauli, Ghazipur district of Uttar Pradesh, India	*Proteobacteria*, *Actinobacteria*	RISA, DGGE	Srivastava et al. (2016)
20.	P. roxburghii forest area, Solan, Himachal Pradesh, India	*Gamma* and *Alpha Proteobacteria*	RISA	Mittal et al. (2019)

typing (Ying et al. 2012). The advantages of ARDRA are quick analysis of variations, multiple strains, and species. The drawback of this method is the inability to provide information about the sample microorganisms and the requirement of large quantities of DNA. Similar to the outcome of ribotyping, this method is faster but is not much sensitive.

9.4.1.2 DGGE

Denaturing gradient gel electrophoresis (DGGE) is another culture-independent method for analysis of microbial community which yields a visual fingerprint representation of the microbial community. The axis of a polyacrylamide gel is exposed to a denaturing gradient. Primers are used for PCR amplification after which the DNA undergoes gel electrophoresis by the denatured gel. Change in melting temperatures due to variations cause different migration ultimately leading to separation. The DNA separates due to the melting domains which are changed from the denaturation whose branching patterns are analyzed. The change in denaturation concentration leads to a change in the migration rate which in turn leads to a banding pattern. By comparing these patterns and known sequences, the various species in the sample can be identified and analyzed (Nimnoi et al. 2011). Digital image analysis can be used to interpret the DGGE profiles. By comparing distances between the migrated and reference strains, species can be identified and classified. This analysis technique poses one drawback as its bands only generate partial sequences. As a molecular tool, this method is often mentioned as a pair with temperature gradient gel electrophoresis (TGGE). Around 95–99% of microbial diversity in a community can be identified through this method.

9.4.1.3 RISA

Ribosomal intergenic spacer analysis (RISA) is an analysis method of community fingerprinting. It involves PCR amplification, electrophoresis, and staining. The region between the two subunits, namely 16S and 23S, is called the intergenic spacer region. The PCR amplification of this region is the main step in RISA (Srivastava et al. 2016). RISA fragments are obtained from specific regions in both the subunits which have been targeted with oligonucleotide primers. Based on the microbial species, tRNAs are encoded by elements of the intergenic spacer region. For the most effective length and sequence heterogeneity, the ISR length ranges from 150 to 500 base pairs. Multiple dominant community members result in a mixed PCR product. This is followed by electrophoresis by polyacrylamide gel. Visualization of the resultant DNA is possible after staining. A complex banding pattern consisting of DNA bands that correspond to a bacterial population gives a community-specific profile. It has been used for soil sustainability studies, species population studies, etc.

9.4.2 Omics-Based Rhizospheric Microbial Diversity

Omics-based approaches involve the use of data available from multiple omics fields including genomics (DNA, genomes), transcriptomics (gene transcripts or mRNA expressions), proteomics (proteins, proteomes), and metabolomics (metabolites, metabolome) for the purpose of microbial diversity analysis for an environmental sample. The following table (Table 9.2) provides a summary of sites where omics-based technologies such as metagenomics, functional metagenomics, metatranscriptomics, metaproteomics, and metabolomics have been used to perform microbial diversity analysis for the rhizosphere. The most common bacteria phyla reported to be present predominantly in these studies are *Proteobacteria*, *Acidobacteria*, and *Firmicutes*.

9.4.2.1 Metagenomics (Using all NGS platforms)

Metagenomics is the study of genetic materials collected immediately from an environment. It is one of the most commonly used culture-independent method and first of the omics methods to be developed for the study of microbial diversity including unculturable soil microbial community. It involved sequencing of genes available directly from environmental samples and analysis of the sequence generated which can then be used for various data analysis and predictions. Sequencing of data is done by the various platforms of NGS, most popular of them being Illumina

Table 9.2 Omics-based rhizospheric microbial diversity analysis

Sl. no.	Approaches	Sample site	Dominant microbial community	Analyzing system	Reference
1.	Metagenomics	Kuwait	*Proteobacteria, Cyanobacteria, Bacteriodetes, Actinobacteria, Acidobacteria, Thermi,* and *Gemmatimonadetes*	Illumina sequencing	Kumar et al. (2018)
2.		Hokkaido, Japan	*Proteobacteria, Fibrobacteres/ Acidobacteria, Actinobacteria, Bacteroidetes/Chlorobi* group	454 sequencing	Unno and Shinano (2012)
3.		Oak forest located in Breuil-Chenue, France	*Acidobacteria, Proteobacteria, Actinobacteria,* and unclassified bacteria	454 sequencing	Uroz et al. (2010)
4.		Loktak lake, Manipur, India	*Proteobacteria, Acidobacteria (Candidatus solibacter), Actinobacteria, Bacteroidetes.*	Illumina sequencing	Puranik et al. (2016)
5.		Mey Blossom Garden, Wuxi city, adjacent to Taihu Lake, in the Yangtze River delta of China	*Proteobacteria, Acidobacteria, Firmicutes, Bacteroidetes*	Illumina sequencing	Wei et al. (2017)
6.		Lhenice, south Bohemia, Czech Republic	*Actinobacteria, Acidobacteria, Proteobacteria*	Pyrosequencing	Ridl et al. (2016)
7.		Golzow, Germany	*Mycobacterium, Pseudomonas, Burkholderia, Caulobacter, Actinomycetales, Rhizobiales,*	Illumina sequencing	Kröber et al. (2014)
8.		Central Rice Research Institute, Cuttack, Odisha, India	*Proteobacteria, Firmicutes, Actinobacteria, Acidobacteria,* Methanogens	Illumina sequencing	Bhattacharyya et al. (2016)
9.		International Rice Research Institute, Los Ban˜ os, Philippines.	*Alpha-, Beta-,* and *Deltaproteobacteria, Firmicutes, Actinobacteria,*	454 sequencing	Knief et al. (2012)

(continued)

Table 9.2 (continued)

Sl. no.	Approaches	Sample site	Dominant microbial community	Analyzing system	Reference
			Gammaproteobacteria, and the *Deinococcus-Thermus* phylum		
10.		The coast of Sa˜o Paulo State, Brazil	*Proteobacteria, Firmicutes, Actinobacteria, Bacteroidetes.* and *Chloroflexi*	Pyrosequencing	Andreote et al. (2012)
11.	Functional metagenomics	Es Trenc saltern, Mallorca, Spain	*Ardenticatena maritima, Cytophagales, Sorangiineae, Bacillus halosaccharovorans, Actinobacteria, Rhodobacteraceae*	454 sequencing	Mirete et al. (2015)
12.		Forest and grassland sites of the German Biodiversity Exploratorien Schorfheide-Chorin and Schwäbische Alb	*Proteobacteria, Actinobacteria, Bacteroidetes,* and *Chloroflexi*	Pyrosequencing	Willms et al. (2019)
13.		Mangrove shore, Thuwal, Saudi Arabia	*Proteobacteria, Bacteriodetes, Firmicutes,* sulfate reducers, methanogens	454 sequencing	Alzubaidy et al. (2016)
14.		Shenzhen, Guangzhou Province, China	*Proteobacteria*	Illumina sequencing	Bai et al. (2017)
15.		Shijiuyang water source ecological wetland in Jiaxing City, China	*Proteobacteria* (mainly *gammaproteobacteria*), *Firmicutes*	Illumina sequencing	Bai et al. (2014)
16.		Southeastern Brazilian Amazon, in the state of Mato Grosso, Brazil	*Proteobacteria, Actinobacteria, Firmicutes, Acidobacteria*	454 sequencing	Mendes et al. (2014)
17.		Meijendel, Netherlands	*Proteobacteria, Actinobacteria, Bacteroidetes, Firmicutes, Acidobacteria, Verrucomicrobia, Planctomycetes*	Illumina sequencing	Yan et al. (2017)

18.	Metatranscriptomics	Agricultural field near Norwich, UK	*Proteobacteria, Actinobacteria, Firmicutes, Acidobacteria, Planctomycetes* and *Bacteroidetes, and Cyanobacteria*	454 sequencing	Turner et al. (2013)
19.		Michigan Extension Station, Benton Harbor, MI	*Acidobacteria, Actinobacteria, Bacteroidetes, Cyanobacteria*	454 sequencing	Chaparro et al. (2014)
20.		Farm in Avon, South Australia, in the southern wheat cropping region of Australia.	*Micrococcaceae* (phylum *Actinobacteria), Pseudomonadaceae, Enterobacteriaceae (Gammaproteobacteria) Nitrososphaeraceae* (phylum *Thaumarchaeota) Ascomycota, Basidiomycota, Glomeromycota*	Illumina sequencing	Hayden et al. (2018)
21.		Hawaii Ocean Time-Series (HOT) in the North Pacific subtropical gyre	*Prochlorococcus, Cyanobacteria, a-Proteobacteria, Bacteroidetes b-Proteobacteria* and *g-Proteobacteria, and Firmicutes*	Pyrosequencing	Poretsky et al. (2009)
22.		Former petroleum refinery site, Montreal, Canada	*Firmicutes, Glomeromycota, Acidobacteria, Verrucomicrobia, Gammaproteobacteria*	Illumina sequencing	Yergeau et al. (2018)
23.		Six distinct mangroves rhizospheres located on the coast of Mauritius	*Proteobacteria, Bacteroidetes*, and *Firmicutes,* with a high abundance of sulfate reducers, nitrogen reducers, and methanogens	Illumina sequencing	Rampadarath et al. (2018)
24.		Greenhouse of the IPK Gatersleben, Germany.	*Proteobacteria, Rhodospirillaceae, Planctomycetes,* and *Verrucomicrobia*	Illumina sequencing	Cao et al. (2015)

(continued)

Table 9.2 (continued)

Sl. no.	Approaches	Sample site	Dominant microbial community	Analyzing system	Reference
25.	Metaproteomics	Vineyard located close to Carpeneto (Italy)	*Streptomyces, Bacillus, Bradyrhizobium, Burkholderia,* and *Pseudomonas.*	MS/MS analysis	Bona et al. (2019)
26.		Mount Prinzera, a Natural Reserve in the Northern Apennines (Italy)	*Methylobacterium mesophylicum, Phyllobacterium, Microbacterium oxydans, Pseudomonas oryzihabitans, Stenotrophomonas rhizophila,* and *Bacillus methylotrophicus*	Illumina sequencing, LCHRMS analysis	Mattarozzi et al. (2017)
27.		Experimental station of Fujian Agriculture and Forestry University, Fuzhou, China	*Proteobacteria, Actinobacteria, Firmicutes,* and *Bacteroidetes*	Tandem mass spectrometry MALDI-TOF system	Chen et al. (2019)
28.		Fuzhou, Fujianin, southeast China; Zhengzhou, Henanin, Central China; Yuxi, Yunnanin southwest China	*Proteobacteria, Actinobacteria, Bacilli, Bacteroidetes, Clostridia*	MALDI-TOF/TOF mass spectrometer	Wang et al. (2011)
29.		Kashimadai experimental field of Tohoku University, Japan	*Methylosinus* and *Methylocystis, Bradyrhizobium* and *Rhodopseudomonas* in *Alphaproteobacteria, Betaproteobacteria, Deltaproteobacteria, Firmicutes, Gammaproteobacteria, Actinobacteria* and Archaea	454 sequencing Nano-liquid chromatography (LC)–electrospray ionization–tandem mass spectrometry (MS/MS) analysis	Bao et al. (2014)

30.		Rice fields at the International Rice Research Institute, Los Baños, Philippines	*Alpha-*, *Beta-*, and *Deltaproteobacteria*, *Firmicutes*, *Actinobacteria*, *Gammaproteobacteria* and the *Deinococcus-Thermus* phylum	454 sequencing Mass spectroscopy	Knief et al. (2012)
31.	Metabolomics	Farms at Sweden; France; Wisconsin, Colorado, OR, USA	*Actinobacteria*, *Proteobacteria*, *Cyanobacteria* and *Firmicutes*	454 pyrosequencing GC-MS	Badri et al. (2013)
32.		France	*Alphaproteobacteria*, *Betaproteobacteria*, *Gammaproteobacteria* and *Acidobacteria*	HPLC/DAD/ESI-MS	Bressan et al. (2009)
33.		Spen Farm, Leeds, UK	*Burkholderiaceae*, *Oxalobacteraceae*, *Pseudomonadaceae*, *Xanthomonadaceae*, *Bradyrhizobiaceae*, *Rhizobiaceae*	Illumina sequencing UPLC Q-TOF mass spectrometry	Pétriacq et al. (2017)
34.		University of California Hopland Research and Extension Center, Hopland, CA, USA	*Acidobacteria*, *Actinobacteria*, *Proteobacteria* (alpha, beta, gamma), *Bacteroidetes*, *Firmicutes*	High throughput sequencing, UPHLC-MS	Zhalnina et al. (2018)
35.		France	*Alphaproteobacteria*, *Betaproteobacteria*, *Gammaproteobacteria* *Acidobacteria*	HPLC-MS	Bressan et al. (2013)

sequencing followed by 454 sequencing and pyrosequencing. Metagenomics approach has been widely used to characterize microbial communities of any environmental habitat (Andreote et al. 2012; Puranik et al. 2016; Wei et al. 2017). Metagenomic analysis has been used in the study of microbiome present in soil contaminations and other aspects related to it (Kumar et al. 2018; Ridl et al. 2016). Whole metagenome sequencing has been used to study the role of rice rhizosphere in the metabolism of methane and nitrogen (Bhattacharyya et al. 2016) and effect of a specific strain of bacillus on lettuce rhizosphere microbial community (Kröber et al. 2014). Studies have also been conducted on bacterial diversity of specific crop species (Knief et al. 2012; Uroz et al. 2010) as well as utilization of compound present in rhizosphere soil (Unno and Shinano 2012) thereby revealing the wide use of this method in diverse fields of microbiology.

9.4.2.2 Functional Metagenomics

Functional metagenomic approach is a type of metagenomic approach which focuses on the studying of gene function from a mixed population of DNA. It involves construction and screening of metagenomic libraries which help in annotations of gene function. It involves the identification of functional gene during screening without using previously retrieved sequenced genes enabling the identification of both novel and known genes (Mirete et al. 2015). This approach has been widely used to study functional traits in rhizosphere community selection, functional potentials of various ecosystems along with determining the community diversity (Alzubaidy et al. 2016; Bai et al. 2014; Mendes et al. 2014; Yan et al. 2017). This approach has been used for discovery of resistance genes in a particular ecological habitat (Mirete et al. 2015; Willms et al. 2019) and study of the effect on pollutant removal in ecological wastewater (Bai et al. 2017).

9.4.2.3 Metatranscriptomics

Metatranscriptomics studies gene expression of microbes (RNA expressions or gene transcripts) within natural environments, i.e., metatranscriptome. Metatranscriptome provides information about the active metabolic processes of the microbiome in a given condition in an environment (Kothari et al. 2017). Metatranscriptomics can retrieve and sequence mRNAs from an environmental microbial community without any previous idea of the genes that might be expressed by the community thereby making it advantageous and less biased (Poretsky et al. 2009). NGS is the preferred technique for sequencing in metatranscriptomics with Illumina sequencing and 454 sequencing being the most popular platforms. Metatranscriptome analysis has been used for studying microbial diversity and host–microbiome interactions of various plant species (Cao et al. 2015; Rampadarath et al. 2018). Effects on the rhizosphere due to various factors of soil, soil contamination, developmental stages of plants have been investigated through this approach (Chaparro et al. 2014;

Yergeau et al. 2018). Comparative studies based on metatranscriptomics have been performed among various samples to reveal the variations and effects of changing factors such as plant species, day timings, or soil types (Hayden et al. 2018; Poretsky et al. 2009; Turner et al. 2013).

9.4.2.4 Metaproteomics

Metaproteomics refers to the study of all proteins and peptides present in environmental samples. Mass spectrometry combined with various chromatography techniques is usually used for extraction and analysis of protein samples and latest advancements in this field has resulted in various high efficiency analyzing systems for protein and peptide identification. Metaproteomics is used for the characterization of soil microbial communities, biological samples based not only on taxonomy but also their functional activity and protein expression. For soil ecosystems, it provides an analysis of the functional proteins and helps in understanding various metabolic processes and signal transductions involved in the soil biotic community. Metaproteomic analysis and characterization have been done in different soil systems and crops giving an insight into the metabolic activities inside the soil by the microbes or the plants, molecular (secreted protein) interactions between microbes and plants and the effect of the proteins on rhizosphere community (Bona et al. 2019; Knief et al. 2012; Mattarozzi et al. 2017; Wang et al. 2011). Apart from characterizing bacterial community, metaproteomics analysis can be used to evaluate the role of particular microbes in specific processes such as nitrogen fixation, methane oxidation (Bao et al. 2014), or effect of fertilizers and related crop yield improvement (Chen et al. 2019).

9.4.2.5 Metabolomics

Metabolomics is the whole-community sampling of all the metabolites (i.e., sugars, lipids) represented within a microbial community, representing the functional substrate and products of metabolomic pathways within an ecosystem (White et al. 2017). Metabolomics uses mass spectroscopy-chromatography-based systems for analyzing metabolites to draw conclusions about different cellular, biochemical processes, interactions with environmental factors, and microbial contributions to metabolic varieties thereby determining the functional diversity of given microbial samples based on their biochemical activity. It has been used for studying the effects of specific exogenous metabolites or its hydrolyzed products produced on the rhizospheric microbial community and also evaluating the effect of engineered plants with altered metabolic profiles (Bressan et al. 2009, 2013). Non-sterile soils have been studied for the presence of plant-derived metabolites and their suitable application for crop agriculture (Pétriacq et al. 2017). Using a combined approach of comparative genomics and exometabolomics, it has been demonstrated that root exudate chemistry during developmental stages of a plant and microbial preferential substrate uptake is correlated and aid in predicting microbial response to root growth apart from determining the rhizospheric diversity (Zhalnina et al. 2018). Along with these, the effect of diverse soil microbiome on leaf metabolome, plant growth patterns and herbivore feeding habits has also been successfully investigated (Badri et al. 2013).

9.5 Pros and Cons of the Omics-Based Approach

Omics-based approaches have gained rapid popularity among researchers as they have made it possible to study entire genome sequences, transcripts, proteins, and metabolites from environmental samples providing a more comprehensive view of genetic diversity of the culture-independent microbiome. They provide more in-depth information about composition, the function of a whole microbial community, their interactions with biotic/abiotic factors, metabolic processes occurring inside the soil microbe community (Lagos et al. 2015). However, these approaches have a few drawbacks such as huge data volume and complex data for data analysis, complex extraction methods of analytes from samples, shorter read length of NGS platforms leading to overestimation of taxonomic classifications, high error rate (Pal et al. 2019), and lack of adequate libraries and databases.

9.6 Future Prospects

Future prospects include advancements in existing techniques with the objective of discovering the lesser-known rhizospheric bacterial communities; predict the responses of various species in their native environment, activities, and prosperity of such communities in soil (Lagos et al. 2015). Advancements in software development are needed for qualitative improvement of data generated from recent NGS platforms, developing robust reference libraries, spectral databases for metaproteomic and metabolomic studies, and adequate data analysis tools to make an integrated multi-omics approach possible (White et al. 2017). An integrated multi-omics study will enable us to look at the complete picture of soil microbiome with all its aspects and will help in deciding and implementing strategies for the selection of native bacterial strains capable of beneficial use, to comprehend the significance and function of the rhizosphere microbial activity to support healthy plant growth, improved yield and much more.

References

Abou-Shanab RA, Van Berkum P, Angle JS, Delorme TA, Chaney RL, Ghozlan HA, Ghanem K, Moawad H (2010) Characterization of Ni-resistant bacteria in the rhizosphere of the hyperaccumulator *Alyssum murale* by 16S rRNA gene sequence analysis. World J Microbiol Biotechnol 26(1):101–108

Ali A, Mustofa M, Asmara W, Rante H, Widada J (2018) Diversity and functional characterization of antifungal-producing Streptomyces-like microbes isolated from the rhizosphere of cajuput plants (*Melaleuca leucodendron* L.). Malaysian J Microbiol 14(7):663–673

Alzubaidy H, Essack M, Malas TB, Bokhari A, Motwalli O, Kamanu FK, Jamhor SA, Mokhtar NA, Antunes A, Simões MF, Alam I (2016) Rhizosphere microbiome metagenomics of gray mangroves (*Avicennia marina*) in the Red Sea. Gene 576(2):626–636

Andreote FD, Jiménez DJ, Chaves D, Dias ACF, Luvizotto DM, Dini-Andreote F, Fasanella CC, Lopez MV, Baena S, Taketani RG, de Melo IS (2012) The microbiome of Brazilian mangrove sediments as revealed by metagenomics. PLoS One 7(6):e38600

Badri DV, Zolla G, Bakker MG, Manter DK, Vivanco JM (2013) Potential impact of soil microbiomes on the leaf metabolome and on herbivore feeding behavior. New Phytol 198 (1):264–273

Bai Y, Liang J, Liu R, Hu C, Qu J (2014) Metagenomic analysis reveals microbial diversity and function in the rhizosphere soil of a constructed wetland. Environ Technol 35(20):2521–2527

Bai Y, Huo Y, Liao K, Qu J (2017) Influence of microbial community diversity and function on pollutant removal in ecological wastewater treatment. Appl Microbiol Biotechnol 101 (19):7293–7302

Bao Z, Okubo T, Kubota K, Kasahara Y, Tsurumaru H, Anda M, Ikeda S, Minamisawa K (2014) Metaproteomic identification of diazotrophic methanotrophs and their localization in root tissues of field-grown rice plants. Appl Environ Microbiol 80(16):5043–5052

Benidire L, Pereira SI, Naylo A, Castro PM, Boularbah A (2020) Do metal contamination and plant species affect microbial abundance and bacterial diversity in the rhizosphere of metallophytes growing in mining areas in a semiarid climate? J Soils Sediments 20(2):1003–1017

Bhattacharyya P, Roy KS, Das M, Ray S, Balachandar D, Karthikeyan S, Nayak AK, Mohapatra T (2016) Elucidation of rice rhizosphere metagenome in relation to methane and nitrogen metabolism under elevated carbon dioxide and temperature using whole genome metagenomic approach. Sci Total Environ 542:886–898

Bona E, Massa N, Novello G, Boatti L, Cesaro P, Todeschini V, Magnelli V, Manfredi M, Marengo E, Mignone F, Berta G (2019) Metaproteomic characterization of the *Vitis vinifera* rhizosphere. FEMS Microbiol Ecol 95(1):204

Bressan M, Roncato MA, Bellvert F, Comte G, el Zahar HF, Achouak W, Berge O (2009) Exogenous glucosinolate produced by *Arabidopsis thaliana* has an impact on microbes in the rhizosphere and plant roots. ISME J 3(11):1243–1257

Bressan M, Achouak W, Berge O (2013) Exogenous glucosinolate produced by transgenic *Arabidopsis thaliana* has an impact on microbes in the rhizosphere and plant roots. Mol Microb Ecol Rhizosphere 1:1173–1179

Cao HX, Schmutzer T, Scholz U, Pecinka A, Schubert I, Vu GT (2015) Metatranscriptome analysis reveals host-microbiome interactions in traps of carnivorous Genlisea species. Front Microbiol 6:526

Chaparro JM, Badri DV, Vivanco JM (2014) Rhizosphere microbiome assemblage is affected by plant development. ISME J 8(4):790–803

Chen J, Arafat Y, Din IU, Yang B, Zhou L, Wang J, Letuma P, Wu H, Qin X, Wu L, Lin S (2019) Nitrogen fertilizer amendment alter the bacterial community structure in the rhizosphere of rice (*Oryza sativa* L.) and improve crop yield. Front Microbiol 10:2623

Da Silva AC, da Costa Rachid CT, de Jesus HE, Rosado AS, Peixoto RS (2017) Predicting the biotechnological potential of bacteria isolated from Antarctic soils, including the rhizosphere of vascular plants. Polar Biol 40(7):1393–1407 de los Reyes AM, Ocampo ET, Manuel MC, Mendoza BC (2020) Analysis of the bacterial and fungal community profiles in bulk soil and rhizospheres of three mungbean [*Vigna radiata* (L.) R. Wilczek] genotypes through PCR-DGGE. Int Lett Nat Sci 77:1–26

Ghosh UD, Singh PK, Ganguli S, Saha C, Chandra A, Seal A, Ghosh MM (2019) Rhizospheric soil of *Typha angustifolia* L. from heavy metal contaminated and free sites: comparative profiling reveals selective abundance of γ-proteobacteria and β-proteobacteria. IJEB 57(10):733–740

Hayden HL, Savin KW, Wadeson J, Gupta VV, Mele PM (2018) Comparative metatranscriptomics of wheat rhizosphere microbiomes in disease suppressive and non-suppressive soils for *Rhizoctonia solani* AG8. Front Microbiol 9:859

Ibekwe AM, Poss JA, Grattan SR, Grieve CM, Suarez D (2010) Bacterial diversity in cucumber (*Cucumis sativus*) rhizosphere in response to salinity, soil pH, and boron. Soil Biol Biochem 42 (4):567–575

Knief C, Delmotte N, Chaffron S, Stark M, Innerebner G, Wassmann R, Von Mering C, Vorholt JA (2012) Metaproteogenomic analysis of microbial communities in the phyllosphere and rhizosphere of rice. ISME J 6(7):1378–1390

Kothari V, Kothari C, Rank J, Joshi A, Singh RP, Kothari R (2017) Metatranscriptomic studies of the plant rhizosphere for finding biological agents. In: Understanding host-microbiome interactions-an omics approach. Springer, Singapore, pp 267–275

Kröber M, Wibberg D, Grosch R, Eikmeyer F, Verwaaijen B, Chowdhury SP, Hartmann A, Pühler A, Schlüter A (2014) Effect of the strain *Bacillus amyloliquefaciens* FZB42 on the microbial community in the rhizosphere of lettuce under field conditions analyzed by whole metagenome sequencing. Front Microbiol 5:252

Kumar V, AlMomin S, Al-Aqeel H, Al-Salameen F, Nair S, Shajan A (2018) Metagenomic analysis of rhizosphere microflora of oil-contaminated soil planted with barley and alfalfa. PLoS One 13 (8):e0202127

Lagos L, Maruyama F, Nannipieri P, Mora ML, Ogram A, Jorquera MA (2015) Current overview on the study of bacteria in the rhizosphere by modern molecular techniques: a mini-review. J Soil Sci Plant Nutr 15(2):504–523

Lopez-Fuentes E, Ruiz-Valdiviezo VM, Martinez-Romero E, Gutierrez-Miceli FA, Dendooven L, Rincon-Rosales R (2012) Bacterial community in the roots and rhizosphere of *Hypericum silenoides* Juss. 1804. Afr J Microbiol Res 6(11):2704–2711

Marquez-Santacruz HA, Hernandez-Leon R, Orozco-Mosqueda MD, Velazquez-Sepulveda I, Santoyo G (2010) Diversity of bacterial endophytes in roots of Mexican husk tomato plants (*Physalis ixocarpa*) and their detection in the rhizosphere. Genet Mol Res 9(4):2372–2380

Mattarozzi M, Manfredi M, Montanini B, Gosetti F, Sanangelantoni AM, Marengo E, Careri M, Visioli G (2017) A metaproteomic approach dissecting major bacterial functions in the rhizosphere of plants living in serpentine soil. Anal Bioanal Chem 409(9):2327–2339

Mendes LW, Kuramae EE, Navarrete AA, Van Veen JA, Tsai SM (2014) Taxonomical and functional microbial community selection in soybean rhizosphere. ISME J 8(8):1577–1587

Mirete S, Mora-Ruiz MR, Lamprecht-Grandío M, de Figueras CG, Rosselló-Móra R, González-Pastor JE (2015) Salt resistance genes revealed by functional metagenomics from brines and moderate-salinity rhizosphere within a hypersaline environment. Front Microbiol 6:1121

Mittal D, Shukla R, Verma S, Sagar A, Verma KS, Pandey A, Negi YS, Saini RV, Saini AK (2019) Fire in pine grown regions of Himalayas depletes cultivable plant growth promoting beneficial microbes in the soil. Appl Soil Ecol 139:117–124

Nimnoi P, Pongsilp N, Lumyong S (2011) Actinobacterial community and diversity in rhizosphere soils of *Aquilaria crassna* Pierre ex Lec assessed by RT-PCR and PCR-DGGE. Biochem Syst Ecol 39(4–6):509–519

Pal S, Roy A, Kazy SK (2019) Exploring microbial diversity and function in petroleum hydrocarbon associated environments through omics approaches. In: Microbial diversity in the genomic era. Academic, Cambridge, pp 171–194

Pétriacq P, Williams A, Cotton A, McFarlane AE, Rolfe SA, Ton J (2017) Metabolite profiling of non-sterile rhizosphere soil. Plant J 92(1):147–162

Poretsky RS, Hewson I, Sun S, Allen AE, Zehr JP, Moran MA (2009) Comparative day/night metatranscriptomic analysis of microbial communities in the North Pacific subtropical gyre. Environ Microbiol 11(6):1358–1375

Puranik S, Pal RR, More RP, Purohit HJ (2016) Metagenomic approach to characterize soil microbial diversity of Phumdi at Loktak Lake. Water Sci Technol 74(9):2075–2086

Rampadarath S, Bandhoa K, Puchooa D, Jeewon R, Bal S (2018) Metatranscriptomics analysis of mangroves habitats around Mauritius. World J Microbiol Biotechnol 34(4):59

Reddy GC, Goyal RK, Puranik S, Waghmar V, Vikram KV, Sruthy KS (2020) Biofertilizers toward sustainable agricultural development. In: Plant microbe symbiosis. Springer, Cham, pp 115–128

Ridl J, Kolar M, Strejcek M, Strnad H, Stursa P, Paces J, Macek T, Uhlik O (2016) Plants rather than mineral fertilization shape microbial community structure and functional potential in legacy contaminated soil. Front Microbiol 7:995

Saccà ML, Manici LM, Caputo F, Frisullo S (2019) Changes in rhizosphere bacterial communities associated with tree decline: grapevine esca syndrome case study. Can J Microbiol 65 (12):930–943

Sánchez-Pérez BN, Zenteno-Rojas A, Rincón-Molina CI, Ruíz-Valdiviezo VM, Gutiérrez-Miceli FA, Vences-Guzmán MA, Villalobos-Maldonado JJ, Rincón-Rosales R (2020) Rhizosphere and endophytic bacteria associated to *Ocimum basilicum* L. with decaclorobiphenyl removal potential. Water Air Soil Pollut 231(3):1–15

Shu W, Pablo GP, Jun Y, Danfeng H (2012) Abundance and diversity of nitrogen-fixing bacteria in rhizosphere and bulk paddy soil under different duration of organic management. World J Microbiol Biotechnol 28(2):493–503

Singh A, Kumar M, Verma S, Choudhary P, Chakdar H (2020) Plant microbiome: trends and prospects for sustainable agriculture. In: Plant microbe symbiosis. Springer, Cham, pp 129–151

Srivastava M, Kaushik MS, Mishra AK (2016) Linking the physicochemical properties with the abundance and diversity of rhizospheric bacterial population inhabiting paddy soil based on a concerted multivariate analysis of PCR-DGGE and RISA. Geomicrobiol J 33(10):894–905

Tanim MT, Chowdhury MM, Bari L, Rahaman MM, Rahman SR, Rahman MM (2019) Genetic diversity of Rhizobiums: isolated from soil samples of Bangladesh. Bangladesh J Microbiol 36 (1):7–10

Tian L, Shi S, Ma L, Zhou X, Luo S, Zhang J, Lu B, Tian C (2019) The effect of *Glomus intraradices* on the physiological properties of *Panax ginseng* and on rhizospheric microbial diversity. J Ginseng Res 43(1):77–85

Turner TR, Ramakrishnan K, Walshaw J, Heavens D, Alston M, Swarbreck D, Osbourn A, Grant A, Poole PS (2013) Comparative metatranscriptomics reveals kingdom level changes in the rhizosphere microbiome of plants. ISME J 7(12):2248–2258

Unno Y, Shinano T (2012) Metagenomic analysis of the rhizosphere soil microbiome with respect to phytic acid utilization. In: Microbes and environments, p ME12181

Uroz S, Buée M, Murat C, Frey-Klett P, Martin F (2010) Pyrosequencing reveals a contrasted bacterial diversity between oak rhizosphere and surrounding soil. Environ Microbiol Rep 2 (2):281–288

Verma P, Yadav AN, Khannam KS, Mishra S, Kumar S, Saxena AK, Suman A (2019) Appraisal of diversity and functional attributes of thermotolerant wheat associated bacteria from the peninsular zone of India. Saudi J Biol Sci 26(7):1882–1895

Wang HB, Zhang ZX, Li H, He HB, Fang CX, Zhang AJ, Li QS, Chen RS, Guo XK, Lin HF, Wu LK (2011) Characterization of metaproteomics in crop rhizospheric soil. J Proteome Res 10 (3):932–940

Wang J, Wu L, Tantai H, Khan MU, Letuma P, Wu H, Zhang S, Chen T, Lin S, Lin W (2019) Properties of bacterial community in the rhizosphere soils of *Achyranthes bidentata* tolerant to consecutive monoculture. Plant Growth Regul 89(2):167–178

Wei Z, Hu X, Li X, Zhang Y, Jiang L, Li J, Guan Z, Cai Y, Liao X (2017) The rhizospheric microbial community structure and diversity of deciduous and evergreen forests in Taihu Lake area, China. PLoS One 12(4):e0174411

White RA, Borkum MI, Rivas-Ubach A, Bilbao A, Wendler JP, Colby SM, Köberl M, Jansson C (2017) From data to knowledge: the future of multi-omics data analysis for the rhizosphere. Rhizosphere 3:222–229

Willms IM, Kamran A, Aßmann NF, Krone D, Bolz SH, Fiedler F, Nacke H (2019) Discovery of novel antibiotic resistance determinants in forest and grassland soil metagenomes. Front Microbiol 10:460

Yan Y, Kuramae EE, de Hollander M, Klinkhamer PG, van Veen JA (2017) Functional traits dominate the diversity-related selection of bacterial communities in the rhizosphere. ISME J 11 (1):56–66

Yergeau E, Tremblay J, Joly S, Labrecque M, Maynard C, Pitre FE, St-Arnaud M, Greer CW (2018) Soil contamination alters the willow root and rhizosphere metatranscriptome and the root–rhizosphere interactome. ISME J 12(3):869–884

Ying YX, Ding WL, Li Y (2012) Characterization of soil bacterial communities in rhizospheric and nonrhizospheric soil of *Panax ginseng*. Biochem Genet 50(11–12):848–859

Zhalnina K, Louie KB, Hao Z, Mansoori N, da Rocha UN, Shi S, Cho H, Karaoz U, Loqué D, Bowen BP, Firestone MK (2018) Dynamic root exudate chemistry and microbial substrate preferences drive patterns in rhizosphere microbial community assembly. Nat Microbiol 3 (4):470–480

Chapter 10
Rhizosphere Fingerprints: Novel Biomolecules Via Meta-Omics Technology

Nidhi Srivastava and Indira P. Sarethy

Abstract The rhizosphere is a highly dynamic environment with complex signaling mechanisms between plant roots and microbes in a constant flux as a consequence of various (bio)-chemical and physical processes. Unraveling the chemical agents or biomolecules within this complex niche environment can be of tremendous value in mapping various functional aspects established between plants and microbes. Recent advances in high-throughput and meta-omics technologies offer opportunities to discover novel molecules in this environment, overcoming the challenges faced by conventional methods. Metabolomics studies in rhizosphere has facilitated the profiling and annotation of metabolites (exo/endo) participating in various physiological processes within biological systems. This chapter will elucidate various aspects of plant–microbe interactions resulting via the production of various signaling molecules, including those of novel bioactive molecules that have been discovered by meta-omics technology.

Keywords Plant–microbe interactions · Meta-omics · Metagenomics · Metatranscriptomics · Metabolite fingerprinting · GC-MS

10.1 Introduction

Soil environments, enriched with the carbon-containing exudates secreted by terrestrial plants, attract microbial populations surrounding the root environment and promote plant–microbe associations (Dennis et al. 2010). The plethora of microbial population is enriched in terms of diversity, with unique metabolic pathways in response to the molecules secreted by the plants in the vicinity. Free-living and endophytic microbial populations (symbiotic/free-living bacteria, actinomycetes, and mycorrhizal fungi) are also affected by the type of soil in that area (Lakshmanan

---

N. Srivastava · I. P. Sarethy (✉)
Department of Biotechnology, Jaypee Institute of Information Technology, Noida, India
e-mail: indirap.sarethy@jiit.ac.in

R. N. Pudake et al. (eds.), *Omics Science for Rhizosphere Biology*, Rhizosphere Biology, https://doi.org/10.1007/978-981-16-0889-6_10

et al. 2014). Plants are affected by the secretory/regulatory chemicals secreted by associated microbes (Turner et al. 2013). The microbial population in the plant vicinity imparts functional diversity to them and is, hence, considered to constitute the "second genome" and "metaorganism" complex (East 2013).

It has been estimated that the global loss of 24 billion acres of fertile land is due to a variety of causes such as deforestation, nutrient cycling (nitrogen) imbalance, dynamic fluctuations in climatic conditions, desertification, changed physicochemical properties of soil (organic carbon/pH), stress caused by physical and biotic environments, and anthropological activities (Dubois 2011). Microbial populations within or in the surroundings of the plants make the plants more resistant/resilient to unfavorable conditions.

Rhizospheric environments are tightly coupled with the metabolic functioning of plants and microbes. Signaling molecules like allelochemicals, plant growth-promoting hormones (PGPR), stress-related bioactive compounds, and chemical release in response to defense (against weeds / insects / pathogens) are the metabolic activities of the plants that occur in the rhizospheres (Berendsen et al. 2012; Gurusinghe et al. 2019). The microbial structural communities accordingly vary with the associated flora and in response, microbes also secrete chemicals (antimicrobials/toxins/enzymes/siderophores) as a result of secondary metabolism by multiple biosynthetic pathways to overcome the competition between them and favor plant growth by enriching the nutrients and other molecules (Raaijmakers and Mazzola 2012; Gurusinghe et al. 2019). However, research in such environments is limited due to less comprehensive high-throughput screening (HTS) methods than conventional approaches associated with the screening and characterization of biodiversity and chemo-diversity (Gurusinghe et al. 2019).

Recent advances in meta-omics technologies such as metagenomic sequencing, proteomics, metabolomics, and better microorganism co-culture techniques have provided more interesting information concerning the microbial diversity of rhizospheric regions, as compared to those by conventional methods. Emergence of high-throughput screening methods with automation concomitant with high resolution in identifying the naturally occurring complex bioactive compounds in a shorter duration, development of bioinformatics tools and software, database availability, low cost of Next-Generation Sequencing (NGS) services, and advances in Mass Spectrometry techniques have speeded up the research in even underexplored habitats. This chapter focuses on various meta-omics technologies that have played an important role in assessing structural and functional diversity of the rhizosphere in the detection of new bioactive compounds (Gurusinghe et al. 2019; Fierer et al. 2012).

10.2 Plant–Microbe Interactions

There are three kinds of microbial interactions—beneficial, pathogenic, and saprophytic—that affect plants. Based on the applications, beneficial microbes can be of varied applications such as biofertilizers, phytostimulators, rhizomediators, and biopesticides. Pathogenic microbes secrete antibiotics/toxins in response to defense or to cause virulence. Saprophytic microbes are well-known key players in balancing ecosystems by composting and elemental/nutrient cycling. The beneficial/pathogenic/saprophytic microbes choose almost similar strategies while interacting with plants. However, the mechanisms behind the interactions are still not well understood (Lugtenberg et al. 2002).

10.3 Key Barriers During Interactions

Innate immunity effectors and modulators from plants enable them to resist herbivory and entrance or invasion of the pathogenic microbial population. Under such conditions, plant immune response is activated, known as microbe-/pathogen-associated molecular pattern (MAMP)-triggered immunity (MTI). MAMP detection occurs via molecular signatures enunciated by extracellular transmembrane receptors or plant pattern recognition receptors (PRRs) (Gao et al. 2013). In response, in some pathogens, effector molecules are secreted, that down-regulate MTI and cause effector-triggered susceptibility (ETS); in turn, the effector molecules are recognized by the plant's resistance proteins, and quick response (hypersensitivity) is activated (effector-triggered immunity—ETI) by activation of the second line of defense (Jones and Dangl 2006). Similar defense mechanisms are employed in all types of microbial populations (pathogenic/beneficiary). Hence, the plant–microbe interaction is a complicated mechanism resulting either in elimination of pathogens or promoting colonization of the beneficial microbes (Hacquard et al. 2017).

10.4 Chemical Communication in Rhizospheres Via Secretion of Signaling Molecules

Rhizospheres are very complex niche habitats with abundance of less diverse microbes. Plants and microbes interact with each other apart from interacting between themselves in sophisticated ways, by secreting small signaling molecules. Various metabolomics (signalomics) studies have helped in detection and quantification of these signaling metabolites involved in chemical communication between various combinations of players—plant–plant, bacteria–bacteria, and plant–bacteria—all in the microbiome assemblage in the rhizosphere (Hacquard et al. 2017; Yang et al. 2017). Figure 10.1 depicts the various ways in which plant–microbe

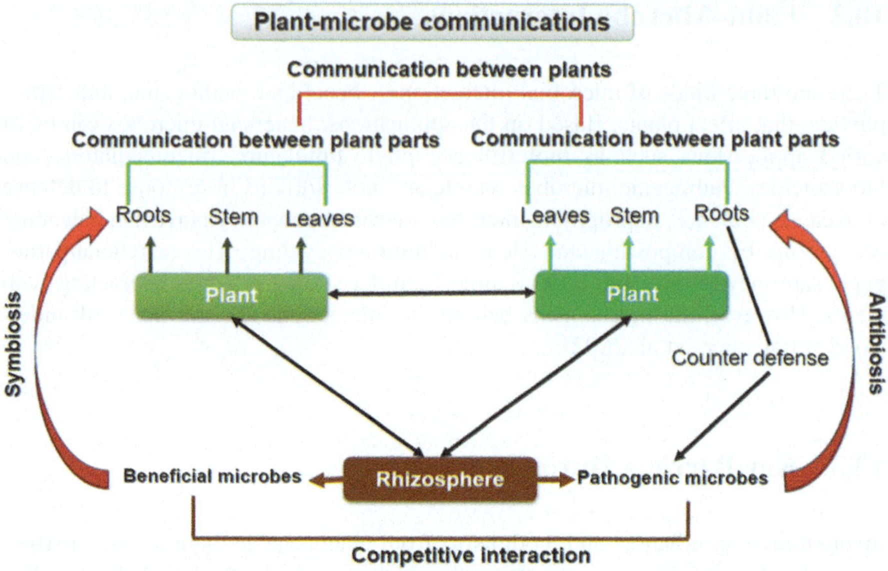

Fig. 10.1 Communications between plants, plants, and microbes, and between beneficial and pathogenic microbes

communications occur, and Fig. 10.2 shows the signaling molecules secreted from plants and microbes establishing their communication system leading to various effects. Some of these interactions and the role by the metabolites are discussed below:

10.4.1 Plant to Plant

Plant communicate with the other plants both in above and under-ground parts via signaling molecules such as phytotoxins and volatile organic carbons (VOCs) showing the effect of allelopathy. The secreted compounds responsible for inter- and intraspecific competition among plants either promote establishment or growth of same/other plant in the vicinity or inhibit it. Allelopathic molecules generally belong to phenolics/benzene-derived compounds/hydroxamic acids/terpenes chemical categories (Massalha et al. 2017).

10.4.2 Bacteria to Bacteria

Bacterial communities in the rhizospheric regions communicate with each other by signaling molecules such as lipophilic VOCs (low molecular weight) as a product

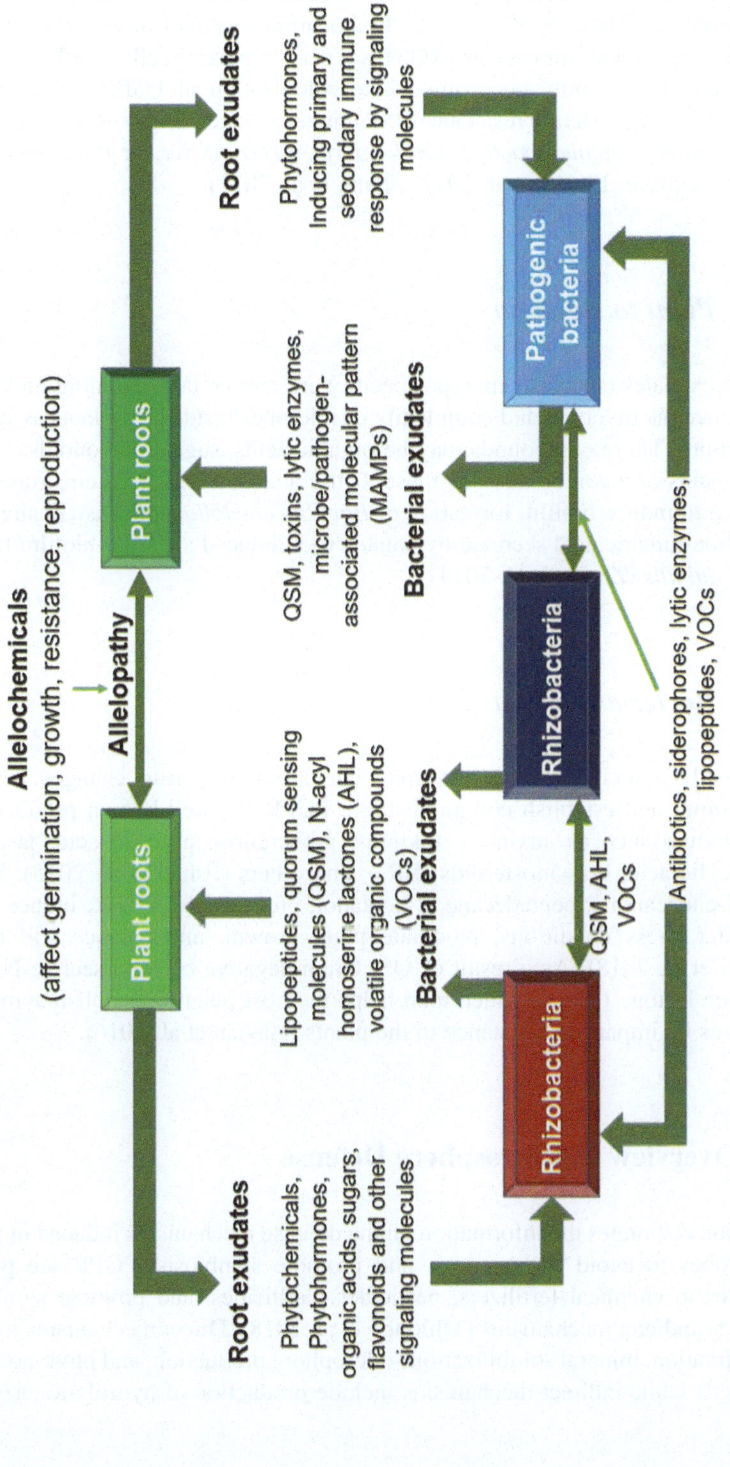

Fig. 10.2 Overview of signaling molecules among rhizospheric plants and associated microbes

synthesized in various metabolic pathways (Kanchiswamy et al. 2015) specific to species/genotype. These VOCs may be hydrocarbons/alcohols/ketones/terpenoids/sulfur compounds. Quorum sensing (QS) is a very important cell-to-cell communication mechanism, mainly occurring where colonization of PGPR [plant growth promoters/induced systemic resistance (ISR) inducers] such as *Bacillus, Pseudomonas, Enterobacter, Acinetobacter, Burkholderia, Arthrobacter,* or *Paenibacillus* is likely to take place (Finkel et al. 2017; Zhang et al. 2017).

10.4.3 Plant to Bacteria

Plant root exudates (species/genotype specific) are also certain signaling molecules that influence the diversity and complexity of microbes inside host plants as well as in the vicinity. These compounds may be organic acids, sugars, flavonoids, VOCs, or amino acids or a combination of these. Citric acid secreted from cucumber root was shown to induce biofilm formation of *Bacillus amyloliquefaciens* (Zhang et al. 2014), while fumaric acid secreted by banana root induced a similar biofilm formation of *B. subtilis* (Zhang et al. 2014).

10.4.4 Bacteria to Plant

Symbiotically associated bacteria secrete chemicals triggering changes in plant transcriptomes and establish communication. PGPR are well-known producers of phytohormones such as auxins, cytokinins, gibberellins, abscisic acid, jasmonic acid, salicylic acid, brassinosteroids (BRs), and others (Fahad et al. 2015). VOCs such as 2-endecanone, pentadecane, 2-heptanol, produced by PGPR, induce resistance under stress conditions, modulate plant growth, and balance soil health (Mhlongo et al. 2018). As a result of QS, Gram-negative bacteria secrete N-acyl-homoserine lactone (AHL), which then helps the host plant to establish symbiotic associations by imparting resistance to the plants (Hassan et al. 2016).

10.5 Overview of Rhizosphere Defense

This section elaborates the information on the defense mechanisms induced in plants and microbes to avoid pathogenesis and promote symbiosis. PGPR are proven alternatives to chemical fertilizers, pesticides, herbicides, and promote symbiosis by direct or indirect mechanisms (Mhlongo et al. 2018). Direct mechanisms include nitrogen fixation, mineral solubilization, siderophore production, and phytohormone biosynthesis while indirect mechanisms include production of hydrolytic enzymes, antibiotics, siderophores, lipopolysaccharides (LPs), and ISR (García-Fraile et al. 2015).

Some bacteria secrete antibiotics and protect the host plant from diseases. Asadhi et al. (2013) reported an antibiotic compound (2,4 diacetylphloroglucinol), produced from *Pseudomonas fluorescens* that exhibited 75% inhibition efficacy to *Sclerotium rolfsii* (Asadhi et al. 2013). Similarly, Lohitha et al. (2016) have reported that Phenazine-1-carboxylic acid antibiotic produced by *P. fluorescens* inhibited *S. rolfsii* and *Gaeumannomyces graminis* and protected the groundnut plant from stem rot. Hydrolytic enzymes (proteases/chitinases/lipases/glucanases/amylases) inhibit fungal growth by degrading many components of the fungal cell wall (Saraf et al. 2014). LPs also are a kind of antibiotics belonging to three different categories (fengycin, iritin, and surfactin) based on their fatty acid branching pattern and play an inhibitory effect on pathogens and activate the plant defense system (Saraf et al. 2014). Recently, a study by Chowdhury et al. (2015) demonstrated that LPs extracted from *B. amyloliquefaciens* protected rhizospheric plants from *Rhizoctonia solani*. Siderophores such as hydrocarboxylic acid, hydroxamic acid, and catechols are not only iron chelators (that provide iron to plants in iron-deprived soil) but also suppress soil-borne diseases as proved by the studies of Tank et al. (2012) and Patil et al. (2014). As reported in Sect. 10.4.4, AHLs are responsible for enhancing plant perception and resistance and induce the elongation of roots. AHLs produced by *Serratia liquefaciens* and *P. putida* have been reported to trigger the defense mechanisms against *Alternaria alternata* in tomato plants (Schuhegger et al. 2006). 3-hydroxydecanoyl-homoserine lactone secreted by *Acidovorax radices* (barley endophytic bacteria), induced resistance in the host by facilitating accumulation of flavonoids (saponarin/lutonarin) (Han et al. 2016). VOCs produced by rhizobacteria also promote plant growth and resistance and inhibit soil-borne pathogens (Song and Ryu 2013). Similarly, VOCs such as (E)-ocimene, 6-methyl-5-hepten-2-one and (E)-(1R, 9S) caryophyllene (Pickett et al., 2007), (Z)-3-hexen-1-ol (Wei and Kang 2011) produced by plants triggered defense response when exposed to insect or herbivory attack.

Several studies using PGPR have revealed ISR and RMPP-associated enhanced gene expressions in case of plant stress, or during pathogens/insects/pests/herbivore attack. An initial subjection to such a smaller scale trigger, known as priming, induces a quick (secondary) defense response by the host plant (Conrath et al. 2009). PGPR are priming agents and due to this ability they can modify cell wall and primary metabolites and induce effector response (defense mechanisms) and biosynthetic pathways associated with the production of secondary metabolites (Conrath et al. 2009). Initial stage priming includes signal transduction through phytohormones acting either synergistically or antagonistically. Primary metabolism during priming helps in biosynthesis or activation of phytohormones/phytoalexins/phytoanticipins, and enhanced gene expression related to defense by providing energy. Secondary metabolites accumulated during initial priming phase become activated when plant is exposed to secondary stress (Mhlongo et al. 2016a, b, 2017).

10.6 Meta-Omics Technology

As we can see from information provided in the previous sections, the rhizosphere region is very complex due to a diverse microbial population and interactions with the various signaling molecules among and within the plant and microbial communities. However, most of the molecules and the mechanisms are still not well understood. Meta-omics technologies incorporating advanced high-throughput screening technologies offer greater resolution, higher sensitivity, robustness, and in combination with better bioinformatics tools and software, database availability, and powerful computers new avenues to explore such understudied niches for the novel chemo-diversity and the mechanisms behind them have opened up (Mhlongo et al. 2018). Details of various meta-omics technologies can be accessed from the comprehensive review by Srivastava et al. (2019). Figure 10.3 presents an overview of meta-omics approaches/technology to assess bio- and chemo-diversity and interaction mechanisms in rhizospheres.

Metabolomics studies facilitate the complete metabolite fingerprinting analysis of the signature molecules and demonstrate the chemistry established or the mechanisms of the reciprocal interactions of these complex niche habitats (Lloyd et al. 2015) and hence provide better understanding of the metabolism in plants and associated microorganisms. These studies are associated with various bio-analytical techniques for metabolite fingerprinting and studying the respective interactions (Naz 2014).

Gas chromatography–mass spectroscopy is one of the most used technologies in metabolomics. A recent study by Miller et al. (2019) has used this technology to detect the metabolites secreted by two genotypically different *Sorghum* roots in sand, clay, and soil and has reported various metabolites belonging majorly to amino acids, carbohydrates, and organic acids. Sixty-eight metabolites have been identified by GC-MS from *Bacillus* sp., *B. amyloliquefaciens*, and *B. thuringiensis* associated with the rhizosphere of Barabara groundnut from an agricultural land of South Africa (Ajilogba and Babalola 2019). Combined metabolomics by GC-MS and transcriptomics studies were conducted on soybean to show the resistance offered to this plant by rhizobacterial *Bacillus simplex* Sneb545 against soybean cyst nematode by over production of palmitic acid, methionine, 4-vinylphenol, and piperine (Kang et al. 2018). Stable isotope probing (SIP) coupled with GC-MS has proven useful in detection of rate determining steps in various metabolic pathways (You et al. 2014). High resolution and efficient separation have been provided using two different stationary phases in GCxGC with advanced mass analyzer [Time-of-flight (TOF)/semi-fasts scan Q] (Jin et al. 2015).

Liquid chromatography–mass spectrometry is another technique to detect bioactive metabolites including hydrophilic components. For better resolution and sensitivity, use of advanced column dimensions, porosity (column particle size) combination of reverse and normal stationary phases has helped better. Ultra-high performance liquid chromatography (UHPLC) is better than high performance liquid chromatography (HPLC) in terms of diameter of the column (1.0–1.2 mm) with

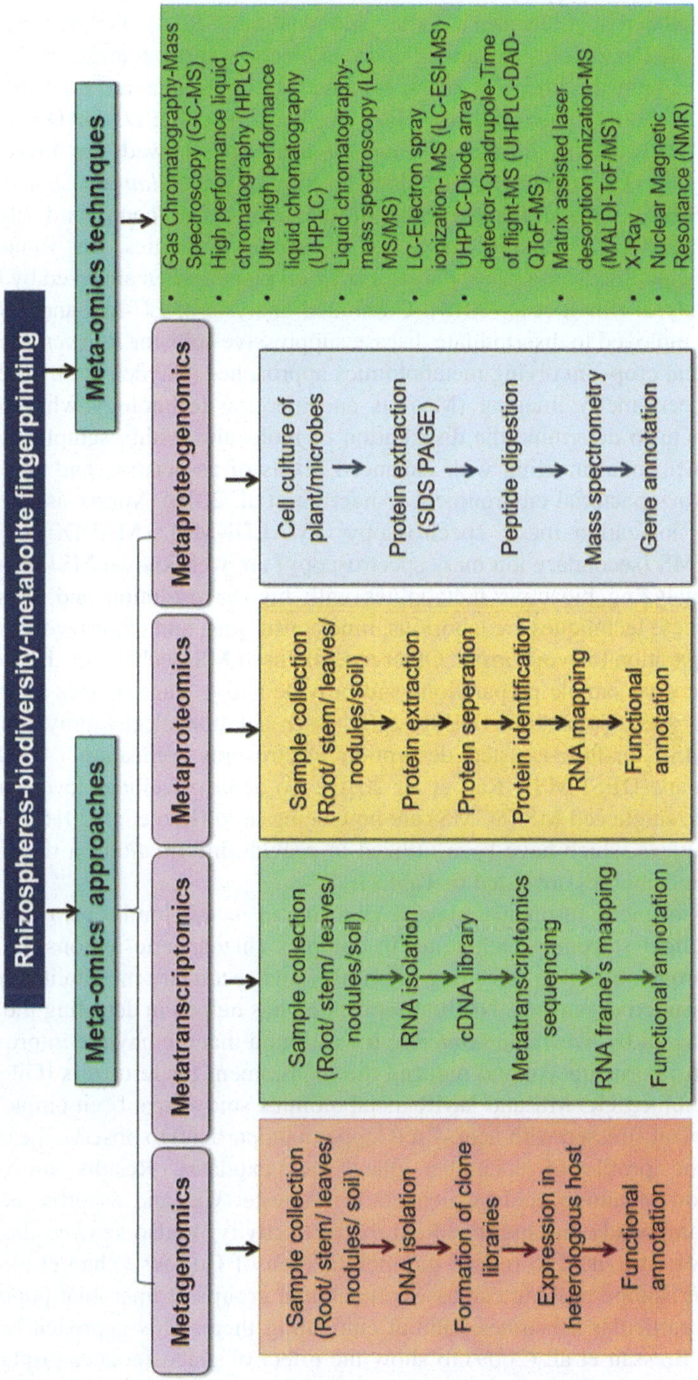

Fig. 10.3 Meta-omics approaches and technologies to assess rhizospheric biodiversity and metabolites

smaller particle size (≤2 μm) and under very high pressure that allows the separation of smaller bioactive molecules (Walter and Andrews 2014). For accurate mass detection, soft ionization mechanism such as electron spray ionization (ESI) that hold and maintain the integrity of the separated biomolecules are now frequently used for metabolite fingerprinting (Hird et al. 2014). UHPLC-DAD-QToF-MS analysis of the culture filtrate of *Acinetobacter* sp. showed the presence of tetracenomycin D1 (Rohr et al. 1988). Similarly, *Williamsia muralis*, an actinobacterium, showed the presence of the antioxidant compound ribesin B (Sasaki et al. 2013). Vicenin-2, with UV absorption properties, was found to be the major metabolite from *Lychnophora salicifolia* plants, when analyzed by UPLC-MS/MS analysis (Silva et al. 2014). Combined analysis of LC-MS and 1H-NMR have been employed to discriminate disease suppressive soils for *Rhizoctonia solani* AG8 in cereal crops involving metabolomics approaches (Hayden et al. 2019).

Mass spectrometry imaging (MSI) is an emerging technology which can be applied in situ to determine the distribution of molecules within samples. It helps in multidimensional imaging with enhanced pixels of associating and co-existing species in rhizobacterial environment (Anderton et al. 2016). Matrix-assisted laser desorption ionization–mass spectroscopy (MALDI-MS), MALDI-TOF, and MALDI-SIMS (secondary ion mass spectroscopy) are well-known MSI techniques used for analyzing bioactive metabolites with higher resolution and sensitivity. However, these techniques are laborious, time consuming, and require very accurate sample preparation to avoid errors. Other recently used MSI techniques that does not require extensive sample preparations and provide native stage images are desorption electro spray ionization (DESI) MSI, laser ablation electrospray ionization (LAESI) MSI, air-flow-assisted desorption electrospray ionization (AFADESI)-MSI, and nano-DESI MSI (Rao et al. 2016). To achieve cellular level chemical information, single cell MS (SCMS) are now being used (Rao et al. 2016). Some of these techniques which have been utilized to provide information on rhizosphere-associated metabolites are listed in Table 10.1.

Nuclear magnetic resonance plays a vital role in metabolomics as a robust and high-throughput screening technique. In complex rhizospheric regions, this technique has proven useful in screening of primary and secondary metabolites secreted by plants or microorganisms. NMR spectroscopy has helped in detecting the role of phytase secreted by *Nicotiana tabacum*. It was found that the phytase improved the organic phosphorus uptake and reduced the requirement for fertilizers (Giles et al. 2017). Combined GC-MS and NMR metabolomics study have been employed on cucumber plant stressed with nano-Cu (Copper nanoparticle) to observe the changes in metabolic profile of cucumber plant root exudates. Results showed the up-regulation of amino acids which could sequester Cu, and ascorbic acid and phenolic compounds that improved antioxidant activity. It also showed the down-regulation of citric acid to reduce the mobilization of Cu ions (Zhao et al. 2016). DNA-stable isotope probing can assess functional groups of microbial populations specific to particular substrates without cultivating them. This approach has been applied by Bressan et al. (2009) to show the effect of glucosinolates produced by *Arabidopsis thaliana* on rhizobacterial and fungal diversity.

Table 10.1 MSI technology for plant-associated rhizobacterial activity analysis

MSI Technology	Rhizospheric activity analysis	Summary of results	References
MALDI-FTICR-MSI	Distribution of metabolites in soybean root nodules involved in regulating symbiosis and nitrogen fixation	Asymmetric distribution of S-adenosylmethionine in nodule affecting plant–microbe interaction in terms of nitrogen fixation	Veličković et al. (2018)
MALDI-FTMS	Bacterial and plant physiology in leaves of *Arabidopsis thaliana*	To assess the shifts in carbohydrate sugars profile on leaf surface after colonization of *Sphingomonas melonis, Pseudomonas syringae* and *Methylobacterium extorquens* in turn altering arginine metabolism and phytoalexin biosynthesis	Ryffel et al. (2016)
MALDI-LTQ-Orbitrap	Study of pesticide (metalaxyl) kinetics under given O_2 and pH conditions in lupine roots	Results indicated positive and negative correlation of pH and O_2 on enantioselective dissipation of the metalaxyl	Rudolph-Mohr et al. (2015)
MALDI-MSI	Interactions between rice–bacterium and soybean–aphid	In soybean–aphid system, isoflavone, and salicylic acid resistance mechanisms and in rice–bacterium interactions, role of antibiotic diterpenoids was visualized	Klein et al. (2015)
MALDI-TOF/TOF	Resistance of banana plant to burrowing nematode	Development of resistance in banana in response to infection by *Radopholus similis by* production of phenylphenalenone anigorufone phytoalexins	Hölscher et al. (2014)
MALDI-TOF/TOF	Antimicrobiome secretion by root bacteria to show plant–microbe interactions	Elicitation of production of antibiotics (macrolactin A, macrolactin D, bacillaene A, difficidin, and oxydifficidin) by *Bacillus amyloliquefaciens* associated with tomato root	Debois et al. (2014)
MALDI-MSI	Detection of surface heterogeneity of plant lipid polymers	Effect of in situ release of the lipid monomers by alkaline hydrolysis on cutin polymers of nectarine, tomato, and apple fruits	Veličković et al. (2014a)
MALDI-MSI	Physiological roles of polysaccharides [Arabinoxylans (AX) and (1→3), (1→4)-β-glucans (BG)] in cell walls of wheat grains and the specificity of the hydrolytic enzymes involved	Assessment of quality and use of wheat grains by observing chemical and distributional heterogeneity of Arabinoxylans and β-glucans after reaction to hydrolytic enzymes	Veličković et al. (2014b)
MALDI-MSI	Metabolite distribution during nitrogen fixation in *Medicago*	Exhibiting distribution of organic acids, amino acids,	Ye et al. (2013)

(continued)

Table 10.1 (continued)

MSI Technology	Rhizospheric activity analysis	Summary of results	References
	truncatula-Sinorhizobium meliloti symbiosis	sugars, lipids, flavonoids between roots and nodules during *Medicago-Sinorhizobium* association	
NanoSIMS	Nutrient uptake (nitrogen assimilation)	Assessment of nitrogen assimilation between plant roots (wheat) of freely draining soil of Australia and associated soil microbial communities	Clode et al. (2009)

Metagenomics studies in rhizospheric regions have provided considerable information about the structural and functional diversity (Wu et al. 2018). It is estimated that only 1% cultivable bacteria have been documented by conventional methods, while metagenome analyses have revealed the presence of microorganisms of a much larger magnitude. Rhizospheres are rich in nutrient contents. By optimizing conditions like temperature, growth media, increased incubation period, and reducing nutrients, the recovery rate may be increased up to 10% (Ling et al. 2015). Metagenomics study has been conducted to observe the change in fungal community (abundance/diversity) in the rhizosphere of sugarcane soil with respect to nitrogen fertilizer dosages (Paungfoo-Lonhienne et al. 2015). Growth-promoting ability of the novel *Burkholderia australis* in nitrogen-deficient soil environment of sugarcane has been investigated by culture-independent methods and bacterial identification (Paungfoo-Lonhienne et al. 2014). Microbial diversity profiling showing their role in carbon and nitrogen cycling has been done in Amazon forest by metagenomics (Fonseca et al. 2018). Recently, the innovative "i chip" technology (in situ culture apparatus) has facilitated the culture recovery up to 50% of cultivable bacteria. This technique can be applied to understand novel bio- and chemo-diversity and associated genes (Ling et al. 2015). This technique has enabled the identification of the novel antibiotic teixobactin from *Eleftheria terrae* (Nichols et al. 2010). Recently, Wu et al. 2018 have used comparative metagenomic analysis to assess the impact of monoculture of *Rehmannia glutinosa* consecutively for 2 years at Jiaozuo City, Henan Province, China on microbial community shifts (increased abundances of *Sphingomonadaceae* and *Streptomycetaceae* and decreased abundances of *Pseudomonadaceae* and *Burkholderiaceae*) with associated functional aspects (production of bioactive molecules and their responses).

Metatranscriptomics is another technology which is helpful in metabolic profiling of complex niche habitats by elucidating the gene expression of microbial communities associated with niche habitats (Chialva et al. 2020). The technique allows the identification of microbes, active genes, their expression levels, and products. It also facilitates the observation of changes in metabolic profiles in different environmental conditions such as under stress (Bashiardes et al. 2016). Metatranscriptomic profiling of 12 rhizospheric areas of Australia to assess microbial and functional diversity in wheat microbiome, which were suppressive and non-suppressive to *Rhizoctonia solani* was carried out. The mRNA data annotation suggested that *Buttiauxella* sp. and *Stenotrophomonas* sp. were the dominant microbes of suppressive samples, while *Pseudomonas* sp. and *Arthrobacter* sp. were dominant in non-suppressive samples. Polyketide cyclase, encoded by *dxs* gene and cold shock proteins (*csp*), were majorly expressed by suppressive samples. *R. solani* infection in suppressive samples of wheat led to the expression of detoxifying genes and antibiotics genes chloroperoxidase (*cpo*), involved in pyrrolnitrin and phenazine biosynthesis family protein F (phzF) and its transcriptional activator protein (phzR) (Hayden et al. 2018). Recently, metatranscriptomics study has been conducted by Chialva et al. (2020) to reveal whether host-targeted transcriptomics (RNA-seq dataset of *Solanum lycopersicum* grown in the native soil of Italy) can detect the taxonomic and functional diversity (plant nutrition/disease resistance/plant development mechanisms) of root microbiota.

Metaproteomics is another high-throughput screening technique that provides information on the protein recovered from the environments and is useful in investigating microbial and functional diversity. This technique has been applied by Gupta et al. (2019) to explore microbial and functional diversity of maize. A total of 696 proteins with different functions were recovered and were shown to belong to 244 genus and 393 species. This technique has also been found useful to detect the microbial flora associated with *Vitis vinifera* rhizospheres by Bona et al. (2019). They found that *Streptomyces, Bacillus, Bradyrhizobium, Burkholderia,* and *Pseudomonas* genera were active in expression of proteins, phosphorus metabolism, and regulation of primary metabolic processes.

Metaproteogenomics approach has also been reported to assess the biodiversity in rhizospheres and phyllosopheres. The technique comprises of combining 16S rDNA and metagenome sequencing with protein identification to provide insight into the associated metabolic pathways and physiological phenomena. This method was employed by Knief et al. (2012) to demonstrate the microbial diversity and the physiological activities of the microbial communities associated with the rhizosphere and phyllosphere of rice fields of Philippines. *Rhizobium*, *Methylobacterium,* and *Agrobacterium* were the dominant genera in the rhizosphere, while *Bradyrhizobium, Rhodopseudomonas, Azospirillum, Methylobacterium, Magnetospirillum,* and *Methylosinus* were dominant in the phyllosphere region. The authors identified 4600 proteins by metaproteomic datasets and concluded that one carbon conversion occurs in both regions. They also found that the identified proteins were involved in methanotrophy and methanogenesis. Response to stress and transport processes and the presence of nifH genes were more noticeable in the phyllosphere than in the rhizosphere.

10.7 Conclusion and Future Aspect

It is evident from the literature and from the quoted examples in this chapter that the meta-omics technologies are not only high-throughput but also provide a huge dataset of information on the activities of microorganisms in the rhizosphere. Each technology is complementary to the other but cannot be supplanted. They provide a piece of information of a large puzzle, which then needs orchestrated assembling to provide a comprehensive picture. The availability of better software and processors would facilitate in easier processing of the information; however, the challenge lies in accurately assembling the information in the correct sequence.

To unravel the complexity and heterogeneity of rhizospheric regions and to facilitate the understanding of plant–microbes interactions by assessing the production of novel natural products and their monitoring to specify their targeted roles and dynamics, the high-throughput meta-omics technologies will play an important role. Accordingly, with increased usage, costs are expected to reduce considerably leading to even more high-throughput data availability.

References

Ajilogba CF, Babalola OO (2019) GC–MS analysis of volatile organic compounds from Bambara groundnut rhizobacteria and their antibacterial properties. World J Microbiol Biotechnol 35 (6):83. https://doi.org/10.1007/s11274-019-2660-7

Anderton CR, Chu RK, Tolić N, Creissen A, Paša-Tolić L (2016) Utilizing a robotic sprayer for high lateral and mass resolution MALDI FT-ICR MSI of microbial cultures. J Am Soc Mass Spectrom 27:556–559

Asadhi S, Bhaskara Reddy BV, Sivaprasad Y, Prathyusha M, Murali Krishna T, Vijay Krishna Kumar K, Raja Reddy K (2013) Characterization, genetic diversity and antagonistic potential of 2,4-diacetylphloroglucinol producing *Pseudomonas fluorescens* isolates in groundnut-based cropping systems of Andhra Pradesh, India. Arch Phytopathol Plant Protect 45:1966–1977. https://doi.org/10.1080/03235408.2013.782223

Bashiardes S, Zilberman-Schapira G, Elinav E (2016) Use of metatranscriptomics in microbiome research. Bioinform Biol Insights 10:19–25. https://doi.org/10.4137/BBI.S34610

Berendsen RL, Pieterse CM, Bakker PA (2012) The rhizospheres microbiome and plant health. Trends Plant Sci 17:478–486

Bona E, Massa N, Novello G, Boatti L, Cesaro P, Todeschini V, Magnelli V, Manfredi M, Marengo E, Mignone F, Berta G (2019) Metaproteomic characterization of the *Vitisvinifera* rhizospheres. FEMS Microbiol Ecol 95(1):fiy204. https://doi.org/10.1093/femsec/fiy204

Bressan M, Roncato MA, Bellvert F, Comte G, el Zahar HF, Achouak W, Berge O (2009) Exogenous glucosinolate produced by *Arabidopsis thaliana* has animpacton microbes in the rhizospheres and plant roots. ISME J 3:1243–1257

Chialva M, Ghignone S, Novero M, Hozzein WN, Lanfranco L, Bonfante P (2020) Tomato RNA-seq data mining reveals the taxonomic and functional diversity of root-associated microbiota. Microorganisms 2020(8):38. https://doi.org/10.3390/microorganisms8010038

Chowdhury SP, Uhl J, Grosch R, Alquéres S, Pittroff S, Dietel K, Schmitt-Kopplin P, Borriss R, Hartmann A (2015) Cyclic lipopeptides of *Bacillus amyloliquefaciens* subsp. plantarum colonizing the lettuce rhizosphere enhance plant defense responses toward the bottom rot pathogen

Rhizoctonia solani. Mol Plant Microbe Interact 28:984–995. https://doi.org/10.1094/MPMI-03-15-0066-R

Clode PL, Kilburn MR, Jones DL, Stockdale EA, Cliff JB, Herrmann AM, Murphy DV (2009) *In Situ* mapping of nutrient uptake in the rhizosphere using nanoscale secondary ion mass spectrometry. Plant Physiol 151:1751–1757

Conrath U, Beckers GJ, Flors V, García-Agustín P, Jakab G, Mauch F, Newman MA, Pieterse CM, Poinssot B, Pozo MJ, Pugin A (2009) Priming: getting ready for battle. Mol Plant-Microbe Interact 19:1062–1071. https://doi.org/10.1094/MPMI-19-1062

Debois D, Jourdan E, Smargiasso N, Thonart P, De Pauw E, Ongena M (2014) Spatiotemporal monitoring of the antibiome secreted by bacillus biofilms on plant roots using MALDI mass spectrometry imaging. Anal Chem 86:4431–4438

Dennis PG, Miller AJ, Hirsch PR (2010) Are root exudates more important than other sources of rhizodeposits in structuring rhizosphere bacterial communities? FEMS Microbiol Ecol 72:313–327

Dubois O (2011) The state of the world's land and water resources for food and agriculture: managing systems at risk. Earthscan, London

East R (2013) Microbiome: soil science comes to life. Nature 501(7468):S18–S19

Fahad S, Hussain S, Bano A, Saud S, Hassan S, Shan D (2015) Potential role of phytohormones and plant growth-promoting rhizobacteria in abiotic stresses: consequences for changing environment. Environ Sci Pollut Res 22:4907–4921. https://doi.org/10.1007/s11356-014-3754-2

Fierer N, Leff JW, Adams BJ, Nielsen UN, Bates ST, Lauber CL, Owens S, Gilbert JA, Wall DH, Caporaso JG (2012) Cross-biome metagenomic analyses of soil microbial communities and their functional attributes. Proc Natl Acad Sci U S A 109:21390–21395

Finkel OM, Castrillo G, Paredes SH, González IS, Dangl JL (2017) Understanding and exploiting plant beneficial microbes. Curr Opin Plant Biol 38:155–163. https://doi.org/10.1016/j.pbi.2017.04.018

Fonseca JP, Hoffmann L, Cabral BC, Dias VH, Miranda MR, de Azevedo Martins AC, Boschiero C, Bastos WR, Silva R (2018) Contrasting the microbiomes from forest rhizosphere and deeper bulk soil from an Amazon rainforest reserve. Gene 642:389–397

Gao X, Chen X, Lin W, Chen S, Lu D, Niu Y, Li L, Cheng C, McCormack M, Sheen J, Shan L (2013) Bifurcation of *Arabidopsis* NLR immune signaling via Ca2+- dependent protein kinases. PLoS Pathog 9:e1003127. https://doi.org/10.1371/journal.ppat.1003127

García-Fraile P, Menéndez E, Rivas R (2015) Role of bacterial biofertilizers in agriculture and forestry. AIMS Bioeng 2:183–205. https://doi.org/10.3934/bioeng.2015.3.183

Giles CD, George TS, Brown LK, Mezeli M, Shand CA, Richardson AE, Mackay R, Wendler R, Darch T, Menezes-Blackburn D, Cooper P (2017) Linking the depletion of rhizosphere phosphorus to the heterologous expression of a fungal phytase in *Nicotiana tabacum* as revealed by enzyme-labile P and solution 31P NMR spectroscopy. Rhizosphere 3(1):82–91

Gupta SK, Rai AK, Sarim KM, Sharma A, Budhlakoti N, Arora D, Verma DK, Singh DP (2019) Metaproteomic data of maize rhizosphere for deciphering functional diversity. Data Brief 27:104574

Gurusinghe S, Brooks TL, Barrow RA, Zhu X, Thotagamuwa A, Dennis PG, Gupta VV, Vanniasinkam T, Weston LA (2019) Technologies for the selection, culture and metabolic profiling of unique rhizosphere microorganisms for natural product discovery. Molecules 24 (10):1955

Hacquard S, Spaepen S, Garrido-Oter R, Schulze-Lefert P (2017) Interplay between innate immunity and the plant microbiota. Annu Rev Phytopathol 55:565–589. https://doi.org/10.1146/annurev-phyto-080516-035623

Han S, Li D, Trost E, Mayer KF, Vlot AC, Heller W, Schmid M, Hartmann A, Rothballer M (2016) Systemic responses of barley to the lactone producing plant beneficial endophyte *Acidovorax radicis* N35. Front Plant Sci 7:1868. https://doi.org/10.3389/fpls.2016.01868

Hassan R, Shaaban MI, Abdel Bar FM, El-Mahdy AM, Shokralla S (2016) Quorum sensing inhibiting activity of *Streptomyces coelicoflavus* isolated from soil. Front Microbiol 7:659. https://doi.org/10.3389/fmicb.2016.00659

Hayden HL, Savin KW, Wadeson J, Gupta VV, Mele PM (2018) Comparative metatranscriptomics of wheat rhizosphere microbiomes in disease suppressive and non-suppressive soils for *Rhizoctonia solani* AG8. Front Microbiol 9:859. https://doi.org/10.3389/fmicb.2018.00859. eCollection 2018.

Hayden HL, Rochfort SJ, Ezernieks V, Savin KW, Mele PM (2019) Metabolomics approaches for the discrimination of disease suppressive soils for Rhizoctonia solani AG8 in cereal crops using 1H NMR and LC-MS. Sci Total Environ 651:1627–1638

Hird SJ, Lau BP, Schuhmacher R, Krska R (2014) Liquid chromatography-mass spectrometry for the determination of chemical contaminants in food. TrAC Trends Anal Chem 59:59–72. https://doi.org/10.1016/j.trac.2014.04.005

Hölscher D, Dhakshinamoorthy S, Alexandrov T, Becker M, Bretschneider T, Buerkert A, Crecelius AC, De Waele D, Elsen A, Heckel DG, Heklau H (2014) Phenalenone-type phytoalexins mediate resistance of banana plants (*Musa* spp.) to the burrowing nematode *Radopholus similis*. Proc Natl Acad Sci U S A 111:105–110

Jin Y, Deng Y, Yue J, Zhao Y, Yu W, Liu Z, Huang H (2015) Significant improvements in the characterization of volatile compound profiles in squid using simultaneous distillation-extraction and GC×GC-TOFMS. CyTA-J Food 13:434–444. https://doi.org/10.1080/19476337.2014.997798

Jones JDG, Dangl JL (2006) The plant immune system. Nature 444:323–329. https://doi.org/10.1038/nature05286

Kanchiswamy CN, Malnoy M, Maffei ME (2015) Chemical diversity of microbial volatiles and their potential for plant growth and productivity. Front Plant Sci 6:151. https://doi.org/10.3389/fpls.2015.00151

Kang W, Zhu X, Wang Y, Chen L, Duan Y (2018) Transcriptomic and metabolomic analyses reveal that bacteria promote plant defense during infection of soybean cyst nematode in soybean. BMC Plant Biol 18(1):86

Klein AT, Yagnik GB, Hohenstein JD, Ji Z, Zi J, Reichert MD, MacIntosh GC, Yang B, Peters RJ, Vela J, Lee YJ (2015) Investigation of the chemical interface in the soybean-aphid and rice-bacteria interactions using MALDI-mass spectrometry imaging. Anal Chem 87:5294–5301

Knief C, Delmotte N, Chaffron S, Stark M, Innerebner G, Wassmann R, von Mering C, Vorholt JA (2012) Metaproteogenomic analysis of microbial communities in the phyllosphere and rhizosphere of rice. ISME J 6(7):1378–1390. https://doi.org/10.1038/ismej.2011.192

Lakshmanan V, Selvaraj G, Bais HP (2014) Functional soil microbiome: belowground solutions to an aboveground problem. Plant Physiol 166(2):689–700

Ling LL, Schneider T, Peoples AJ, Spoering AL, Engels I, Conlon BP, Mueller A, Schäberle TF, Hughes DE, Epstein S, Jones M (2015) A new antibiotic kills pathogens without detectable resistance. Nature 517:455–459

Lloyd N, Johnson DL, Herderich MJ (2015) Metabolomics approaches for resolving and harnessing chemical diversity in grapes, yeast and wine. Aust J Grape Wine Res 21:723–740. https://doi.org/10.1111/ajgw.12202

Lohitha SR, Bhaskara RB, Sivaprasad Y, Prathyusha M, Sujitha A, Krishna T, Reddy KR (2016) Molecular characterization and antagonistic potential of phenazine-1-carboxylic acid producing *Pseudomonas fluorescens* isolates from economically important crops in South India. Int J Clin Biol Sci 1(2):30–40

Lugtenberg B, Chin-A-Woeng T, Bloemberg G (2002) Microbe-plant interactions: principles and mechanisms. Antonie Van Leeuwenhoek 81:373–383

Massalha H, Korenblum E, Tholl D, Aharoni A (2017) Small molecules below-ground: the role of specialized metabolites in the rhizosphere. Plant J 90:788–807. https://doi.org/10.1111/tpj.13543

Mhlongo MI, Piater LA, Madala NE, Steenkamp PA, Dubery IA (2016a) Phenylpropanoid defences in *Nicotiana tabacum* cells: overlapping metabolomes indicate common aspects to priming responses induced by lipopolysaccharides. PLoS One 11:e0151350. https://doi.org/10.1371/journal.pone.0151350

Mhlongo MI, Steenkamp PA, Piater LA, Madala NE, Dubery IA (2016b) Profiling of altered metabolomic states in *Nicotiana tabacum* cells induced by priming agents. Front Plant Sci 7:1527. https://doi.org/10.3389/fpls.2016.01527

Mhlongo MI, Tugizimana F, Piater LA, Steenkamp PA, Madala NE, Dubery IA (2017) Untargeted metabolomics analysis reveals dynamic changes in azelaic acid- and salicylic acid derivatives in LPS-treated *Nicotiana tabacum* cells. Biochem Biophys Res Commun 482:1498–1503. https://doi.org/10.1016/j.bbrc.2016.12.063

Mhlongo MI, Piater LA, Madala NE, Labuschagne N, Dubery IA (2018) The chemistry of plant–microbe interactions in the rhizosphere and the potential for metabolomics to reveal signaling related to defense priming and induced systemic resistance. Front Plant Sci 9:112. https://doi.org/10.3389/fpls.2018.00112

Miller SB, Heuberger AL, Broeckling CD, Jahn CE (2019) Non-targeted metabolomics reveals sorghum rhizosphere-associated exudates are influenced by the belowground interaction of substrate and sorghum genotype. Int J Mol Sci 20:431. https://doi.org/10.3390/ijms20020431

Naz S (2014) Analytical protocols based on LC–MS, GC–MS and CE–MS for non-targeted metabolomics of biological tissues. Bioanalysis 6:1657–1677. https://doi.org/10.4155/BIO.14.119

Nichols D, Cahoon N, Trakhtenberg EM, Pham L, Mehta A, Belanger A, Kanigan T, Lewis K, Epstein SS (2010) Use of ichip for high-throughput in situ cultivation of "uncultivable" microbial species. Appl Environ Microbiol 76:2445–2450

Patil S, Bheemaraddi MC, Shivannavar CT (2014) Biocontrol activity of siderophore producing Bacillus subtilis CTS-G24 against wilt and dry root rot causing fungi in Chickpea. J Agric Vet Sci 7:63–68

Paungfoo-Lonhienne C, Lonhienne TGA, Yeoh YK, Webb RI, Lakshmanan P, Chan CX, Lim P-E, Ragan MA, Schmidt S, Hugenholtz P (2014) A new species of burkholderia isolated from sugarcane roots promotes plant growth. Microb Biotechnol 7:142–154. https://doi.org/10.1111/1751-7915.12105

Paungfoo-Lonhienne C, Yeoh YK, Kasinadhuni NR, Lonhienne TG, Robinson N, Hugenholtz P, Ragan MA, Schmidt S (2015) Nitrogen fertilizer dose alters fungal communities in sugarcane soil and rhizosphere. Sci Rep 5:8678. https://doi.org/10.1038/srep08678

Raaijmakers JM, Mazzola M (2012) Diversity and natural functions of antibiotics produced by beneficial and plant pathogenic bacteria. Annu Rev Phytopathol 50:403–424

Rao W, Pan N, Yang Z (2016) Applications of the single-probe: mass spectrometry imaging and single cell analysis under ambient conditions. J Vis Exp 112:e53911. https://doi.org/10.3791/53911

Rohr J, Eick S, Zeeck A, Reuschenbach P, Zähner H, Fiedler HP (1988) Metabolic products of microorganisms- 249. J Antibiot 8:1066–1073

Rudolph-Mohr N, Gottfried S, Lamshöft M, Zühlke S, Oswald SE, Spiteller M (2015) Non-invasive imaging techniques to study O-2 micro-patterns around pesticide treated lupine roots. Geoderma 239:257–264

Ryffel F, Helfrich EJ, Kiefer P, Peyriga L, Portais JC, Piel J, Vorholt JA (2016) Metabolic footprint of epiphytic bacteria on *Arabidopsis thaliana* leaves. ISME J 10:632–643

Saraf M, Pandya U, Thakkar A (2014) Role of allelochemicals in plant growth promoting rhizobacteria for biocontrol of phytopathogens. Microbiol Res 169:18–29. https://doi.org/10.1016/j.micres.2013.08.009

Sasaki T, Li W, Zaike S, Asada Y, Li Q, Ma F, Zhang Q, Koike K (2013) Antioxidant lignoids from leaves of *Ribes nigrum*. Phytochemistry 95:333–340

Schuhegger R, Ihring A, Gantner S, Bahnweg G, Knappe C, Vogg G, Hutzler P, Schmid M, Van Breusegem F, Eberl LE, Hartmann A (2006) Induction of systemic resistance in tomato by

N-acyl-L-homoserine lactone-producing rhizosphere bacteria. Plant Cell Environ 29:909–918. https://doi.org/10.1111/j.1365-3040.2005.01471.x

Silva DB, Turatti IC, Gouveia DR, Ernst M, Teixeira SP, Lopes NP (2014) Mass spectrometry of flavonoid vicenin-2, based sunlight barriers in *Lychnophora* species. Sci Rep 4:4309. https://doi.org/10.1038/srep04309

Song GC, Ryu CM (2013) Two volatile organic compounds trigger plant self-defense against a bacterial pathogen and a sucking insect in cucumber under open field conditions. Int J Mol Sci 14:9803–9819

Srivastava N, Gupta B, Gupta S, Danquah MK, Sarethy IP (2019) Analyzing functional microbial diversity: an overview of techniques. In: Microbial diversity in the genomic era. Academic, Cambridge, pp 79–102

Tank N, Rajendran N, Patel B, Saraf M (2012) Evaluation and biochemical characterization of a distinctive pyoverdin from a Pseudomonas isolated from chickpea rhizosphere. Braz J Microbiol 43(2):639–648

Turner TR, James EK, Poole PS (2013) The plant microbiome. Genome Biol 14(6):209

Veličković D, Herdier H, Philippe G, Marion D, Rogniaux H, Bakan B (2014a) Matrix-assisted laser desorption/ionization mass spectrometry imaging: a powerful tool for probing the molecular topology of plant cutin polymer. Plant J 80:926–935

Veličković D, Ropartz D, Guillon F, Saulnier L, Rogniaux H (2014b) New insights into the structural and spatial variability of cell-wall polysaccharides during wheat grain development, as revealed through MALDI mass spectrometry imaging. J Exp Bot 65:2079–2091

Veličković D, Agtuca BJ, Stopka SA, Vertes A, Koppenaal DW, Paša-Tolić L, Stacey G, Anderton CR (2018) Observed metabolic asymmetry within soybean root nodules reflects unexpected complexity in rhizobacteria-legume metabolite exchange. ISME J 12:2335–2338. https://doi.org/10.1038/s41396-018-0188-8

Walter TH, Andrews RW (2014) Recent innovations in UHPLC columns and instrumentation. Trends Anal Chem 63:14–20

Wei J, Kang L (2011) Roles of (Z) -3-hexenol in plant-insect interactions. Plant Signal Behav 6:369–371. https://doi.org/10.4161/psb.6.3.14452

Wu L, Wang J, Wu H, Chen J, Xiao Z, Qin X, Zhang Z, Lin W (2018) Comparative metagenomic analysis of rhizosphere microbial community composition and functional potentials under *Rehmannia glutinosa* consecutive monoculture. Int J Mol Sci 19(8):2394. https://doi.org/10.3390/ijms19082394

Yang Y, Wang N, Guo X, Zhang Y, Ye B (2017) Comparative analysis of bacterial community structure in the rhizospheres of maize by high throughput pyrosequencing. PLoS One 12(5): e178425. https://doi.org/10.1371/journal.pone.0178425

Ye H, Gemperline E, Venkateshwaran M, Chen R, Delaux PM, Howes-Podoll M, Ané JM, Li L (2013) MALDI mass spectrometry-assisted molecular imaging of metabolites during nitrogen fixation in the *Medicago truncatula-Sinorhizobium meliloti* symbiosis. Plant J 75:130–145

You L, Zhang B, Tang YJ (2014) Application of stable isotope-assisted metabolomics for cell metabolism studies. Metabolites 4(2):142–165

Zhang N, Wang D, Liu Y (2014) Effects of different plant root exudates and their organic acid components on chemotaxis, biofilm formation and colonization by beneficial rhizosphere-associated bacterial strains. Plant Soil 374:689–700. https://doi.org/10.1007/s11104-0131915-6

Zhang X, Zhang R, Gao J, Wang X, Fan F, Ma X, Yin H, Zhang C, Feng K, Deng Y (2017) Thirty-one years of rice-rice-green manure rotations shape the rhizosphere microbial community and enrich beneficial bacteria. Soil Biol Biochem 104:208–217. https://doi.org/10.1016/j.soilbio.2016.10.023

Zhao L, Huang Y, Hu J, Zhou H, Adeleye AS, Keller AA (2016) 1H NMR and GC-MS based metabolomics reveal defense and detoxification mechanism of cucumber plant under nano-Cu stress. Environ Sci Technol 50(4):2000–2010

Chapter 11
Rhizosphere Virology and Plant Health

Shikha Thakur, Lisha Khungar, Yashika Dhingra, and Rahul Arora

Abstract Plants are at the interface of above- and below-ground interactions. The diverse group of microbes interacts with the plants in both beneficial and detrimental manner. Plants have evolved a number of different strategies to overcome these attacks. Viruses make an important group of organisms that interact with the plant to use the host molecular machinery to replicate its own genome and propagate further. Plants have developed defence strategies involving electrical signalling, using reactive oxygen species and involve the role of different plant hormones in a signalling cascade. This prevents the plant from further damage and also intimates its neighbouring plant to prepare themselves for a forthcoming attack. Volatile organic compounds play an important role in such phenomenon.

Keywords Plant–virus interaction · Rhizosphere · SAR · Chaperons · Phytohormones · VOCs · NGS · Signalling

11.1 Introduction

Plant which belongs to the kingdom *plantae* may broadly be defined as any multicellular eukaryotes characteristically carrying out photosynthesis for nutrition wherein chemical energy is produced from water, minerals, and carbon dioxide, aided by chlorophyll and solar radiations. They often show continuous growth at localized regions. They have cells with cellulose in their walls, providing them the

S. Thakur
CSIR-National Chemical Laboratory, Pune, India

L. Khungar · Y. Dhingra
Department of Plant Molecular Biology, University of Delhi, Delhi, India

R. Arora (✉)
Division of Biosciences, University College London, London, UK

The Francis Crick Institute, London, UK
e-mail: rahul.arora.15@ucl.ac.uk

R. N. Pudake et al. (eds.), *Omics Science for Rhizosphere Biology*, Rhizosphere Biology, https://doi.org/10.1007/978-981-16-0889-6_11

essential rigidity to stand without any external support. Plants lack organs for locomotion, thus making them sessile in nature. Plants however are not seen in the light of being an isolated entity, rather they are recognized as meta-organism which possesses a distinct microbiome showing symbiosis with the co-inhabiting microbes (Mendes et al. 2013). These microbes reside in the area around the plant root known as the rhizosphere. The rhizosphere can formally be defined as the plant–root interface. The rhizosphere is inhabited by the distinct microorganisms the composition of which is dictated by chemicals secreted by the plant roots. The rhizosphere, however well understood, is not a region which can be well defined with its shape or size; in fact, it may consist of various strata based on the varying chemical, biological, and physical properties of the plant root system (McNear Jr. 2013). It is one important niche for the microbial communities to dwell upon. The number and the great diversity of microbes living in these niches is a result of the rhizodeposits in this region. A deep knowledge of the plant–microbe interaction in the rhizosphere will lead to sustainable food, fuel, and fibre production, sufficient to sustain the ever-increasing world population (Berg and Smalla 2009). The under-ground microbes are important for both, the plant as well as soil ecology. These microbes take active part in soil structure determination, organic matter decomposition, removing toxins, and the recycling of some important elements. Besides, the microbes also play monumental role in keeping at bay, the soil-borne plant diseases. They help promote plant growth and the changing vegetation types (Doran et al. 1996).

The microbes colonizing the rhizosphere range from bacterium-fixing nitrogen for the plants to mycorrhizal fungi helping in the uptake of certain minerals in close association with the roots. The rhizosphere also hosts a huge range of viruses. Viruses exist where any other form of life exists. They are the most abundant biological entities prevailing on Earth. Recent work shows that viruses can play important role in association with plants; they confer the plant resistance to extreme environmental conditions. The viruses, besides being a therapeutic tool and model system for biological enquiries, act as the major drivers of biogeochemical cycles. They also play a pivotal role in shaping the rhizospheres' microbial diversity. Viruses tend to be the greatest reservoirs of gene helping in the widely occurring horizontal gene transfers (Kimura et al. 2008). Viruses cannot function outside the host cells and can only carry out the processes for its replication and integration for further transmission only when they are inside a host cell. This unique property of the viruses, wherein it does not identify with the living organisms renders them a unique position in the classification system of organisms. They show intrinsic properties, which is a factor of its size and shape, as well as relational properties, which is dictated by the type of host they are being harboured in (Regenmortel 2000). The viruses present in the rhizosphere are of great significance due to their influence on other soil biological communities through gene transfer from one host to another host and the death of the microbes which occurs as a consequence of the transmission of these particles (Emerson 2019).

Viruses follow a similar pathway for infection. They can be lytic, lysogenic, or pseudo lysogenic in nature. A temperate phage in the lysogenic state is known as a prophage. Viral lysogeny involves integration of the phage genome into the host cell chromosome as a prophage. The production of a temperate phage in the lysogenic state does not depend on the host cell density. In contrast, a virulent virus in the lytic state redirects the host metabolism towards the production of new virus particles, resulting in lysis of the cells. As a consequence, the production and survival of a virulent virus require frequent, successful host–phage encounters to exceed the rate of phage destruction and inactivation. Viral lysogeny posits advantage over the virulence for phage and its hosts where long inactive periods of hosts are survived by the resident phages (Kimura et al. 2008; Mann 2003).

The plant viruses are sub-microscopic in nature and show a great diversity in their shape, ranging from spherical, rod-shaped to filamentous, which may further contain different kinds of genomes. A major chunk of these virus genome happens to be single-stranded RNA (ssRNA), while some of those even might have double-stranded RNA (dsRNA), single-stranded DNA (ssDNA), and double-stranded DNA (dsDNA) (Gaffar and Koch 2019; Matthews' Plant Virology 2002). The viral genetic material is encapsulated by a coat or capsid and the infection is mediated by vectors. Virus multiplication within the host cell works in an intricate manner. This also needs the systemic movement of viruses via the plasmodesmata and the phloem. The plant however has evolved a complex mechanism to deal with various strains of different viruses it encounters in the rhizosphere (Gallet et al. 2018).

It is important to study the rhizosphere viruses to determine plant health and ultimately the ecosystem health. An approach that works best to study these complexes below-ground interactions is metagenomics. Microbial metagenomics for soil bacterium has been widely used and standardized using the 16srRNA sequencing, but the same does not hold true in case of viruses. Though there are methods like plaque formation studies, but they impose a number of limitations. Metagenomic sequencing of viruses is limited because of the lack of universal markers and the phylogenetic means for defining viral populations. This probably is one of the plausible reasons for high local viral diversity as opposed to the low global viral diversity. Modern techniques involving single molecule real-time sequencing (SMRT) will help in the improved identification by stratifying different viral variants which otherwise would have got flattened out in the short reads (Breitbart and Rohwer 2005; Rhoads and Au 2015). This would also enhance the traceability for molecular epidemiology and ecology emitted by the viruses. This in turn would help establish virus-associated risk analysis and the viral durability in the proximal area (Jones 2014). The plant–virus interactions will continue to harbour attention in future as well; the viruses will continue to invade both cultivated and wild plants and thus, the new plant–virus interactions would keep evolving. The accelerated rate of plant–virus interactions and the emergence of new virus-borne diseases in a wide variety of crop plants offer a great potential for future research in this domain.

An insight into the rhizosphere's viral diversity would help to develop different strategies for sustainable agricultural practices and to mitigate the food security challenge. Apart from impacting the plant health in a deteriorating manner, these virus particles also provide resistance to their host plants against a number of different biotic and abiotic stresses (Roossinck 2015). The ability of viruses to confer beauty in the ornamental plants, and tolerance in several different crops, which has long been overlooked can be put as an interesting research domain. The interplay of rhizosphere inhabitants needs to be thoroughly investigated to look for the enhanced beneficial functionalities (Berendsen et al. 2012). Furthermore, a systems approach will facilitate designing of robust rhizosphere engineering platforms. The immense impact of this will revolutionize the approaches towards agricultural practices.

11.2 Virus Particles Enter the Host Plants Through Their Roots

Most plant viruses enter the host through its aerial parts, but there are some classes which invade plants via roots with the help of various soil-inhabiting vectors. The first ever root invading plant virus identified was soil-borne wheat mosaic virus (SBWMV) which causes mosaic disease in wheat (Hamm et al. 2007). Since then various soil-borne viruses have been identified. The viruses have single-stranded RNA (ssRNA) genome and belong to 17 different genera from 8 virus families (Table 11.1). The viruses are transmitted by soil-inhabiting vectors belonging to three different groups, namely, *plasmodiophorids* (Prostita), *Olpidium* spp. (Fungi), and nematodes (Invertebrates). *Plasmodiophorids* transmit viruses from *Potyviridae, Benyviridae,* and *Virgaviridae*, *Olpidium* spp. transmit *Alphaflexiviridae*, *Tombusviridae*, *Rhabdoviridae,* and *Ophioviridae*, while nematodes transmit viruses from *Secoviridae*, *Virgaviridae,* and *Tombusviridae*. The virions may be icosahedral, rod-shaped, or filamentous. Moreover, the structure of virus has no association

Table 11.1 Classification of soil-borne viruses

S. no.	Family	Genus	Virus representative
1	Benyviridae	Benyvirus	Beet necrotic yellow vein virus
2	Virgaviridae	Furovirus	Soil-borne wheat mosaic virus
		Pecluvirus	Peanut clump virus
		Pomovirus	Potato mop-top virus
		Tobravirus	Tobacco rattle virus
3	Potyviridae	Bymovirus	Barley yellow mosaic virus
4	Ophioviridae	Ophiovirus	Mirafiori lettuce big vein virus
5	Rhabdoviridae	Varicosavirus	Lettuce big vein-associated virus
6	Alphafexiviridae	Potexvirus	Pepino mosaic virus
7	Tombusviridae	Tombusvirus	Cucumber necrosis virus
		Gamma-carmovirus	Melon necrotic spot virus
		Dianthovirus	Red clover necrotic mosaic virus
		Alpha-/Beta-necrovirus	Tobacco necrotic virus
8	Secoviridae	Nepovirus	Tomato ringspot virus
		Cheravirus	Cherry rasp leaf virus

with the type of vector mediating its transmission (Kormelink et al. 2011; Verchot-Lubicz 2003). In icosahedral virions, the coat protein (CP) plays a significant part in the transmission process. It adsorbs the virus either on to the surface of vector zoospores or to nematode feeding apparatus while in rod-shaped or filamentous virions some other proteins are also involved in the transmission in addition to the coat protein (Adams et al. 2001; Bragard et al. 2013; MacFarlane 2003).

11.2.1 Nematode-Transmitted Viruses

Nematodes generally transmit viruses belonging to *Nepovirus, Tobravirus, and Dianthovirus.* Movement proteins in *Nepoviruses* form a hollow tubular structure inside the infected cell which extends and connects the neighbouring cells. The tubular structures contain viral movement proteins and form the channel for virion transmission to adjacent cells and its spread to other plant areas (Wieczorek and Sanfaçon 1993). *Tobraviruses* move from one cell to the other without any coat protein. The movement protein binds in a cooperative manner to the RNA genome forming a ribonucleoprotein complex which moves from one to the other through plasmodesmatal junctions. Whereas the *Tobraviruses* move from one cell to the other without any coat protein (Carrington et al. 1996).

11.2.2 Fungi-Transmitted Viruses

Olpidium spp. transmits viruses belonging to *Dianthovirus, Necrovirus, Tombusvirus,* and *Varicosavirus. Tombusvirus* and *Dianthovirus* do not need coat proteins for cell-to-cell movement. The movement protein of Red Clover Necrotic Mosaic Virus (RCNMV; *Dianthovirus*) has a special RNA-binding property. It moves the viral RNA through plasmodesmatal pores by inducing their expansion (Fujiwara et al. 1993; Giesman-Cookmeyer and Lommel 1993).

11.2.3 Plasmodiophorids-Transmitted Viruses

It includes viruses from *Benyvirus, Furovirus, Pecluvirus,* and *Pomovirus* (Mayo 1999). *Beny-, Pomo-,* and *Pecluviruses* encode three proteins, namely, TGBp1, TGBp2, and TGBp3, also known as triple gene block proteins. TGBp1 is an RNA helicase and associates itself with plasmodesmata while TGBp2 and TGBp3 are small membrane-binding proteins and are essential in cell-to-cell viral movements (Torrance and Mayo 1997).

11.2.4 Viral Transport to Aerial Parts

After entering the host plant through roots, these soil-borne viruses generally spread their invasion to the aerial parts via a vascular system of the plant, specifically the xylem. Immunogold-labelling studies and transmission electron microscopic studies have revealed the invasion of beet necrotic yellow vein virus (BNYVV) and SBWMV in root xylem parenchyma and vessels of infected plants (Chiba et al. 2008; Verchot et al. 2001). Rice yellow mottle virus (RYMV) particles move from one cell to another into immature xylem elements where they also undergo replication. The virus particles are released into the xylem upon programmed cell death and travel upward to the aerial parts and subsequently, generate symptoms of infection (Opalka et al. 1998; Verchot et al. 2001).

Cellular level damage and molecular signalling triggered by viruses upon entering the plant forms the basis of plant defence mechanisms. Once a virus enters the plant system, it starts using the host plant's cellular machinery to replicate itself. Plants develop immune responses against the invading viruses to restrict the putative damage by the virus. They either employ RNAi-based signalling or resistance (R) gene-based defence mechanisms.

11.3 RNAi-Based Signalling Mechanisms

Viruses that infect plant systems generally have an RNA genome. Therefore, RNAi- or RNA-induced silencing-based immune responses are an important basal defence mechanism (Ding and Voinnet 2007; Vaucheret 2006). The RNA genome comprises of various stem-loop structures, which are replicated by virus-encoded RNA-dependent RNA polymerases (RDRPs) to form double-stranded RNA (dsRNA). Plant dicer-like (DCL) ribonucleases act on these dsRNA structures to generate 21–24 nucleotide short interfering RNAs (siRNAs). The siRNAs are taken up by RISC, RNA-induced silencing complex, processed into guide RNAs (giRNAs) and then targeted to the foreign RNA molecules with complementary sequences (Ruiz-Ferrer and Voinnet 2009). Host Argonaut (AGO) family proteins degrade these targeted viral RNA molecules to impede the spread of viral infection (Vaucheret 2008). The RNAi-based signalling mechanism tends to limit the viral invasion, but viruses also have a counter-mechanism to combat this layer of plant defence. Viruses encode anti-silencing proteins, also known as viral suppressors of RNAi (VSRs) which interfere with host RNA silencing machinery (Ding and Voinnet 2007). Generally, all plant–virus families encode VSRs. These VSRs are diverse in sequence, evolutionary origin, competency, and mode of action (Bazzini et al. 2007). VSRs undertake diverse mechanisms to suppress plant RNAi pathway, i.e. they interfere with the cellular RNAi pathway at various steps (Fig. 11.1). For example, they may either impede the initiation of RNAi response by sequestering dsRNA or DCL like Potyviral HcPro and beet yellow virus P21 protein (Carbonell

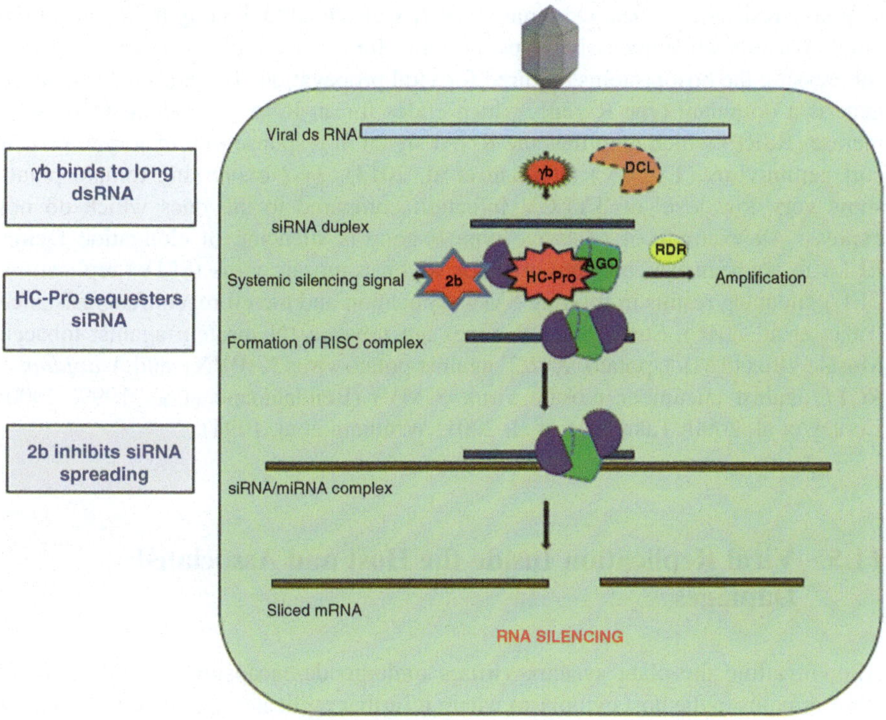

Fig. 11.1 VSR-mediated suppression of plant RNAi signalling

and Carrington 2015; Goto et al. 2007; Haas et al. 2008; Lakatos et al. 2006) or they may interact with AGO proteins resulting in mis-assembly of RISC like cucumber mosaic virus 2b (Carbonell and Carrington 2015; Csorba et al. 2015; Duan et al. 2012). They may even suppress RDRPs or altogether block any other downstream RNAi pathway (Glick et al. 2008). In addition to performing RNA silencing suppressor activities, these viral proteins also participate in other viral processes like replication, assembly, or movement of these virus particles.

11.4 R-Gene-Mediated Defence Responses

The RNAi-based mechanism restricts the entry of viral pathogen inside the plant host while the R-gene-mediated defence limits viral replication inside the host cells and its escalation to other plant areas (Nakahara and Masuta 2014; Verlaan et al. 2013; Zhou and Chai 2008). R proteins generally belong to nucleotide-binding site–leucine-rich repeat (NBS–LRR) class of proteins which trigger broad spectrum defence processes against various pathogens including bacteria, fungi, and viruses (Jones and Dangl 2006; Kachroo et al. 2006). R-genes are either dominant or recessive resistance genes. Dominant type R-genes tend to directly fight against the viral invasion while recessive type proteins limit viral replication by indirectly suppressing the host proteins required for viral propagation. Tomato *Ty-1* resistance gene is a dominant type R-gene, which codes for an RNA-dependent RNA polymerase (RdRP) which amplifies the RNAi signal in response to tomato yellow leaf curl geminivirus (TYLCV) (Verlaan et al. 2013). *Ty-1* expressing tomato plants show very low level of TYLCV infection compared to the ones which do not express. An example of recessive type R-gene is silencing of elongation factor, EF1A in soybean which is required for soybean mosaic virus (SMV) replication. EF1A silencing results in limiting SMV replication and therefore, confers resistance (Luan et al. 2016). Some other R-genes are tobacco 'N' protein against tobacco Mosaic virus (TMV), potato '*Rx1/2*' against potato virus X (PVX), and *Arabidopsis RCY1* against cucumber mosaic virus (CMV) (Bendahmane et al. 1999, 2000; Cooley et al. 2000; Takahashi et al. 2001; Whitham et al. 1994).

11.5 Viral Replication Inside the Host and Associated Damages

After invading the plant system, viruses undergo de-encapsidation and the viral RNA remains in the host cytoplasm where it further replicates and traverses to either cells through plasmodesmata (PD). PDs are continuous intercellular channels among neighbouring plant cells. Viral particles replicate inside viral replication complexes (VRCs) which comprise of host and viral proteins required for viral replication and the viral genome template. The *cis*-acting elements in the viral genome guide the initiation of replication process, recognition of the template by RNA-dependent RNA polymerase, synthesis of complementary RNA strands, VRC formation, and activation. VRCs are highly organized structures and associate with host cells' membranes. They tether to modified membranes of various cellular organelles like endoplasmic reticulum (ER) or mitochondria.

The steps of viral replication depend on the genome type of the virus. The (+) RNA viruses first translate the genomic RNA to form the required viral proteins for replication and then, undergo the actual replication process. The (+) strand is first copied into (−) RNA strands which undergo subsequent rounds of replication to form (+) strands that form the genome of daughter virions. In (−) RNA viruses, the RDRPs are already present along with the genomic RNA which copies the template to form a (+) strand. These (+) strands are again replicated into (−) strands which may undergo subsequent rounds of replication before finally serving as the template for new viruses. In dsRNA viruses, the RDRP synthesizes the (+) strand which is forced out the virion. This (+) strand undergoes translation to form the viral proteins like CP, RDRP which assemble to form an immature virion. The (+) strand is then subsequently copied into the (−) strand to form the complete dsRNA genome

(Gelderblom 1996; International Committee on Taxonomy of Viruses 2012; King et al. 2012).

Viral replication and spread in the plant system are a co-ordinated process. The VRCs are anchored to cellular membranes and therefore, are in contact with the host cytoskeleton. The transport of newly synthesized virions via PDs is mediated by the virion-encoded movement proteins (MPs). MPs are known to bind to both, the viral ribonucleoprotein complex and the PDs. MPs interaction with PDs involves PD dilation easing the virion traversing between adjacent cells (Lucas 2006). MPs are essential for viral transport as viruses deficient in MP do replicate but are unable to spread their infection. Complementation of such viruses with functionally related MP results in viral movement to systemic tissues (Deom et al. 1987; Holt and Beachy 1991; Morozov et al. 1997; Niehl et al. 2014). Moreover, MPs bind to the ribonucleoprotein complexes in a sequence-independent manner and associate themselves with the viral RNA in their vicinity. TMV MP co-localizes with the viral RNA irrespective of being synthesized during the infection or introduced exogenously (Boutant et al. 2010; Heinlein et al. 1998). MPs have an intrinsic mechanism of nucleic acid identification and binding and therefore have a significant role in viral transport.

11.6 Signalling Events in Plant Defence Against Viral Attack

In an attempt to fight against pathogens, plants have various layers of defence strategies. However, before mounting an efficient response against these potential threats, plants must possess the ability of differentiating between different pathogens. This is achieved by recognizing various elicitors like PAMPs (pathogen-associated molecular patterns) and MAMPs (microbe-associated molecular patterns) which further leads to PTI (PAMP-triggered immunity) responses. To this perspective, whether viruses encode for such molecular patterns is still not clear. Nevertheless, upon viral attack, PTI has been reported to be generated in response to intracellular dsRNAs, thus they can be referred to as viral PAMPs (Nicaise 2014; Niehl et al. 2016). After pathogen recognition, typical PTI responses are generated in plants which include crosstalk between electrical signalling, Ca^{2+}, and the reactive oxygen species (ROS) waves, callose deposition, MAPK, and phytohormones signalling. Plants also respond to viral attack by triggering such orchestrated set of processes and signalling events. In case of a viral attack, PTI was reported to be accompanied with similar processes like SA accumulation which triggers a relay of events associated with downstream signalling and differential regulation of pathogen defence genes. The early response involves ROS production and activation of MAPKs (Bigeard et al. 2015). For example, resistance against various viruses was achieved by signalling cascades involving RLKs (receptor-like kinases) in *Arabidopsis* (Gou et al. 2010; Kørner et al. 2013). Likewise, NIKI- a RLK was found to be involved in *Arabidopsis* defence against *CaLCuv* (cabbage leaf curl virus) (Fontes et al. 2004). Various studies suggest that among a plethora of elements, Ca^{2+} ions are the earliest doers of plant defence response. Various Ca^{2+} sensors like CaM (calmodulin), CMLs (calmodulin-like protein), and CDPKs (Ca^{2+}-dependent protein kinases) are involved in plant immunity against pathogens. In context to viruses, a tobacco CML was seen to play a crucial role in plant defence by regulating virus-induced gene silencing (Anandalakshmi et al. 2000). It binds and controls degradation of silencing suppressors of viruses, thus, exerting its antiviral activity. Similarly, CML role has also been described in tobacco against CMV (cucumber mosaic virus) (Jeon et al. 2017). CML was found to be involved in triggering SAR (systemic acquired resistance) by acting as a receptor which further leads to downstream SA (salicylic acid) signalling. Here, CML was proposed to perform dual function as it perceives Ca^{2+} influx as well as viral suppressors for RNA silencing.

Another early plant defence response is an increase in reactive oxygen species (ROS). The literature carries various examples where in response to viral attack, an increase of ROS was observed. For instance, enhancement in level of ROS was seen after infection by cucumber mosaic virus (Riedle-Bauer 2000), clover mosaic virus (Clarke et al. 2002), and plum pox virus (Díaz-Vivancos et al. 2008; Hernández et al. 2006). Thus, an increase in ROS levels seems to be a common antiviral defence strategy adopted by plants. Apart from ROS and Ca^{2+} signalling, NO (nitric oxide) signalling also plays a positive role in antiviral defence. Induction in NOS (nitric oxide synthase) activity was seen upon TMV (tobacco mosaic virus) infection. NO also induces SA accumulation which further induces pathogenesis-related (PR) gene expression mediating N-gene resistance against TMV (Durner et al. 1998). Overall, it can be concluded that plants activate similar signalling pathways in defence against viruses as seen in the case of non-viral pathogens.

11.6.1 Contribution of Cyclins, CDKs, Chaperones, and Other Enzymes During Viral Infection

Modulation of plant's cell cycle progression is very common during viral attack. Alteration in the expression levels of cell cycle-associated genes was observed upon infection of CaLCuV (cabbage leaf curl virus) in *Arabidopsis* (Ascencio-Ibáñez et al. 2008). In parallel, upregulation of genes like *CDKs* (cyclin-dependent kinases), *CYC* (cycloidea) and *PCNA* (proliferating cell nuclear antigen), and downregulation of genes like *RBR1* (retinoblastoma-related gene) and *CDK* inhibitor was reported upon BSCTV (beet severe curly top virus) infection in Arabidopsis (Park et al. 2010). In another study, upon infection with RBSDV (rice black-streaked dwarf virus), upregulation of *cdc2* gene (encodes a CDK) was reported in maize plants (Shen et al. 2016). Accumulating evidences have shown that activation of these core cell cycle genes and interactions between viral and cell cycle proteins lead to perturbations in various cell cycle controls, which in turn facilitates viral infection (Jing et al. 2019; Mills-Lujan and Deom 2010; Shen et al. 2016). One such regulatory switch for transition from G1 phase to S phase in cell cycle is E2F/RBR (Sidle et al. 1996). Interactions between E2F/RBR facilitates control of being in G_0 state (Harbour and Dean 2000; Lavia and Jansen-Dürr 1999); however, cyclin D-dependent CDK-mediated phosphorylation of RBR disrupts its interaction with E2F leading to expression of S phase genes (Kaelin 1999). Some viruses smartly interacts with RBR and escapes this phosphorylation prerequisite, which in turn leads to uncontrolled DNA replication (Jansen-Dürr 1996; Niculescu et al. 1998). It was observed that overexpression of cyclin D (CYCD3:1) leads to no symptom development upon CaLCuV infection in *Arabidopsis* whereas mutations in CYCD (3:1, 3:2 and 3:3) leads to severe symptoms (Ascencio-Ibáñez et al. 2008). Moreover, high titre of viral DNA was detected in mutants of *cycd3*, in contrast to very less amount in CYCD3 overexpression lines, howbeit, highlighting an important defence approach against viral attacks.

Viruses also reconfigure the chaperone machinery of plants to support their own replication. Subversion of key chaperones like HSP70 and HSP101 from their regular function into viral protein complexes is one of the common strategies employed by viruses to amplify their genomes (Aparicio et al. 2005; Carr et al. 2006). Nevertheless, plants have also developed some mechanisms to counteract this viral strategy. HSP90 is one of the key chaperones involved in NLR-mediated defence response as reduction in the defence was seen due to mutations in HSP90. Various studies have reported that HSP90 along with Rar1 (a resistance signalling gene) and SGT1 (suppressor of G2 allele of skp1) forms a complex to achieve N- or Rx-mediated resistance against viruses like TMV and PVX (potato virus X) (Botër et al. 2007; Takabatake et al. 2007). HSP40 might also be playing an important role in plant immunity as susceptibility to SMV (soybean mosaic virus) was enhanced upon silencing of HSP40 in soybean (Liu and Whitham 2013). In another study, ER residing chaperones were also seen to affect antiviral defence responses (Caplan et al. 2009). Calreticulin (CRT)—an ER chaperone was found to be involved in N-mediated resistance against TMV by directly interacting with TMV movement proteins. When CRT was overexpressed, movement of TMV between cells was interfered and moving proteins was directed to microtubules for degradation (Chen et al. 2005). Other than CRT, BiP (binding immunoglobin protein), and PDIs (protein disulphide isomerases) including ERp57 were upregulated during N-mediated resistance to TMV (Caplan et al. 2009). Silencing ERp57, CRT2, and CRT3 in N-gene expressing *N. benthamiana* led to partial restoration of systemic accumulation lending further support to earlier reports that upregulating CRT blocks TMV movement. It is highly probable that ERQC (endoplasmic reticulum quality control) might be identifying viral proteins as foreign and directing them for degradation via proteasome machinery, thus forming an important control in antiviral immune responses.

11.6.2 Systemic Acquired Resistance in Plant Against Viral Infection

Cellular defence responses elicited at the site of infection are passed to distant non-infected cells, ensuing decreased virulence in these regions. This kind of response is termed as SAR (systemic acquired resistance). It is crucial for plants as it limit the spread of virus progeny to only infected cells and can remain sustained for several weeks (Fu and Dong 2013). Generally, SAR is initiated during an incompatible reaction between R/Avr at the site of infection. This further results in an increase of endogenous levels of SA which in turn activates expression of defence-related genes (Tsuda et al. 2008; Yi et al. 2014). A crucial aspect of SAR response is how the immune signal travels to the distant parts of plants. Several molecules like G3P (Glycerol 3 phosphate), MeSA (Methylated SA, a derivative of salicylic acid), glycerolipids, azelaic acid, and indole derivatives act as putative signals during non-viral and viral pathogens (Chanda et al. 2011; Jung et al. 2009). From studies in tobacco infected with TMV, MeSA was found to be involved in mounting SAR. It binds to lipid-transport proteins and its derivatives and travels through phloem to other parts of the plants (Dempsey and Klessig 2012; Park et al. 2007).

How SAR signals remain sustained for a long duration is still ambiguous but epigenetic changes like chromatin remodelling and DNA modifications seem to be essential in maintaining them (Spoel and Dong 2012). Moreover, it has been reported that SAR can be successfully passed to next generation, despite no pathogen attack (Luna et al. 2012). As such, with respect to viral attacks, molecular determinants of SAR and how it will be successfully forwarded to progeny still requires investigation. However, in one interesting example, it was seen that upon TMV infection, frequency of DNA homologous recombination was increased in infected as well as non-inoculated distant leaves. This increased frequency of recombination was seen to transmit in the progeny of TMV-infected plants. The increased rate leads to more DNA rearrangements and less methylation of *LRR* (leucine-rich repeat) gene loci which is homologous to N resistance gene (Boyko et al. 2007). Likewise, the progeny also showed more tolerance to TMV (Kathiria et al. 2010). It is quite an interesting aspect which upon exploration will uncover many crucial information on SAR defence and its transmission against viral attacks.

Apart from SAR, some host proteins and immunity molecules are also involved in mediating resistance against viruses. For instance, a mutation in TOM1 and TOM2a (tonoplast-localized transmembrane proteins) resulted in reduced infection against TMV in *Arabidopsis* (Ishikawa et al. 1993; Tsujimoto et al. 2003). Similarly, mutations in translation initiation factors (eIF4E, eIF4G) impart resistance against viruses in *Arabidopsis*, pepper, tomato, pea, barley, and rice (Albar et al. 2006; Lellis et al. 2002; Ruffel et al. 2002, 2005). In addition, lectin proteins are also involved in antiviral defence. Resistance against TEV (tobacco etch virus) was achieved due to RTM1 (restricted TEV movement1)—a jacalin-type lectin in *Arabidopsis* (Chisholm et al. 2000). Likewise, resistance against *potexviruses* was mediated due to JAX1 proteins (Yamaji et al. 2012). Lectin-mediated resistance is different than other antiviral responses as it doesn't involve SAR or HR responses, neither it changes level of SA nor it incorporates typical defence genes alterations. In summary, it involves a yet unidentified antiviral defence strategy that needs to be determined in future.

11.6.3 Phytohormones-Mediated Defence Against Viral Attack

Being in constant threat of potential viral attacks, plants depend upon phytohormones to fine tune their antiviral defence mechanisms. SA (salicylic acid) is one of the major phytohormones involved in mounting an effective response against viral pathogenicity. The literature carries various examples of role of SA in mediating R-gene resistance, basal immune processes, and SAR. For instance, accumulation of SA was significantly increased after TMV infection in tobacco. Similarly, SA-mediated defence response was generated to restrict tomato ringspot virus (ToRSV) infection in tobacco (Baebler et al. 2014). Moreover, despite of the presence of appropriate R-genes, mutations in SA pathway still made plants more susceptible to viral attacks (Baebler et al. 2014; Dinesh-Kumar et al. 2000; Lewsey et al. 2009; Takahashi et al. 2004). In another similar case of depletion of SA by using NahG (salicylate hydrolase), transgene decreased immunity against potato virus Y in potato Ny-1 R-gene-resistant lines (Baebler et al. 2014). In parallel, overexpression of SA pathway genes and/or spraying plants with SA and its analogues lead to a further delay in virulence and viral disease establishment (Ishihara et al. 2008; Mayers et al. 2005; Peng et al. 2013). Generally, upon SA biosynthesis, signalling events are initiated which leads to the production of ROS, callose deposition, formation of PR proteins, and induction of HR and SAR responses (Jones and Dangl 2006; Torres et al. 2006; Vlot et al. 2009). All these events lead to further downstream activation of other defence genes which helps plant to decrease susceptibility to viral attacks.

Other than SA, phytohormones like JA (jasmonic acid) and ethylene are also involved in plant–virus interactions. On exogenous application of JA, reduction in geminivirus infection was reported in *Arabidopsis* (Lozano-Durán et al. 2011). Similarly, endogenous levels of JA was seen to enhance incompatible plant–virus interactions in potato and tobacco (Dhondt et al. 2000; Kovač et al. 2009). Few studies about the involvement of ethylene in plant–virus interactions have also been reported. It has been observed that spraying plants with ACC (immediate precursor of ethylene) prior to infection leads to a reduced viral pathogenicity. Furthermore, spraying of SA and JA also aided in reducing viral titres (Clarke et al. 1998). Although in plants, it has been seen that during viral attacks, endogenous levels of Et and JA have antagonistic consequences on defence responses mounted via SA, howbeit, this finding implies that the timing of treatment is crucial for an adequate defence response. Moreover, it is also becoming progressively clear that a crosstalk among these hormones exist which regulates the defence responses against viral attacks. A modulation of JA and SA was seen to be crucial in mounting HR responses initiated during Avr-R reactions (Nicaise 2014). It was concluded that a fine balance between JA and SA is required to determine the level of resistance, similar to pathosystems involved in non-viral pathogens (Thaler et al. 2012). Recent researches also indicate a potential role of ABA (Chen et al. 2013) and brassinosteroids (Ali et al. 2014) in antiviral mechanisms; however, the exact mechanisms and their functions still remain unclear. Nevertheless, future researches (along with current findings) would help to gain a better understanding of plant–virus interactions mediated by phytohormones which might encourage researchers to achieve robust resistance phenotypes against viral infections.

11.6.4 Volatile Organic Compounds and Plant–Virus Interactions

Another line of defence system includes utilization of VOCs (volatile organic compounds) for communication among plants and/or between distinct divisions of same plant. Various diverse molecules like fatty acids and amino acid derivatives, benzenoids, phenylpropanoids, and terpenes have been seen to function as plant VOCs (Pichersky et al. 2006). For instance, tobacco plants emits methyl salicylate (a derivative of salicylic acid) in response to TMV infection which leads to increased resistance against TMV (Shulaev et al. 1997; Vlot et al. 2008). Likewise, accumulation of MeSA was also reported in leaves of tomato plants upon TMV infection (Deng et al. 2004). A model has been suggested where upon TMV infection, accumulated SA is converted to MeSA which further travels to distant leaves via phloem transport. Here, SA-binding protein 2 converts MeSA to SA by its methyl esterase activity (Forouhar et al. 2005) which in turn mounts a SAR response in distant tissues (Park et al. 2007). VOCs are also involved in a process known as priming of plants which serves as an alarm for the adjacent plants to construct their defence responses. In a study using tobacco plants (*Nicotiana tabacum* cv. *Xanthi* nc), it was reported that under laboratory conditions, the emitted MeSAs from TMV-infected plants was adequate to induce resistance in the neighbouring tobacco plants (Shulaev et al. 1997).

Apart from VOCs, synthetic elicitors also trigger defence reactions in plants against pathogens. A number of synthetic elicitors like polyacrylic acid, salicylic acid, chitosan, ozone, paraquat, and many more compounds have been shown to induce defence responses against bacteria, fungi, and viruses. Gianinazzi and Kassanis (1974) demonstrated that application of polyacrylic acid (PA) either by spraying on tobacco leaves or by watering the plants with it leads to resistance against TMV and TNV. Similarly, acetylsalicylic acid and 2, 6-dichloro-isonicotinic acid was shown to provide resistance against TMV in tobacco (Uknes et al. 1992; White 1979). Resistance against turnip crinkle virus (TCV) was achieved in

Arabidopsis by treating them with a novel chemical activator benzothiadiazole (BTH) (Lawton et al. 1996). BTH leads to activation of SAR signalling pathway which further provides disease control against various pathogens (bacteria, fungi, and viruses) in crops like rice, lettuce, tobacco, and wheat (Friedrich et al. 1996; Görlach et al. 1996). Burketová et al. (1999) showed that BTH in sugar beets induces synthesis of β-1, 3-glucanase, and chitinase to provide resistance against TNV. Taken together, it can be concluded that a complex cocktail of VOCs and chemical elicitors are required in generating antiviral responses in plants which can further be exploited to combat agricultural loses due to viral attack.

11.7 Conclusion

Plants and virus particles have been evolving simultaneously over the past many years. Where virus particles have developed various strategies of its transmission from one host to another and to survive by means of its own alterations in the genetic material, plants on the other hands have also come up with various defence strategies. Viruses have adapted to the complex multi-dimensional interactions both under and above the ground with its host plants and with the other microbes. It is interesting to note that despite showing parallel evolution, plants have emerged to be more smart than its pathogens. The plants have developed three distinct layers of defence response with early signalling mechanisms involving the participation of ROS and electrical signalling followed with a more chemically and energetically demanding phytohormone signalling. The third layer of fine defence strategy includes the response elicited by the secondary metabolites. The VOCs seems to play a very intelligent role by informing the neighbour plants of a potential attack so that they can decide the fate of signalling and can divert the resources adequately. It would be interesting to see the degree of variation in gene expression of defence-associated genes in plants under control conditions with no viral pathogen versus a viral-attacked plant versus a plant-allocating resource to prepare for a viral attack when informed by a neighbouring plant emitting VOCs.

Future research avenues might look into the integration of proteomic and transcriptomic studies to understand the role of differential transcription and translation and up- and downregulation of the metabolome to have a better insight onto how interaction of a diverse variety of microbes and herbivores in the rhizosphere alter the plant–virus interactions and what all factors contribute towards the synergistic and/or antagonistic nature of the molecular machinery involved. A tremendous improvement over the years in next-generation sequencing can help decipher the role of epigenetic transfer of SAR from one to another generation. Therefore, to have a better insight and a comprehensive understanding of the molecular biology involved in plant defence strategies, an understanding of interactome is very essential.

References

Adams MJ, Antoniw JF, Mullins JGL (2001) Plant virus transmission by plasmodiophorid fungi is associated with distinctive transmembrane regions of virus-encoded proteins. Arch Virol 146 (6):1139–1153. https://doi.org/10.1007/s007050170111

Albar L, Bangratz-Reyser M, Hébrard E, Ndjiondjop MN, Jones M, Ghesquière A (2006) Mutations in the eIF(iso)4G translation initiation factor confer high resistance of rice to rice yellow mottle virus. Plant J 47(3):417–426. https://doi.org/10.1111/j.1365-313X.2006.02792.x

Ali SS, Gunupuru LR, Kumar GBS, Khan M, Scofield S, Nicholson P, Doohan FM (2014) Plant disease resistance is augmented in uzu barley lines modified in the brassinosteroid receptor BRI1. BMC Plant Biol 14(1):227. https://doi.org/10.1186/s12870-014-0227-1

Anandalakshmi R, Marathe R, Ge X, Herr J, Mau C, Mallory A et al (2000) A calmodulin-related protein that suppresses posttranscriptional gene silencing in plants. Science 290 (5489):142–144. https://doi.org/10.1126/science.290.5489.142

Aparicio F, Thomas CL, Lederer C, Niu Y, Wang D, Maule AJ (2005) Virus induction of heat shock protein 70 reflects a general response to protein accumulation in the plant cytosol. Plant Physiol 138(1):529–536. https://doi.org/10.1104/pp.104.058958

Ascencio-Ibáñez JT, Sozzani R, Lee TJ, Chu TM, Wolfinger RD, Cella R, Hanley-Bowdoin L (2008) Global analysis of Arabidopsis gene expression uncovers a complex array of changes impacting pathogen response and cell cycle during geminivirus infection. Plant Physiol 148 (1):436–454. https://doi.org/10.1104/pp.108.121038

Baebler Š, Witek K, Petek M, Stare K, Tušek-Žnidarič M, Pompe-Novak M et al (2014) Salicylic acid is an indispensable component of the Ny-1 resistance-gene-mediated response against Potato virus y infection in potato. J Exp Bot 65(4):1095–1109. https://doi.org/10.1093/jxb/ert447

Bazzini AA, Mongelli VC, Hopp HE, Del Vas M, Asurmendi S (2007) A practical approach to the understanding and teaching of RNA silencing in plants. Electron J Biotechnol 10(2):178–190. https://doi.org/10.2225/vol10-issue2-fulltext-11

Bendahmane A, Kanyuka K, Baulcombe DC (1999) The Rx gene from potato controls separate virus resistance and cell death responses. Plant Cell 11(5):781–791. https://doi.org/10.1105/tpc.11.5.781

Bendahmane A, Querci M, Kanyuka K, Baulcombe DC (2000) Agrobacterium transient expression system as a tool for the isolation of disease resistance genes: application to the Rx2 locus in potato. Plant J 21(1):73–81. https://doi.org/10.1046/j.1365-313X.2000.00654.x

Berendsen RL, Pieterse CMJ, Bakker PAHM (2012) The rhizosphere microbiome and plant health. Trends Plant Sci 17(8):478–486. https://doi.org/10.1016/j.tplants.2012.04.001

Berg G, Smalla K (2009) Plant species and soil type cooperatively shape the structure and function of microbial communities in the rhizosphere. FEMS Microbiol Ecol 68(1):1–3. https://doi.org/10.1111/j.1574-6941.2009.00654.x

Bigeard J, Colcombet J, Hirt H (2015) Signaling mechanisms in pattern-triggered immunity (PTI). Mol Plant 8(4):521–539. https://doi.org/10.1016/j.molp.2014.12.022

Botër M, Amigues B, Peart J, Breuer C, Kadota Y, Casais C et al (2007) Structural and functional analysis of SGT1 reveals that its interaction with HSP90 is required for the accumulation of Rx, an R protein involved in plant immunity. Plant Cell 19(11):3791–3804. https://doi.org/10.1105/tpc.107.050427

Boutant E, Didier P, Niehl A, Mély Y, Ritzenthaler C, Heinlein M (2010) Fluorescent protein recruitment assay for demonstration and analysis of in vivo protein interactions in plant cells and its application to tobacco mosaic virus movement protein. Plant J 62(1):171–177. https://doi.org/10.1111/j.1365-313X.2010.04126.x

Boyko A, Kathiria P, Zemp FJ, Yao Y, Pogribny I, Kovalchuk I (2007) Transgenerational changes in the genome stability and methylation in pathogen-infected plants: (virus-induced plant genome instability). Nucleic Acids Res 35(5):1714–1725. https://doi.org/10.1093/nar/gkm029

Bragard C, Caciagli P, Lemaire O, Lopez-Moya JJ, MacFarlane S, Peters D et al (2013) Status and prospects of plant virus control through interference with vector transmission. Annu Rev Phytopathol 51:177–201. https://doi.org/10.1146/annurev-phyto-082712-102346

Breitbart M, Rohwer F (2005) Here a virus, there a virus, everywhere the same virus? Trends Microbiol 13(6):278–284. https://doi.org/10.1016/j.tim.2005.04.003

Burketová L, Šindelářová M, Šindelář L (1999) Benzothiadiazole as an inducer of β-1,3-glucanase and chitinase isozymes in sugar beet. Biol Plant 42(2):279–287. https://doi.org/10.1023/A:1002172921426

Caplan JL, Zhu X, Mamillapalli P, Marathe R, Anandalakshmi R, Dinesh-Kumar SP (2009) Induced ER chaperones regulate a receptor-like kinase to mediate antiviral innate immune response in plants. Cell Host Microbe 6(5):457–469. https://doi.org/10.1016/j.chom.2009.10.005

Carbonell A, Carrington JC (2015) Antiviral roles of plant Argonautes. Curr Opin Plant Biol 27:111–117. https://doi.org/10.1016/j.pbi.2015.06.013

Carr T, Wang Y, Huang Z, Yeakley JM, Fan JB, Whitham SA (2006) Tobamovirus infection is independent of HSP101 mRNA induction and protein expression. Virus Res 121(1):33–41. https://doi.org/10.1016/j.virusres.2006.03.013

Carrington JC, Kasschau KD, Mahajan SK, Schaad MC (1996) Cell-to-cell and long-distance transport of viruses in plants. Plant Cell 8(10):1669. https://doi.org/10.1105/tpc.8.10.1669

Chanda B, Xia Y, Mandal MK, Yu K, Sekine KT, Gao QM et al (2011) Glycerol-3-phosphate is a critical mobile inducer of systemic immunity in plants. Nat Genet 43(5):421–427. https://doi.org/10.1038/ng.798

Chen MH, Tian GW, Gafni Y, Citovsky V (2005) Effects of calreticulin on viral cell-to-cell movement. Plant Physiol 138(4):1866–1876. https://doi.org/10.1104/pp.105.064386

Chen L, Zhang L, Li D, Wang F, Yu D (2013) WRKY8 transcription factor functions in the TMV-cg defense response by mediating both abscisic acid and ethylene signaling in Arabidopsis. Proc Natl Acad Sci U S A 110(21):E1963–E1971. https://doi.org/10.1073/pnas.1221347110

Chiba S, Miyanishi M, Andika IB, Kondo H, Tamada T (2008) Identification of amino acids of the beet necrotic yellow vein virus p25 protein required for induction of the resistance response in leaves of Beta vulgaris plants. J Gen Virol 89(5):1314–1323. https://doi.org/10.1099/vir.0.83624-0

Chisholm ST, Mahajan SK, Whitham SA, Yamamoto ML, Carrington JC (2000) Cloning of the Arabidopsis RTM1 gene, which controls restriction of long-distance movement of tobacco etch virus. Proc Natl Acad Sci U S A 97(1):489–494. https://doi.org/10.1073/pnas.97.1.489

Clarke SF, Burritt DJ, Jameson PE, Guy PL (1998) Influence of plant hormones on virus replication and pathogenesis-related proteins in Phaseolus vulgaris L. infected with white clover mosaic potexvirus. Physiol Mol Plant Pathol 53(4):195–207. https://doi.org/10.1006/pmpp.1998.0175

Clarke SF, Guy PL, Burritt DJ, Jameson PE (2002) Changes in the activities of antioxidant enzymes in response to virus infection and hormone treatment. Physiol Plant 114(2):157–164. https://doi.org/10.1034/j.1399-3054.2002.1140201.x

Cooley MB, Pathirana S, Wu HJ, Kachroo P, Klessig DF (2000) Members of the Arabidopsis HRT/RPP8 family of resistance genes confer resistance to both viral and oomycete pathogens. Plant Cell 12(5):663–676. https://doi.org/10.1105/tpc.12.5.663

Csorba T, Kontra L, Burgyán J (2015) Viral silencing suppressors: tools forged to fine-tune host-pathogen coexistence. Virology 479:85–103. https://doi.org/10.1016/j.virol.2015.02.028

Dempsey DA, Klessig DF (2012) SOS—too many signals for systemic acquired resistance? Trends Plant Sci 17(9):538–545. https://doi.org/10.1016/j.tplants.2012.05.011

Deng C, Zhang X, Zhu W, Qian J (2004) Gas chromatography-mass spectrometry with solid-phase microextraction method for determination of methyl salicylate and other volatile compounds in leaves of Lycopersicon esculentum. Anal Bioanal Chem 378(2):518–522. https://doi.org/10.1007/s00216-003-2240-3

Deom CM, Oliver MJ, Beachy RN (1987) The 30-kilodalton gene product of tobacco mosaic virus potentiates virus movement. Science 237(4813):389–394. https://doi.org/10.1126/science.237.4813.389

Dhondt S, Geoffroy P, Stelmach BA, Legrand M, Heitz T (2000) Soluble phospholipase A2 activity is induced before oxylipin accumulation in tobacco mosaic virus-infected tobacco leaves and is contributed by patatin-like enzymes. Plant J 23(4):431–440. https://doi.org/10.1046/j.1365-313X.2000.00802.x

Díaz-Vivancos P, Clemente-Moreno MJ, Rubio M, Olmos E, García JA, Martínez-Gómez P, Hernández JA (2008) Alteration in the chloroplastic metabolism leads to ROS accumulation in pea plants in response to plum pox virus. J Exp Bot 59(8):2147–2160. https://doi.org/10.1093/jxb/ern082

Dinesh-Kumar SP, Tham WH, Baker BJ (2000) Structure-function analysis of the tobacco mosaic virus resistance gene N. Proc Natl Acad Sci U S A 97(26):14789–14794. https://doi.org/10.1073/pnas.97.26.14789

Ding SW, Voinnet O (2007) Antiviral immunity directed by small RNAs. Cell 130(3):413–426. https://doi.org/10.1016/j.cell.2007.07.039

Doran JW, Sarrantonio M, Liebig MA (1996) Soil health and sustainability. Adv Agron 56:1–54. https://doi.org/10.1016/S0065-2113(08)60178-9

Duan CG, Fang YY, Zhou BJ, Zhao JH, Hou WN, Zhu H et al (2012) Suppression of Arabidopsis Argonaute1-mediated slicing, transgene-induced RNA silencing, and DNA methylation by distinct domains of the cucumber mosaic virus 2b protein. Plant Cell 24(1):259–274. https://doi.org/10.1105/tpc.111.092718

Durner J, Wendehenne D, Klessig DF (1998) Defense gene induction in tobacco by nitric oxide, cyclic GMP, and cyclic ADP-ribose. Proc Natl Acad Sci U S A 95(17):10328–10333. https://doi.org/10.1073/pnas.95.17.10328

Emerson JB (2019) Soil viruses: a new hope. mSystems 4(3):e00120–e00119. https://doi.org/10.1128/msystems.00120-19

Fontes EPB, Santos AA, Luz DF, Waclawovsky AJ, Chory J (2004) The geminivirus nuclear shuttle protein is a virulence factor that suppresses transmembrane receptor kinase activity. Genes Dev 18(20):2545–2556. https://doi.org/10.1101/gad.1245904

Forouhar F, Yang Y, Kumar D, Chen Y, Fridman E, Park SW et al (2005) Structural and biochemical studies identify tobacco SABP2 as a methyl salicylate esterase and implicate it in plant innate immunity. Proc Natl Acad Sci U S A 102(5):1773–1778. https://doi.org/10.1073/pnas.0409227102

Friedrich L, Lawton K, Ruess W, Masner P, Specker N, Rella MG et al (1996) A benzothiadiazole derivative induces systemic acquired resistance in tobacco. Plant J 10(1):61–70. https://doi.org/10.1046/j.1365-313X.1996.10010061.x

Fu ZQ, Dong X (2013) Systemic acquired resistance: turning local infection into global defense. Annu Rev Plant Biol 64:839–863. https://doi.org/10.1146/annurev-arplant-042811-105606

Fujiwara T, Giesman-Cookmeyer D, Biao D, Lommel SA, Lucas WJ (1993) Cell-to-cell trafficking of macromolecules through plasmodesmata potentiated by the red clover necrotic mosaic virus movement protein. Plant Cell 5(12):1783–1794. https://doi.org/10.1105/tpc.5.12.1783

Gaffar FY, Koch A (2019) Catch me if you can! RNA silencing-based improvement of antiviral plant immunity. Viruses 11(7):673. https://doi.org/10.3390/v11070673

Gallet R, Michalakis Y, Blanc S (2018) Vector-transmission of plant viruses and constraints imposed by virus–vector interactions. Curr Opin Virol 33:144–150. https://doi.org/10.1016/j.coviro.2018.08.005

Gelderblom HR (1996) Structure and classification of viruses. In: Medical microbiology, 4th edn. University of Texas Medical Branch at Galveston

Gianinazzi S, Kassanis B (1974) Virus resistance induced in plants by polyacrylic acid. J Gen Virol. https://doi.org/10.1099/0022-1317-23-1-1

Giesman-Cookmeyer D, Lommel SA (1993) Alanine scanning mutagenesis of a plant virus movement protein identifies three functional domains. Plant Cell 5(8):973–982. https://doi.org/10.1105/tpc.5.8.973

Glick E, Zrachya A, Levy Y, Mett A, Gidoni D, Belausov E et al (2008) Interaction with host SGS3 is required for suppression of RNA silencing by tomato yellow leaf curl virus V2 protein. Proc Natl Acad Sci U S A 105(1):157–161. https://doi.org/10.1073/pnas.0709036105

Görlach J, Volrath S, Knauf-Beiter G, Hengy G, Beckhove U, Kogel KH et al (1996) Benzothiadiazole, a novel class of inducers of systemic acquired resistance, activates gene expression and disease resistance in wheat. Plant Cell 8(4):629–643. https://doi.org/10.2307/3870340

Goto K, Kobori T, Kosaka Y, Natsuaki T, Masuta C (2007) Characterization of silencing suppressor 2b of cucumber mosaic virus based on examination of its small RNA-binding abilities. Plant Cell Physiol 48(7):1050–1060. https://doi.org/10.1093/pcp/pcm074

Gou X, He K, Yang H, Yuan T, Lin H, Clouse SD, Li J (2010) Genome-wide cloning and sequence analysis of leucine-rich repeat receptor-like protein kinase genes in Arabidopsis thaliana. BMC Genomics 11(1):19. https://doi.org/10.1186/1471-2164-11-19

Haas G, Azevedo J, Moissiard G, Geldreich A, Himber C, Bureau M et al (2008) Nuclear import of CaMV P6 is required for infection and suppression of the RNA silencing factor DRB4. EMBO J 27(15):2102–2112. https://doi.org/10.1038/emboj.2008.129

Hamm PB, Gieck SL, David NL, Hunger RM (2007) First report of soilborne wheat mosaic virus on wheat (Triticum aestivum) in the Columbia Basin of Oregon. Plant Dis 91(11):1513. https://doi.org/10.1094/pdis-91-11-1513c

Harbour JW, Dean DC (2000) The Rb/E2F pathway: expanding roles and emerging paradigms. Genes Dev 14(19):2393–2409. https://doi.org/10.1101/gad.813200

Heinlein M, Padgett HS, Gens JS, Pickard BG, Casper SJ, Epel BL, Beachy RN (1998) Changing patterns of localization of the tobacco mosaic virus movement protein and replicase to the endoplasmic reticulum and microtubules during infection. Plant Cell 10(7):1107–1120. https://doi.org/10.1105/tpc.10.7.1107

Hernández JA, Díaz-Vivancos P, Rubio M, Olmos E, Ros-Barceló A, Martínez-Gómez P (2006) Long-term plum pox virus infection produces an oxidative stress in a susceptible apricot, Prunus armeniaca, cultivar but not in a resistant cultivar. Physiol Plant 126(1):140–152. https://doi.org/10.1111/j.1399-3054.2005.00581.x

Holt CA, Beachy RN (1991) In vivo complementation of infectious transcripts from mutant tobacco mosaic virus cDNAs in transgenic plants. Virology 181(1):109–117. https://doi.org/10.1016/0042-6822(91)90475-Q

International Committee on Taxonomy of Viruses (2012) Virus taxonomy: classification and nomenclature of viruses. In: Ninth report of the international committee on taxonomy of viruses. https://doi.org/10.1016/B978-0-12-384684-6.00057-4

Ishihara T, Sekine KT, Hase S, Kanayama Y, Seo S, Ohashi Y et al (2008) Overexpression of the Arabidopsis thaliana EDS5 gene enhances resistance to viruses. Plant Biol 10(4):451–461. https://doi.org/10.1111/j.1438-8677.2008.00050.x

Ishikawa M, Naito S, Ohno T (1993) Effects of the tom1 mutation of Arabidopsis thaliana on the multiplication of tobacco mosaic virus RNA in protoplasts. J Virol 67(9):5328–5338. https://doi.org/10.1128/jvi.67.9.5328-5338.1993

Jansen-Dürr P (1996) How viral oncogenes make the cell cycle. Trends Genet 12(7):270–275. https://doi.org/10.1016/0168-9525(96)81455-7

Jeon EJ, Tadamura K, Murakami T, Inaba J, Kim BM, Sato M et al (2017) Rgs-CaM detects and counteracts viral RNA silencing suppressors in plant immune priming. J Virol 91(19):e00761–e00717. https://doi.org/10.1128/jvi.00761-17

Jing C, Li P, Zhang J, Wang R, Wu G, Li M et al (2019) The malvastrum yellow vein virus c4 protein promotes disease symptom development and enhances virus accumulation in plants. Front Microbiol 10:2425. https://doi.org/10.3389/fmicb.2019.02425

Jones RAC (2014) Plant virus ecology and epidemiology: historical perspectives, recent progress and future prospects. Ann Appl Biol 164(3):320–347. https://doi.org/10.1111/aab.12123

Jones JDG, Dangl JL (2006) The plant immune system. Nature 444(7117):323–329. https://doi.org/10.1038/nature05286

Jung HW, Tschaplinski TJ, Wang L, Glazebrook J, Greenberg JT (2009) Priming in systemic plant immunity. Science 24(5923):89–91. https://doi.org/10.1126/science.1170025

Kachroo P, Chandra-Shekara AC, Klessig DF (2006) Plant signal transduction and defense against viral pathogens. Adv Virus Res 66:161–191. https://doi.org/10.1016/S0065-3527(06)66004-1

Kaelin WG (1999) Functions of the retinoblastoma protein. BioEssays 21(11):950–958. https://doi.org/10.1002/(SICI)1521-1878(199911)21:11<950::AID-BIES7>3.0.CO;2-D

Kathiria P, Sidler C, Golubov A, Kalischuk M, Kawchuk LM, Kovalchuk I (2010) Tobacco mosaic virus infection results in an increase in recombination frequency and resistance to viral, bacterial, and fungal pathogens in the progeny of infected tobacco plants. Plant Physiol 153(4):1859–1870. https://doi.org/10.1104/pp.110.157263

Kimura M, Jia ZJ, Nakayama N, Asakawa S (2008) Ecology of viruses in soils: past, present and future perspectives. Soil Sci Plant Nutr 54(1):1–32. https://doi.org/10.1111/j.1747-0765.2007.00197.x

King AMQ, Lefkowitz E, Adams MJ, Carstens EB (2012) Virus taxonomy. In: Virus taxonomy, pp 486–7. https://doi.org/10.1016/b978-012465330-6/50022-1

Kormelink R, Garcia ML, Goodin M, Sasaya T, Haenni AL (2011) Negative-strand RNA viruses: the plant-infecting counterparts. Virus Res 162(1–2):184–202. https://doi.org/10.1016/j.virusres.2011.09.028

Kørner CJ, Klauser D, Niehl A, Domínguez-Ferreras A, Chinchilla D, Boller T et al (2013) The immunity regulator BAK1 contributes to resistance against diverse RNA viruses. Mol Plant-Microbe Interact 26(11):1271–1280. https://doi.org/10.1094/MPMI-06-13-0179-R

Kovač M, Müller A, Milovanovič Jarh D, Milavec M, Düchting P, Ravnikar M (2009) Multiple hormone analysis indicates involvement of jasmonate signalling in the early defence of potato to potato virus YNTN. Biol Plant 53(1):195–199. https://doi.org/10.1007/s10535-009-0034-y

Lakatos L, Csorba T, Pantaleo V, Chapman EJ, Carrington JC, Liu YP et al (2006) Small RNA binding is a common strategy to suppress RNA silencing by several viral suppressors. EMBO J 25(12):2768–2780. https://doi.org/10.1038/sj.emboj.7601164

Lavia P, Jansen-Dürr P (1999) E2F target genes and cell-cycle checkpoint control. BioEssays 21(3):221–230. https://doi.org/10.1002/(SICI)1521-1878(199903)21:3<221::AID-BIES6>3.0.CO;2-J

Lawton KA, Friedrich L, Hunt M, Weymann K, Delaney T, Kessmann H et al (1996) Benzothiadiazole induces disease resistance in Arabidopsis by activation of the systemic acquired resistance signal transduction pathway. Plant J 10(1):71–82. https://doi.org/10.1046/j.1365-313X.1996.10010071.x

Lellis AD, Kasschau KD, Whitham SA, Carrington JC (2002) Loss-of-susceptibility mutants of Arabidopsis thaliana reveal an essential role for eIF(iso)4E during potyvirus infection. Curr Biol 12(12):1046–1051. https://doi.org/10.1016/S0960-9822(02)00898-9

Lewsey M, Palukaitis P, Carr JP (2009) Plant-virus interactions: defence and counter-defence. In: Annual plant reviews, molecular aspects of plant disease resistance, pp 134–76. https://doi.org/10.1002/9781444301441.ch6

Liu JZ, Whitham SA (2013) Overexpression of a soybean nuclear localized type-III DnaJ domain-containing HSP40 reveals its roles in cell death and disease resistance. Plant J 74(1):110–121. https://doi.org/10.1111/tpj.12108

Lozano-Durán R, Rosas-Díaz T, Gusmaroli G, Luna AP, Taconnat L, Deng XW, Bejarano ER (2011) Geminiviruses subvert ubiquitination by altering CSN-mediated derubylation of SCF E3 ligase complexes and inhibit jasmonate signaling in Arabidopsis thaliana. Plant Cell 23(3):1014–1032. https://doi.org/10.1105/tpc.110.080267

Luan H, Shine MB, Cui X, Chen X, Ma N, Kachroo P et al (2016) The potyviral P3 protein targets eukaryotic elongation factor 1A to promote the unfolded protein response and viral pathogenesis. Plant Physiol 172(1):221–234. https://doi.org/10.1104/pp.16.00505

Lucas WJ (2006) Plant viral movement proteins: agents for cell-to-cell trafficking of viral genomes. Virology 344(1):169–184. https://doi.org/10.1016/j.virol.2005.09.026

Luna E, Bruce TJA, Roberts MR, Flors V, Ton J (2012) Next-generation systemic acquired resistance. Plant Physiol 158(2):844–853. https://doi.org/10.1104/pp.111.187468

MacFarlane SA (2003) Molecular determinants of the transmission of plant viruses by nematodes. Mol Plant Pathol 4(3):211–215. https://doi.org/10.1046/j.1364-3703.2003.00164.x

Mann NH (2003) Phages of the marine cyanobacterial picophytoplankton. FEMS Microbiol Rev 27 (1):17–34. https://doi.org/10.1016/S0168-6445(03)00016-0

Matthews' Plant Virology (2002) In: Matthews' plant virology. https://doi.org/10.1016/b978-0-12-361160-4.x5050-6

Mayers CN, Lee KC, Moore CA, Wong SM, Carr JP (2005) Salicylic acid-induced resistance to cucumber mosaic virus in squash and Arabidopsis thaliana: contrasting mechanisms of induction and antiviral action. Mol Plant-Microbe Interact 18(5):428–434. https://doi.org/10.1094/MPMI-18-0428

Mayo MA (1999) Developments in plant virus taxonomy since the publication of the 6th ICTV report. Arch Virol 144(8):1659–1666. https://doi.org/10.1007/s007050050620

McNear DH Jr (2013) The rhizosphere—roots, soil and everything in between. Nature Education Knowledge 4(3):1

Mendes R, Garbeva P, Raaijmakers JM (2013) The rhizosphere microbiome: significance of plant beneficial, plant pathogenic, and human pathogenic microorganisms. FEMS Microbiol Rev 37 (5):634–663. https://doi.org/10.1111/1574-6976.12028

Mills-Lujan K, Deom CM (2010) Geminivirus C4 protein alters Arabidopsis development. Protoplasma 239(1–4):95–110. https://doi.org/10.1007/s00709-009-0086-z

Morozov SY, Fedorkin ON, Jüttner G, Schiemann J, Baulcombe DC, Atabekov JG (1997) Complementation of a potato virus X mutant mediated by bombardment of plant tissues with cloned viral movement protein genes. J Gen Virol 78(8):2077–2083

Nakahara KS, Masuta C (2014) Interaction between viral RNA silencing suppressors and host factors in plant immunity. Curr Opin Plant Biol 20:88–95. https://doi.org/10.1016/j.pbi.2014.05.004

Nicaise V (2014) Crop immunity against viruses: outcomes and future challenges. Front Plant Sci 5:660. https://doi.org/10.3389/fpls.2014.00660

Niculescu AB, Chen X, Smeets M, Hengst L, Prives C, Reed SI (1998) Effects of p21Cip1/Waf1 at both the G1/S and the G2/M cell cycle transitions: pRb is a critical determinant in blocking DNA replication and in preventing Endoreduplication. Mol Cell Biol 18(1):629–643. https://doi.org/10.1128/mcb.18.1.629

Niehl A, Pasquier A, Ferriol I, Mély Y, Heinlein M (2014) Comparison of the oilseed rape mosaic virus and tobacco mosaic virus movement proteins (MP) reveals common and dissimilar MP functions for tobamovirus spread. Virology 456:43–54. https://doi.org/10.1016/j.virol.2014.03.007

Niehl A, Wyrsch I, Boller T, Heinlein M (2016) Double-stranded RNAs induce a pattern-triggered immune signaling pathway in plants. New Phytol 211(3):1008–1019. https://doi.org/10.1111/nph.13944

Opalka N, Brugidou C, Bonneau C, Nicole M, Beachy RN, Yeager M, Fauquet C (1998) Movement of rice yellow mottle virus between xylem cells through pit membranes. Proc Natl Acad Sci U S A 95(6):3323–3328. https://doi.org/10.1073/pnas.95.6.3323

Park SW, Kaimoyo E, Kumar D, Mosher S, Klessig DF (2007) Methyl salicylate is a critical mobile signal for plant systemic acquired resistance. Science 318(5847):113–116. https://doi.org/10.1126/science.1147113

Park J, Hwang HS, Buckley KJ, Park JB, Auh CK, Kim DG et al (2010) C4 protein of beet severe curly top virus is a pathomorphogenetic factor in Arabidopsis. Plant Cell Rep 29 (12):1377–1389. https://doi.org/10.1007/s00299-010-0923-8

Peng H, Li S, Wang L, Li Y, Li Y, Zhang C, Hou X (2013) Turnip mosaic virus induces expression of the LRR II subfamily genes and regulates the salicylic acid signaling pathway in non-heading Chinese cabbage. Physiol Mol Plant Pathol 82:64–72. https://doi.org/10.1016/j.pmpp.2013.01.006

Pichersky E, Noel JP, Dudareva N (2006) Biosynthesis of plant volatiles: Nature's diversity and ingenuity. Science 311(5762):808–811. https://doi.org/10.1126/science.1118510

Regenmortel MHV (2000) Introduction to the species concept in virus taxonomy. In: Virus taxonomy. Seventh report of the international committee on taxonomy of viruses. Biological centrifugation, pp 3–16

Rhoads A, Au KF (2015) PacBio sequencing and its applications. Genom Proteom Bioinform 13 (5):278–289. https://doi.org/10.1016/j.gpb.2015.08.002

Riedle-Bauer M (2000) Role of reactive oxygen species and antioxidant enzymes in systemic virus infections of plants. J Phytopathol 148(5):297–302. https://doi.org/10.1046/j.1439-0434.2000.00503.x

Roossinck MJ (2015) A new look at plant viruses and their potential beneficial roles in crops. Mol Plant Pathol 16(4):331. https://doi.org/10.1111/mpp.12241

Ruffel S, Dussault MH, Palloix A, Moury B, Bendahmane A, Robaglia C, Caranta C (2002) A natural recessive resistance gene against potato virus Y in pepper corresponds to the eukaryotic initiation factor 4E (elF4E). Plant J 32(6):1067–1075. https://doi.org/10.1046/j.1365-313X.2002.01499.x

Ruffel S, Gallois JL, Lesage ML, Caranta C (2005) The recessive potyvirus resistance gene pot-1 is the tomato orthologue of the pepper pvr2-eIF4E gene. Mol Genet Genom 274(4):346–353. https://doi.org/10.1007/s00438-005-0003-x

Ruiz-Ferrer V, Voinnet O (2009) Roles of plant small RNAs in biotic stress responses. Annu Rev Plant Biol 60:485–510. https://doi.org/10.1146/annurev.arplant.043008.092111

Shen J, Chen X, Chen J, Sun L (2016) A phloem-limited fijivirus induces the formation of neoplastic phloem tissues that house virus multiplication in the host plant. Sci Rep 6(1):1–1. https://doi.org/10.1038/srep29848

Shulaev V, Silverman P, Raskin I (1997) Airborne signalling by methyl salicylate in plant pathogen resistance. Nature 385(6618):718–721. https://doi.org/10.1038/385718a0

Sidle A, Palaty C, Dirks P, Wiggan O, Kiess M, Montgomery Gill R et al (1996) Activity of the retinoblastoma family proteins, pRB, p107, and p130, during cellular proliferation and differentiation. Crit Rev Biochem Mol Biol 31(3):237–271. https://doi.org/10.3109/10409239609106585

Spoel SH, Dong X (2012) How do plants achieve immunity? Defence without specialized immune cells. Nat Rev Immunol 12(2):89–100. https://doi.org/10.1038/nri3141

Takabatake R, Ando Y, Seo S, Katou S, Tsuda S, Ohashi Y, Mitsuhara I (2007) MAP kinases function downstream of HSP90 and upstream of mitochondria in TMV resistance gene N-mediated hypersensitive cell death. Plant Cell Physiol 48(3):498–510. https://doi.org/10.1093/pcp/pcm021

Takahashi H, Suzuki M, Natsuaki K, Shigyo T, Hino K, Teraoka T et al (2001) Mapping the virus and host genes involved in the resistance response in cucumber mosaic virus-infected Arabidopsis thaliana. Plant Cell Physiol 42(3):340–347. https://doi.org/10.1093/pcp/pce039

Takahashi H, Kanayama Y, Zheng MS, Kusano T, Hase S, Ikegami M, Shah J (2004) Antagonistic interactions between the SA and JA signaling pathways in Arabidopsis modulate expression of defense genes and gene-for-gene resistance to cucumber mosaic virus. Plant Cell Physiol 45 (6):803–809. https://doi.org/10.1093/pcp/pch085

Thaler JS, Humphrey PT, Whiteman NK (2012) Evolution of jasmonate and salicylate signal crosstalk. Trends Plant Sci 17(5):260–270. https://doi.org/10.1016/j.tplants.2012.02.010

Torrance L, Mayo MA (1997) Proposed re-classification of furoviruses. Arch Virol 142(2):435–439

Torres MA, Jones JDG, Dangl JL (2006) Reactive oxygen species signaling in response to pathogens. Plant Physiol 141(2):373–378. https://doi.org/10.1104/pp.106.079467

Tsuda K, Sato M, Glazebrook J, Cohen JD, Katagiri F (2008) Interplay between MAMP-triggered and SA-mediated defense responses. Plant J 53(5):763–775. https://doi.org/10.1111/j.1365-313X.2007.03369.x

Tsujimoto Y, Numaga T, Ohshima K, Yano MA, Ohsawa R, Goto DB et al (2003) Arabidopsis tobamovirus multiplication (TOM) 2 locus encodes a transmembrane protein that interacts with TOM1. EMBO J 22(2):335–343. https://doi.org/10.1093/emboj/cdg034

Uknes S, Mauch-Mani B, Moyer M, Potter S, Williams S, Dincher S et al (1992) Acquired resistance in arabidopsis. Plant Cell 4(6):645–656. https://doi.org/10.2307/3869523

Vaucheret H (2006) Post-transcriptional small RNA pathways in plants: mechanisms and regulations. Genes Dev 20(7):759–771. https://doi.org/10.1101/gad.1410506

Vaucheret H (2008) Plant ARGONAUTES. Trends Plant Sci 13(7):350–358. https://doi.org/10.1016/j.tplants.2008.04.007

Verchot J, Driskel BA, Zhu Y, Hunger RM, Littlefield LJ (2001) Evidence that soilborne wheat mosaic virus moves long distance through the xylem in wheat. Protoplasma 218(1–2):57–66. https://doi.org/10.1007/BF01288361

Verchot-Lubicz J (2003) Soilborne viruses: advances in virus movement, virus induced gene silencing, and engineered resistance. Physiol Mol Plant Pathol 62(2):55–63. https://doi.org/10.1016/S0885-5765(03)00040-7

Verlaan MG, Hutton SF, Ibrahem RM, Kormelink R, Visser RGF, Scott JW et al (2013) The tomato yellow leaf curl virus resistance genes Ty-1 and Ty-3 are allelic and code for DFDGD-class RNA-dependent RNA polymerases. PLoS Genet 9(3):e1003399. https://doi.org/10.1371/journal.pgen.1003399

Vlot AC, Klessig DF, Park SW (2008) Systemic acquired resistance: the elusive signal(s). Curr Opin Plant Biol 11(4):436–442. https://doi.org/10.1016/j.pbi.2008.05.003

Vlot AC, Dempsey DA, Klessig DF (2009) Salicylic acid, a multifaceted hormone to combat disease. Annu Rev Phytopathol 47:177–206. https://doi.org/10.1146/annurev.phyto.050908.135202

White RF (1979) Acetylsalicylic acid (aspirin) induces resistance to tobacco mosaic virus in tobacco. Virology 99(2):410–412. https://doi.org/10.1016/0042-6822(79)90019-9

Whitham S, Dinesh-Kumar SP, Choi D, Hehl R, Corr C, Baker B (1994) The product of the tobacco mosaic virus resistance gene N: similarity to toll and the interleukin-1 receptor. Cell 78(6):1101–1115. https://doi.org/10.1016/0092-8674(94)90283-6

Wieczorek A, Sanfaçon H (1993) Characterization and subcellular localization of tomato ringspot nepovirus putative movement protein. Virology 194(2):734–742. https://doi.org/10.1006/viro.1993.1314

Yamaji Y, Maejima K, Komatsu K, Shiraishi T, Okano Y, Himeno M et al (2012) Lectin-mediated resistance impairs plant virus infection at the cellular level. Plant Cell 24(2):778–793. https://doi.org/10.1105/tpc.111.093658

Yi SY, Shirasu K, Moon JS, Lee SG, Kwon SY (2014) The activated SA and JA signaling pathways have an influence on flg22-triggered oxidative burst and callose deposition. PLoS One 9(2):e88951. https://doi.org/10.1371/journal.pone.0088951

Zhou JM, Chai J (2008) Plant pathogenic bacterial type III effectors subdue host responses. Curr Opin Microbiol 11(2):179–185. https://doi.org/10.1016/j.mib.2008.02.004

Chapter 12
Long Sequencing Tools for Rhizosphere Study

Debasish Pati and Binod Bihari Sahu

Abstract At present, the rhizosphere study is gaining huge scientific attention worldwide. Plant and microbes are interlinked by a thin networking system comprising bacteria, archaea, fungi, picoeukaryotes, and phage, aggregated with in the narrow zone within close proximity of plant roots. The fine tuning of the microbial community depends upon the plant species, texture of soil, and nature of root secretion, in response to the microbiota directly/indirectly regulates plant growth, metabolism, nutrient cycling, and survival under stress condition. The gain in research interest of rhizosphere is majorly due to the recent rapid development of NGS platform in the last decade. Development and evolution of NGS from the time of Sanger's sequencing is now facilitating researchers with low cost, high throughput, longer read length, and lesser technical complexity in sample processing. With the progress of NGS, huge genomic data has been generating that force the parallel innovation in bioinformatics tools for data processing and storing. In this chapter, we focus on evolution of NGS platform and their applications in rhizosphere study. Rhizosphere study have the immense possibilities towards world food security, improving nutrient cycling in agricultural field, engineering microbial community for plant growth and higher productivity as well have applicable in preserving natural plant diversity of local ecosystem.

Keywords Metagenomics · Meta-omics · Next-generation sequencing · Rhizosphere

D. Pati · B. B. Sahu (✉)
Department of Life Science, National Institute of Technology, Rourkela, Odisha, India
e-mail: sahub@nitrkl.ac.in

R. N. Pudake et al. (eds.), *Omics Science for Rhizosphere Biology*, Rhizosphere Biology, https://doi.org/10.1007/978-981-16-0889-6_12

12.1 Introduction to Rhizosphere

The rhizosphere can be known as the confederate platform between a plant roots and the surrounding soil. The term was first introduced in 1904 by Lorenz Hiltner, where he postulated that a plant's nutrition is remarkably guided and modulated by the structure of its rhizosphere microbiota (Philippot et al. 2013) and vice versa. On the other hand, the activity of a rhizosphere's microbiome is governed through the nature, composition, and characteristics of the plant's root exudates (Doornbos et al. 2012). All of the above mechanism of symbiosis/ mutualism is known as plant effect/rhizosphere effect. It can be evaluated in terms of rhizosphere ratio, i.e., (R:S), where "R" stands for sum of microbe number in the rhizosphere in comparison to the corresponding number in the bulk soil "S" (Antoun and Prévost 2006).

The rhizosphere microbiome is the orchestra of microbial community, which is plumed within the several millimeters of radius around the root. Structurally, the ecosystem can be broadly categorized into three major regions (Prashar et al. 2014):

- Endorhizosphere: Consists of the root outer most layers, the endodermis, and cortical layer.
- Rhizoplane: Consists of root exterior where soil particle adheres and microbiome flourish, consists of epidermis, cortex, and mucilaginous polysaccharide layer.
- Ectorhizosphere: Consists of soil in immediate vicinity to root.

The rhizosphere is a cluster of various microbial lifeform like bacteria, archaea, fungi, picoeukaryotes, and phages. The rhizosphere comprises of microbiome community very specific to the plant species and genotype (White III et al. 2017). The microbe community largely are regulated and characterized by the plant root environment, plant species, and surrounding soil texture/quality. Below is a diagrammatic representation of rhizosphere functional anatomy, represented in Fig. 12.1. The term "Rhizodeposition" was coined by Whipps and Lynch in 1985, which denotes the bulk of organic compounds released by living root to its surrounding. It includes low molecular weight compounds like simple carbohydrate, amino acid, vitamins, plant hormones, phenols, organic acids, sugar phosphate esters, and many other carbon containing secondary metabolites, as well contains higher molecular weight compounds like enzymes, protein, and mucilage. In response to these compounds, the microbial diversity gets regulated, for example, *Agrobacterium tumefaciens,* a bacterial pathogen gets allured towards specific phenolic compound (acetosyringones) secreted from injured plant. Oomycetes and other pathogens can take over symbiotic signaling molecules, such as cutin monomers to initiate infection. Likewise, indigenous secondary metabolites like pyrrolizidine alkaloids can influence the rhizosphere microbial community by favoring resistant/tolerant microorganisms or, in contrast microorganism that utilize these compounds.

The rhizosphere microbiota acts in a species-specific manner, diverse among each other as well from each other. The community itself gets regulated by its host in response to the microbiome that modulates host behavior directly or indirectly. It has been reported that the rhizosphere microbiota is responsible for plant defense against

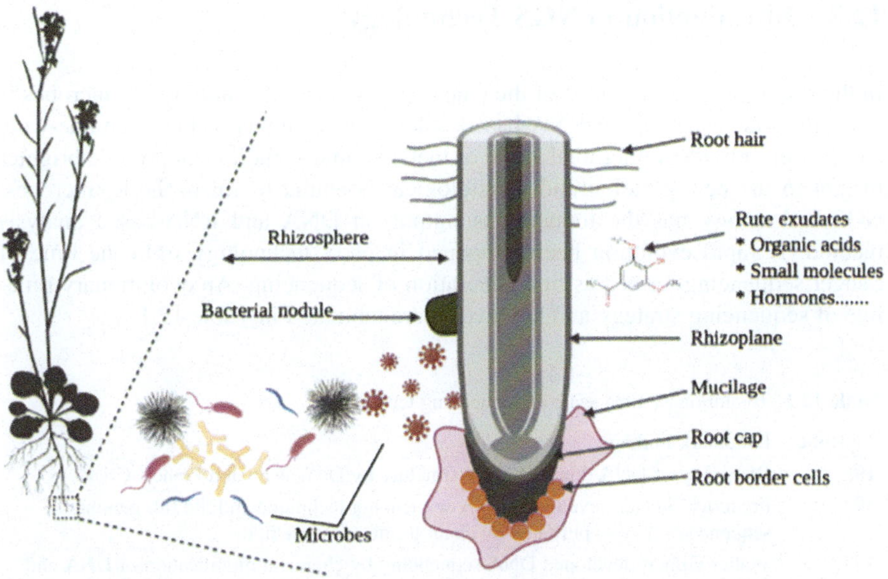

Fig. 12.1 The Rhizosphere ecology: adapted and modified from Philippot et al. (2013). Diagrammatic representation of narrow zone of rhizosphere consists of saprophytic and symbiotic bacteria, archaea, fungi, virus, and picoeukaryotes, influencing and interacting with each other in a microenvironment

soil-mediated pathogen, and the phenomenon is best known as the so-called disease-suppressive soil. The rhizosphere is a versatile entity that has wider applicability in plant metabolism, growth regulation, defense mechanism, and survival, yet many more to find the hidden attributes. Most of the findings emerged from the studies on *Medicago truncatula* and *Arabidopsis thaliana*. However, significant advancement in revealing the non-cultivated plant species rhizosphere microbial ecology is also accelerating at present scenario. To answer some of the key question like:

- How the microbes modulate growth and survival of host plant?
- What are the parameters for microbes diversity and abundances?
- How can the rhizosphere be engineered to enhance food security?

A better understanding of key contender and process that regulates in rhizosphere need to be discovered, for the purpose an array of molecular techniques applied, such as stable isotope- probing and metagenomics. Among all, next-generation sequencing (NGS) technology has magnificently hastened findings in plant–microbe interaction study since last two decade by facilitating the generation of huge sequence data at a significantly reduced price per base.

12.2 Introduction to NGS Technology

In the rhizosphere study, most of the challenges arise as vast number of microbes is nonculturable where DNA/RNA-based studies have higher value. Genome-wide analysis of individual microbial strain or metagenomics studies can provide brighter insight to the configuration and physiological potential of rhizospheric microbes. NGS technology has the tremendous output on DNA and RNA-based analysis method. A rapid evolution been happened in NGS technology from the time of Sanger sequencing termed as first generation of sequencing. An evolutionary time-line of sequencing strategy and key events are tabulated in Table 12.1.

Table 12.1 Evolutionary time scale of sequencing technology

Timeline	Landmark findings
1953	Discovery of DNA double helical structure by D. Watson and Francis Crick
1977	Frederick Sanger develops a DNA sequencing technique and did full genome sequence of a virus phiX174 by chain-termination method
1977	Walter Gilbert developed DNA sequencing by chemical modification of DNA and subsequent cleavage at specific base
1983	The polymerase chain reaction (PCR) is developed by Dr. Kary Mullis
1987	Applied Biosystems incorporated markets the first automated DNA sequencer (AB370)
1990	Human genome project was launched
1995	The first bacterium genome (*Haemophilus influenza*) sequenced
1996	Complete sequencing the genome of yeast (*Saccharomyces cerevisiae*)
1998	Published genome of the nematode worm *C. elegans*
2000	454 Life Science Corporation was founded
2000	*Drosophila melanogaster* full genome sequenced
2001	First draft of the Human Genome sequence released
2002	First mammal-the mouse full genome sequenced
2003	Human Genome project completed
2004	Pacific Bioscience founded
2005	Oxford Nanopore Technologies founded
2006	Illumina launched new Genome analyser
2007	Sequencing by Oligo Ligation Detection (SOLD) launched by Applied Biosystems
2008	1000 Genome project launched/NGS result in dramatic drop in sequencing cost
2009	Pacific Bioscience introduced single molecule Real Time (SMRT) DNA sequencing technique
2011	Major three new sequencing platforms (Ion Torrent's personal genome machine/Paci Biosciences/Illumina Miseq) are released
2014	The average cost for a whole genome sequencing is less than US $10,000
2015	The pocket-sized MinION-Oxford Nanopore Technology
2017	SeqLL release the true single molecule sequencing platform

12.2.1 The First Generation of Sequencing

At the pioneering stage, Sanger introduced sequencing by base termination. They use chain termination method by incorporating dideoxynucleotide to determine the sequence of nucleotide in a DNA strand. Nucleotides get incorporated to the 3′ end of the primer with the targeted DNA as template. The dideoxynucleotide added to respective DNA template while amplifying terminates the primer activity by getting incorporated at the growing chain end. This generates DNA fragments of various lengths. The fragments are discrete by gel-electrophoresis and the bands generated in the autoradiograph of separating gel can be used for direct sequence prediction, starting from the bottom (5′ end). Later on, the above method got automated by the use of unique fluorescent-labeled nucleotides as per four nucleotide bases.

12.2.2 The Next Generation of Sequencing

By the introduction of Solexa in 2005, the next generation of sequencing (NGS) technology became marketable. Then, after a slew of sequencing platforms have been invented, all of which were gradually evolving and being refined at an amazing pace. Gradual development of NGS technologies leads to a drastic reduction in per base sequencing cost, as represented in Fig. 12.2a we can observe by the evolution of sequencing platform, the landmark "US $ 1000" genome goal was achieved in 2015 for human genome sequencing as compared to the first ever human genome project did cost about ~$300 million worldwide. The NGS platforms are calibrated on the basis of two parameters, i.e., the total throughput and the read length. a comparative graphical representation for both the parameter of widely used NGS platforms are illustrated below in Fig. 12.2b, c. These techniques can be grouped into three major categories depending upon the standard chemistry used for sequencing as—"sequencing by synthesis, sequencing by ligation and single-molecule sequencing" (Egan et al. 2012). Each platform has its own advantages and specificity as compared to other approaches represented in Table 12.2.

12.2.2.1 Sequencing by Synthesis

NGS technique uses the basic principle of Sanger sequencing; base incorporation is determined by recording the chemiluminescence from respective nucleotides added during DNA polymerase-mediated complementary DNA strand synthesis in the above approach. DNA is cleaved to proper length, attached with adaptor sequence, and clonally polymerized to amplify the signal generated by chemical reaction or fluorescent. Templates are then separated and immobilized in preparation for flow cell cycle. Roche-455/Illumina and ion torrent use the above principle, among all in general the detection procedure for chemical or fluorescence shift during nucleotide

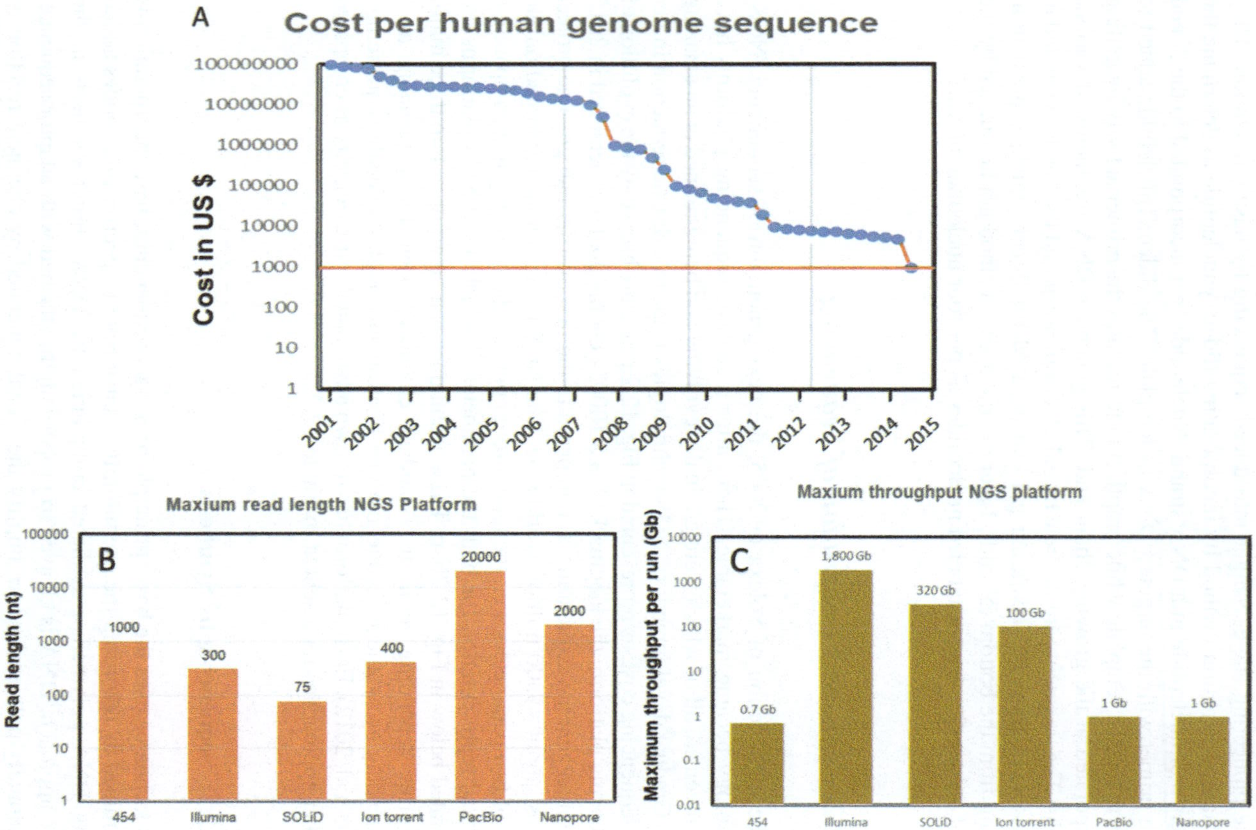

Fig. 12.2 Evolution of NGS platforms in terms of cost, reads length, and throughput. (**a**) represents the feasibility of NGS cost of human genome sequencing from 2001 until 2015, when the landmark "US $ 1000" genome goal achieved. A sharp decrease in cost observed as a result of innovation and upgradation of NGS platforms. (**b**) Total maximum read length achieved by commercially available sequencing platform like 454, Illumina/Solexa, SOLiD, Ion Torrent, Pacific

Bioscience (PacBio), and Nanopore. **(c)** Maximum throughput achieved by commercially available platforms, adapted and modified from van Dijk et al. (2014). *(Note: Maximum throughput and read length are not necessarily obtained with one and same instrument)*

Table 12.2 Comparative representation of widely used next-generation sequencing platforms according to their performance and advantages. Adapted and modified from Egan et al. (2012)

NGS platforms/ company	No. reads per run	Time (h or days)	Raw error rate (%)	Cost per 10^6 bases	Platform cost (USD approx.)	Chemistry	Advantages	Drawbacks
First generation								
Sanger/Life Technologies	1	2 h	0.3	2400	95,000	Dideoxy termination	Higher quality, longer read length	High cost, low throughput, practically limited applicable
Second generation								
454 GS FLX +/Roche	1×10^6	24/48 h	1	10	5,00,000	Pyrosequencing	Larger read length suitable for de novo genome assemblies	Relatively low throughput, high error in homopolymer repeats, high reagent cost
HiSeq/ Illumina	5×10^9	27/ 240 h	0.8	0.1	7,50,000	Reversible termination	High speed and low cost, higher accuracy	Relatively short reads, long run time
MiSeq/Illumia	3×10^8	27 h	0.8	0.13	1,25,000	Reversible termination		
SoLID/Life Technologies	1×10^9	14 days	0.01	0.13	3,50,000	Ligation	Higher resolution, larger dynamic range	Short reads, requires sequence complexity, challenging sample loading technique
Ion Proton/ Life Technologies	6×10^7	2–5 h	1.7	1	2,15,000	Proton detection	High speed and low cost, rapid run time, free from fluorescence detector and camera scanning	High error rate in homopolymers
Third generation								
SMRT/Pac Bio	1×10^6	1–2 h	12.9	2	7,50,000	Real-time SMS	Extremely longer reads, eliminate PCR-based bias	Lowest throughput of all plat-form, >15% error rate, much more expensive, limited range of application
Nanopore/ Oxford Nanopore Technologies	6×10^4	48/72 h	34	<1	1000	Real-time SMS		

addition is carried out by the sequential washing of nucleotide and is common in all procedures (Egan et al. 2012).

12.2.2.2 Roche 454 Pyrosequencing

In sequencing by Roche 454 a single, primed DNA template is attached on a microbead and polymerized by emulsion PCR. Each bead represents a homogeneous genomic unit which is then placed in a well of PicoTiter Plate, then subjected to a flow well for incubation with the mixture of apyrase, luciferase, DNA polymerase, and ATP sulfurylase, along with the adenosine 5′-phosphosulfate (ASP) and luciferin as substrates. It applies the basic principle of DNA synthesis, where DNA polymerase incorporates correct nucleotide into the strand, and a single phosphate molecule gets released. In the presence of ASP, the phosphate molecule utilizes for ATP (adenosine triphosphate) synthesis. ATP catalyzes the conversion of luciferase to oxyluciferin, in the process fluorescent light is released in equal proportion to ATP generated during the reaction (Nyrén 2007). The process is then followed by the removal of unused ATP and nucleotide by apyrase. A new set of reaction mixture is flooded over the template. The same procedure is cycled, until the whole template is elongated. The light and its intensity is recorded by camera for further processing, i.e., for DNA sequence evaluation

12.2.2.3 Illumina

It was preliminarily introduced by Solexa where the Illumina platform uses solid-phase bridge PCR principle, where the solid substrate contains complementary known sequence in response to the adaptor ligated to the DNA library. Both 5′ and 3′ adaptors are ligated to substrate that forms a bridge supplemented with forward and reverse primers. The DNA remains attached to the base and amplified with the primers and form a cluster of amplicons, the cluster helps in intensifying the chemiluminescence. Numerous identical clusters are generated within each channel of the flow cell. The amplicons are denatured, primed as well as elongated sequentially. The base incorporation takes place cyclically multiple times with repeated flooding of mixture of all the nucleotide; each nucleotide is labeled with a different fluorophore. Each fluorophore is modified with reversible terminator (3′-O-azidomethyl) (Bentley et al. 2008). The emitted fluorescence captured for imaging; sequencing is restored by addition of free 3′OH group with the help of tris (2-carboxyethyl) phosphine. The overall process is repeated cyclically until the desired length of DNA is synthesized.

12.2.2.4 Ion Torrent

The ion torrent system uses the measurement of pH shift due to liberation of hydrogen ion in DNA polymerization is a unique approach irrespective of other NGS technologies (Rothberg et al. 2011). The microwell containing DNA templates is washed with deoxyribonucleotide triphosphate (dNTP). By the incorporation of correct dNTP, it releases one H^+ that signals an ISFET (ion-sensitive field-effect transistor) ion detector. In the presence of homopolymer repeats in the template, multiple dNTP gets incorporated in response multiple hydrogen ion will produce leading to higher intensity.

12.2.2.5 Sequencing by Ligation

DNA polymerase was used in sequencing by synthesis process as the elongation factor, in contrast the mismatch sensitivity feature of DNA ligase is used in sequencing by ligation approach to determine the sequence (Landegren et al. 1988). Depending on the nucleotide(s) to be determined, random length of fluorescently tagged oligonucleotide probes are used. The cleaved DNA templates are ligated with a short, known anchor sequence, then DNA ligase is added to the flow cell, which hybridize the fluorescently labeled probe to the primer and template. To determine the incorporated probe, fluorescence imaging is performed. To assess the sequence of nucleotides, the above procedure is cycled with different sets of probes.

12.2.2.6 SOLiD

From the sequencing industry leader, Life Technologies/Applied Biosystems comes the support oligonucleotide ligation detection (SOLiD) system, that is a revolutionary new platform with built-in scalability delivering ultra-high-throughput with greater accuracy. At first, the desired DNA is sheared to a particular length and resulting fragments are ligated with adaptors. The microbeads subjected to clonal amplification containing single adaptor ligated molecule through emulsion PCR reaction. Microbeads attached to glass slide; the templated beads are combined with a universal sequencing primer ligase and a large pool of dye-based probes, which are fluorescently labeled with four dye, each dye represents four of 16 possible dye nucleotide sequences. The first two positions are specific to the fluorophore followed by the hybridized primer. Bases location of three to five in the primer are degenerate bases separated from bases of six to eight place, made up of universal inosine bases, by a phosphorothiolate linkage (Mckernan et al. 2012). The complementary probe gets hybridized to the template sequence and then the signature fluorescence is recorded after which the dye is cleaved off leaving 5′ free end available for further addition. This process is repeated for several cycles until the desired length is achieved. After that, the whole amplified strand is replaced with a

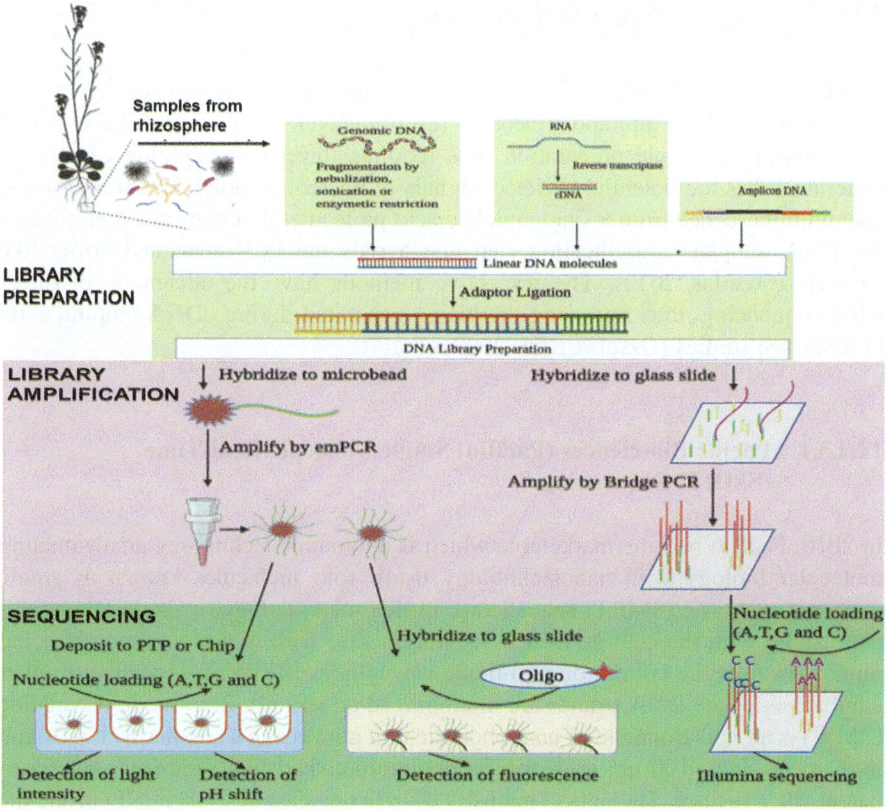

Fig. 12.3 Overview of the library preparation and sequencing strategy of the most commonly used NGS platforms: adapted and modified from Knief (2014)

new primer that is incorporated at n-1 base position and the successive recurrent cycle of ligation takes place. To enhance the accuracy, the primer's reset process is cycled five times more, providing a dual measurement of each base separated by several rounds (Ashelford et al. 2011).

A generalized work flow for the so-called second generation of sequencing has been drawn below in Fig. 12.3, representing the outlook of library preparations, amplification, and sequencing procedure for above-mentioned platforms.

All of the genomic molecules are collected from rhizosphere's microbes, cleaved, and converted into double-stranded DNA, conjugated with adaptors for library preparation. Adaptors are specific for sequencing platform and enable the binding of the library molecules to the surface of beads or a flow cell depending on the strategy, where they are amplified prior to sequencing by emulsion PCR/ bridge PCR, respectively. Sequencing is either by synthesis process like in illumine, 454, and ion torrent sequencing or by sequencing by ligation process in SOLiD.

12.2.3 Single-Molecule Sequencing

Third-generation sequencing, often termed as single-molecule sequencing (SMS), can tackle few of the limitation, faced by rest of the NGS technologies. This method uses simplified sample preparation thus requires lower concentration of starting material. It has the potential to detect signals of nucleotide polymerization through chemiluminescence from a single nucleic acid molecule, by the means it eliminates the DNA template amplification step and avoids the PCR-generated errors like GC-bias (Ozsolak 2012). The respective methods have the advantage of direct RNA sequencing, thus removing the biases generated during cDNA amplification in RNA-seq studies (Ozsolak et al. 2009).

12.2.3.1 Pacific Biosciences (PacBio) Single Molecule Real-Time (SMRT) Sequencing

In 2010, PacBio became marketable which is a dynamic technology amalgamating molecular biology with nanotechnology using solo molecules known as single molecule real-time (SMRT) sequencing. In this methodology, a single DNA template along with the φ29 derived DNA polymerase with higher accuracy is being immobilized, at the bottom of zero-mode waveguides (ZMW) irrespective of other approaches where DNA strands are immobilized on surface. Zero-mode waveguide (ZMW) is an imprisonment of nanophotonic unit possessing a circular well (~70 nm in diameter and ~100 nm in depth) in aluminum-coated film sediment on a transparent silica surface. The optical intensity of light decreases exponentially inside the chamber when it moves through the narrow aperture. The observation volume within an illuminated ZMW is ~20 zeptoliters ($20*10^{-21}$ L). With this characteristic, single nucleotide incorporation by DNA polymerase can readily be detected. The DNA polymerase adds the labeled nucleotide to the immobilized template, induces signals, which excite the instrument optics, continuously monitoring the enzymes active site in ZMW. The overall strategy adapted by PacBio platform is represented schematically in Fig. 12.4. The reads length of sequence in PacBio can range up to ~40Kb, but with >15% error rate, which is the limiting factor as compared to second generation of sequencing (Mardis 2013).

12.2.3.2 Nanopore DNA Sequencing

Nanopore technology is the most recent single molecule third-generation sequencer developed by Oxford Nanopore Technologies, UK. This technology uses the characteristic of biological nanopores lodged in the polymer membrane. Biological nanopores are either set up from engineered proteins like α-hemolysin MspA (*Mycobacterium smegmatis* porin A protein) or derived from *Staphylooccus aureus*, or are entirely synthetic, e.g., graphene (Knief 2014). An ionic solution is maintained

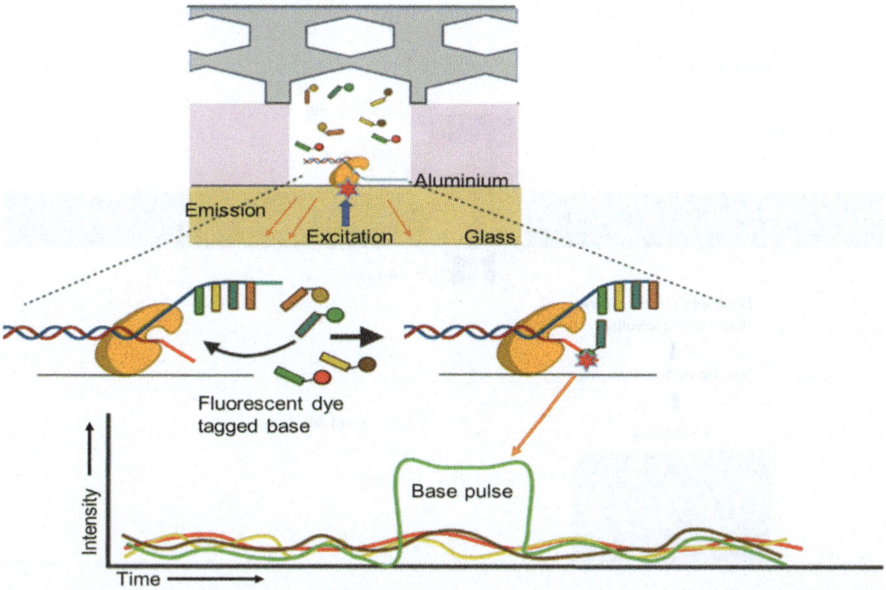

Fig. 12.4 Sequencing by PacBio through light pulse detection: adapted and modified from Mardis (2013. The hairpin adaptor ligated double-stranded DNA molecule (SMARTbell) diffuse into ZMW; the adaptor binds to the polymerase immobilized at the bottom, which carries out template polymerization. To differentiate all four nucleotides according to their emission spectrum, each labeled with different fluorescent dye. A fluorescently labeled nucleotide is incorporated with the help of DNA polymerase. The fluorescent output of the color corresponding to the incorporated base is elevated, which can be represented in intensity vs. time plot

across the nanopore. The nanopore consists of 1 nm diameter pore coupled with adaptors. The adaptor can detect a simple alteration in electrical density, each nucleotide possesses its characteristic property, thus each base that can be detected according to its signature electrical hindrance pattern (Hart et al. 2010) is represented in Fig. 12.5. The table-free, portable sequencing device MinION was introduced in the year 2004 and has been commercially in May 2015. The performance of MinION nanopore sequencer has been calibrated by using M13 genomic DNA and shows 99% sequence similarity when compared to reference genome (Hayden 2012). The sequencer is like a USB chip and can be used only once. One can avoid sample collection from site and carry them to lab which may cause component degradation or change in status of sample. Instead of that, the sequencers can readily be taken to field and sequencing will be done on spot directly, which will further reduce the cost and efforts in sequencing significantly and help in maintaining sample integrity (Jain et al. 2015).

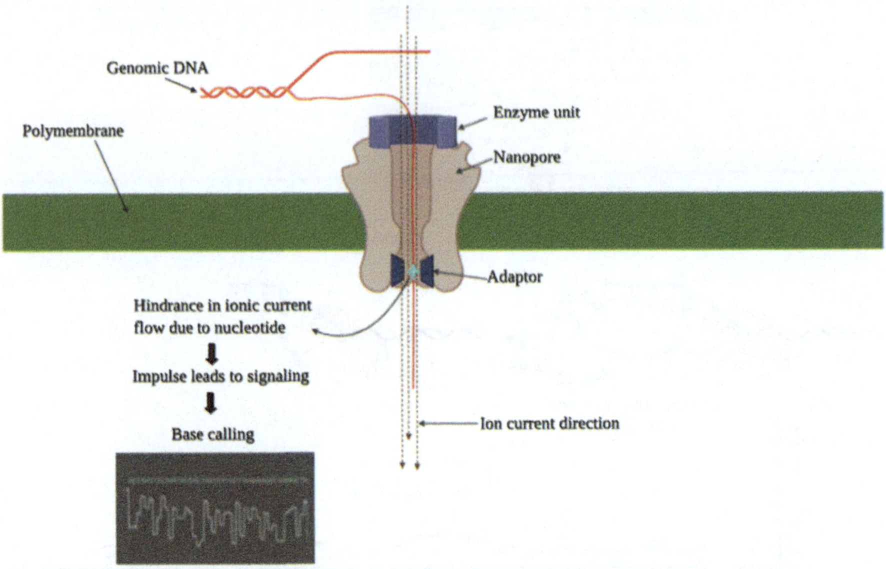

Fig. 12.5 Diagrammatic representation of a nanopore sequencing during base calling: adapted and modified from Ambardar et al. (2016)

12.3 Application of NGS in Rhizosphere Study

NGS is one of the holistic approaches that can revolutionize the rhizosphere study, as about ~98% of microbe cannot be cultured in vitro that limits its detailed study. In particular, sequence-based study is an indispensable approach as well as challenging due to the uneven composition of microbial communities that creates convolution. Due to new platform invention and upgradation, NGS is now paving path for high throughput, cost- and time-effective analysis like microbial diversity and abundance evaluation, their direct-indirect tunning with particular host species, microbes impact on plant growth during stress condition and many more. The whole applicability of NGS technology towards rhizosphere study can be summed in single term, i.e., "meta-omic," it includes metagenomics, metatranscriptomics, and metabolomics study in a single sphere, the various aspects of meta-omic been diagrammatically represented in Fig. 12.6. Meta-omics provides a major tool for studying evolution within microbial communities, which can occur due to random mutation during convergent evolution (Philippot et al. 2013).

High-throughput sequencing-mediated microbiome study can be achieved by two main approaches (Roumpeka et al. 2017)

- By marker gene studies: PCR amplification of a particular gene is done by designed inclusive primer, (e.g., *rbcl* and *matk* for plants, 16s rRNA for prokaryotes or ribosomal ITS for fungi and *cox1* for animals), from whole genomes present in a sample then, the resultant product is sequenced. The sequences are

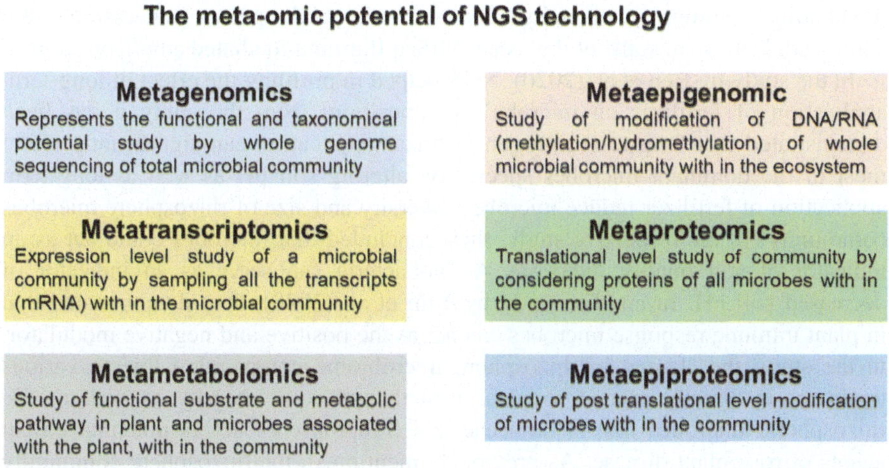

Fig. 12.6 Image representing and dictating the application spectrum of meta-omic studies. Modified from White et al. (2017)

clustered into operational taxonomic units (OTUs) and these are compared across the rhizosphere samples.

- By shotgun metagenomics: whole microbial DNA is extracted from rhizosphere and subject to collective functional analysis.

In particular, NGS has been used to deduce rhizosphere microbial community anatomy, diversity and metabolic potential by metagenomics microbial community in response to chemical treatment and plant–microbe interactions. Many other researches have furnished with mine of information applicable to study the structure, abundance, distribution pattern, and key members of rhizosphere. Studies of the rhizosphere using NGS technologies have mainly focused on model plants such as *Arabidopsis thaliana* (Bulgarelli et al. 2012), legumes (e.g., soybeans and peas) and cereals (e.g., corn, oats, and wheat) (Newman et al. 2016; Turner et al. 2013).

Amplicon sequence study of microbe's genes through NGS platform is a dynamic approach towards characterization of the rhizospheric community. At preliminary stage, majority of the studies used the Roche 454 sequencing technology as it was the first introduced sequencing platform. Only few used the Ion PGM platform, but recently the Illumina MiSeq have been used widely, as Illumina provides conjoining insert libraries with paired end reads, which denotes its accuracy (Knief 2014). For example, in study (Joshi and Chitanand 2020) it has been observed that microbes community helps directly in plant growth mechanism, they observed that a strain of *Pseudomonas aeruginosa* possesses traits for plant growth as well as has antifungal activity studied through Illumina sequencing.

Majority of amplicon sequencing of rhizosphere research was focused to evaluate and identify the plant-associated microbial communities; later on, the studies that were made on the climatic adaptation, topological factors, role of host taxon in modulating community diversity, and the microbial species succession were addressed. Below are some of the recent MiSeq Illumina-mediated amplicon studies:

In the study by Ren et al. (2020), NGS helped in profiling the effect of long-term application of fertilizer on microbe's composition; from the study it has been demonstrated that long-term nitrogen fertilizer application can significantly affect most of the dominant microbes species by altering soil pH as well as long-term application of fertilizer reduce microbes diversity and size of rhizosphere microbes community. In the respective study, they concluded that microbes could act as an indicator of soil quality index, as Actinobacteria can serve as an indicator of decreased soil pH. In another study by Yim et al. (2020), they demonstrated that in plant immune response microbes can act as the positive and negative modulator, in the study the change in rhizosphere microbiome composition due to various treatment is directly related to plant growth and root integrity as well as the rhizospheric microbes like Nectriaceae and Streptomyces act as potential casual agents of rose plant disease. As previously mentioned, the rhizosphere community depends upon the species genotype; in the study by Chen et al. (2020), plant itself regulates its rhizosphere as an umbrella species, and the Miscanthus cultivation significantly alters the bacterial and fungal community composition and reduces bacterial fungal diversity. They also observed concentration of organic matter and nitrogen that are the key regulators for microbial diversity and composition in the rhizosphere community.

Metagenomics-based taxonomic analysis has advantages over amplicon studies as it provides results in less bias manner, also facilitates direct analysis of metabolic dynamic of a rhizosphere ecosystem. De novo assembly of metagenomics data has been featured like added length assembly during alignment to reference databases, which provides error rectification, linking genes responsible for protein coding and more robust functional and taxonomical assignment. By the above approach, one can reconstruct a complete genome of novel, unculturable phyla (Knief 2014). To strengthen the metabolomic analysis, technology like metatranscriptomic have generated enormous possibilities, as one can deduce any of metabolic gene expression at the time of sampling. Both 454 and illumine technology have been revolutionaries in the research area. NGS has potential advantages over its nearby competitive technologies as it does not rely upon prior knowledge of organism's sequence, as well in some studies the transcriptome of the plant and the microbe were analyzed in parallel (Zhuang et al. 2012). The first metatranscriptomic studies by Chaparro et al. (2014) in *Arabidopsis thaliana* they analyzed microbial metatranscriptome and concluded that genes participating in microbial metabolism changes their expression level in relation to root exudate composition at various time point of growth, as well the root secretion itself alter accordingly. Turner et al. (2013) in there study they performed rRNA sequencing instead of mRNA of functional microbiota in the rhizosphere of some crop species and found profound difference in the composition of whole microbial community in bulk soil comparison as well as between the different plant species.

Overall, the second-generation sequencing technologies have superiority over first-generation sequencing; some of their drawbacks like short-read length make them less applicable for some studies, including determination and assembly of critical genome region, gene isoform identification, and epigenetic modification detection. Single molecule real-time sequencing by PacBio provides an unparallel opportunity to carry over major challenges faced by the second-generation sequencing, as it enables shorter run time, wide information on kinetic variation and longer read length. However, the technology requires improvement, such as the high error rate of raw single pass data which can be tackled by hybrid sequencing technology in near future (Rhoads and Au 2015). The other de novo sequence assembler is the nanopore sequencing platform which has maximum advantages over other technologies; it can generate read length >10Kb appropriate for genomic assembly study, as well as can help in-depth study of telomeric region sequences. By the direct RNA sequencing application, no prior sequence knowledge is required, also major metatranscriptomic analysis can be achieved. The epigenetic modification like methylation and hydroxy methylation can be identified through the SMRT approach thus, key regulatory functional genes of rhizosphere can be assessed, by the development of table-free portable MinION platform one can analyze the sequence on the sport avoiding sample alteration or degradation while transporting to laboratory for sequencing. The major drawback of high error percentage can be tackled, and it is not going to be the limiting factor in recent future.

12.4 Computational Tools for Metagenomics Study

The metagenomics sequencing analysis produces massive amount of data, which need to be processed, analyzed, and stored to get significant output. Irrespective of the NGS platform, the overall goal of each analysis is basically the same; however, each platform has its own identities and specificities. Many metagenomics computational/ statistical analysis procedure and databases have been evolved due to an increase in computational capabilities as well as algorithms and applications to assist all the required steps. It will be beyond the scope of this study to discuss the whole analytical procedure, bioinformatics pipeline, and software available for metagenomics analysis. Thus, below is a generalized representation of workflow and some of the tools required for metagenomics analysis being diagrammatically represented in Fig. 12.7.

12.5 Future Aspects

Study of the rhizosphere ecology of both cultivated and non-cultivated plant species has created awareness about the conservation of biodiversity and the ecological succession depend on plant–microbes interaction linked by a narrow, complex, and unrevealed networking system. A critical thinking is required diverging from the boundaries of mutualism—competition prospective, as well major studies are

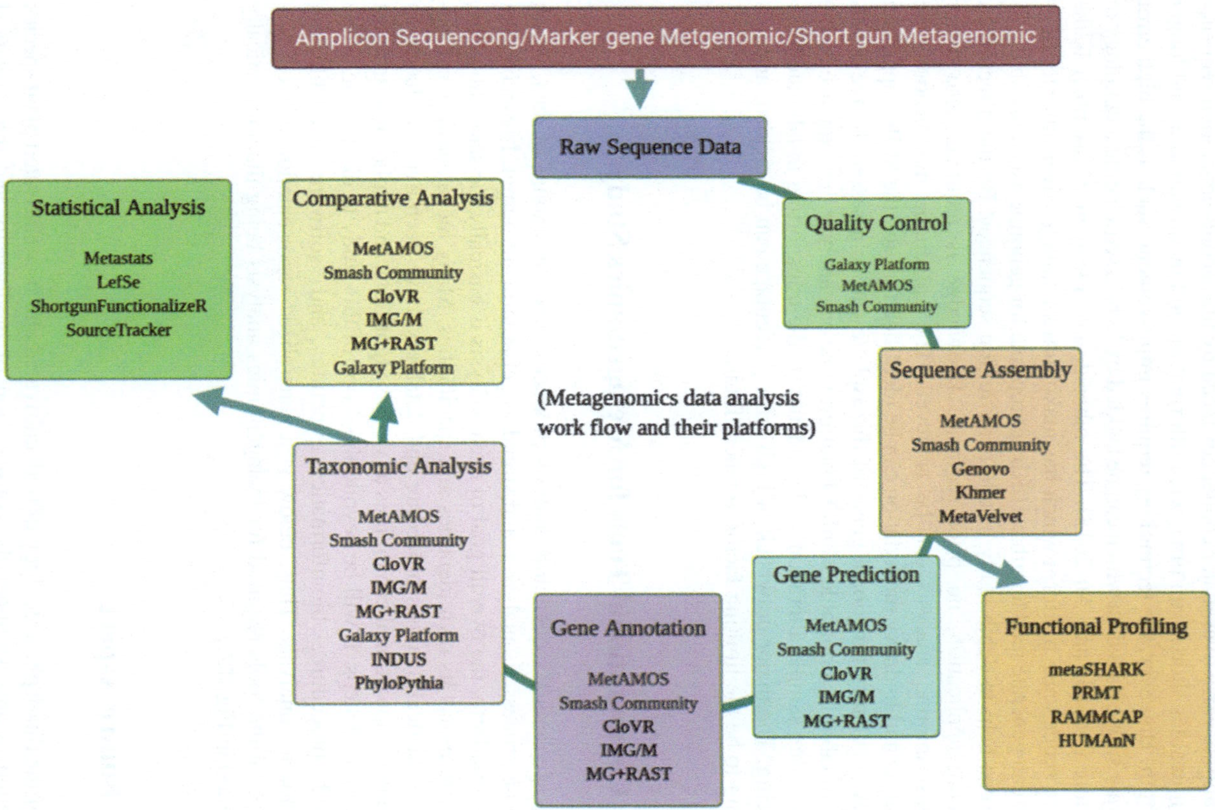

Fig. 12.7 Diagrammatic representation of generalize workflow and bioinformatics tools for metagenomics data analysis (adapted and modified from Kumar and Dubey 2020; Roumpeka et al. (2017))

required to reveal the effect of climatic and edaphic factor on rhizosphere community. As climate and geography of plant directly influence the composition of root exudates, which in turn influence the microbiota. In terms of crop improvement, restoration of natural species and to remove exotic species' root microbial composition have pivotal role. The sustainability of any species of agricultural practice depends on reduced input of mineral nutrients and pesticides. Thus, we have to focus on genotypes, which can take advantage of biotic and abiotic soil resources at a time. To ease the above critical problem and eliminate major limitations faced by rhizosphere study, NGS technology paves a wider path in particular direction by metagenomics studies. The high ratio of organisms' unannotated genes is one of the major limitations in metagenomics sequencing for which no homology is found in public database. Interlinking of genes and genome sequencing of representative pure culture of strain will remain an important task to overcome major challenges in sequencing.

Using specialized sequence data analysis method can compensate limitation of NGS such as short-read length and higher sequencing error rate. Sequence information is not going to be further limiting factors as sequencing of uncultivated microorganisms is now possible due to the innovative technologies like single cell genomic sequencing. The complementation of metagenomics with meta-omics data will be one of the major aims to obtain more detailed view of rhizosphere activity. By the hand-on-hand approach of NGS, rhizosphere study has wider application towards sustainable food crop improvement, natural species conservation, bioenergy crop improvement, and engineered microbial community development that will possibly accelerate in the near future.

References

Ambardar S, Gupta R, Trakroo D, Lal R, Vakhlu J (2016) High throughput sequencing: an overview of sequencing chemistry. Indian J Microbiol 56:394–404. https://doi.org/10.1007/s12088-016-0606-4

Antoun H, Prévost D (2006) Ecology of plant growth promoting rhizobacteria, *In* PGPR: Biocontrol and biofertilization, pp. 1-38. Springer, Dordrecht. https://doi.org/10.1007/1-4020-4152-7_1

Ashelford K et al (2011) Full genome re-sequencing reveals a novel circadian clock mutation in Arabidopsis. Genome Biol 12:R28. https://doi.org/10.1186/gb-2011-12-3-r28

Bentley DR et al (2008) Accurate whole human genome sequencing using reversible terminator chemistry. Nature 456:53–59. https://doi.org/10.1038/nature07517

Bulgarelli D et al (2012) Revealing structure and assembly cues for Arabidopsis root-inhabiting bacterial microbiota. Nature 488:91–95. https://doi.org/10.1038/nature11336

Chaparro JM, Badri DV, Vivanco JM (2014) Rhizosphere microbiome assemblage is affected by plant development. ISME J 8:790–803. https://doi.org/10.1038/ismej.2013.196

Chen Y et al (2020) Miscanthus cultivation shapes rhizosphere microbial community structure and function as assessed by Illumina MiSeq sequencing combined with PICRUSt and FUNGUIld analyses. Arch Microbiol 202:1157–1171. https://doi.org/10.1007/s00203-020-01830-1

Doornbos RF, van Loon LC, Bakker PAHM (2012) Impact of root exudates and plant defense signaling on bacterial communities in the rhizosphere. A review. Agron Sustain Dev 32:227–243. https://doi.org/10.1007/s13593-011-0028-y

Egan AN, Schlueter J, Spooner DM (2012) Applications of next-generation sequencing in plant biology. Am J Bot 99:175–185. https://doi.org/10.3732/ajb.1200020

Hart C, Lipson D, Ozsolak F, Raz T, Steinmann K, Thompson J, Milos PM (2010) Chapter 19—Single-molecule sequencing: sequence methods to enable accurate quantitation. In: Walter NG (ed) Methods in enzymology, vol 472. Academic, Cambridge, pp 407–430. https://doi.org/10.1016/S0076-6879(10)72002-4

Hayden E (2012) Nanopore genome sequencer makes its debut. Nature 482. https://doi.org/10.1038/nature.2012.10051

Jain M, Fiddes IT, Miga KH, Olsen HE, Paten B, Akeson M (2015) Improved data analysis for the MinION nanopore sequencer. Nat Methods 12:351–356. https://doi.org/10.1038/nmeth.3290

Joshi A, Chitanand M (2020) Complete genome sequence of plant growth promoting Pseudomonas aeruginosa AJ D 2 an isolate from monocropic cotton rhizosphere. Genomics 112:1318. https://doi.org/10.1016/j.ygeno.2019.07.022

Knief C (2014) Analysis of plant microbe interactions in the era of next generation sequencing technologies. Front Plant Sci 5:216. https://doi.org/10.3389/fpls.2014.00216

Kumar A, Dubey A (2020) Rhizosphere microbiome: engineering bacterial competitiveness for enhancing crop production. J Adv Res 24:337–352. https://doi.org/10.1016/j.jare.2020.04.014

Landegren U, Kaiser R, Sanders J, Hood L (1988) A ligase-mediated gene detection technique. Science 241:1077–1080. https://doi.org/10.1126/science.3413476

Mardis ER (2013) Next-generation sequencing platforms. Annu Rev Anal Chem 6:287–303. https://doi.org/10.1146/annurev-anchem-062012-092628

Mckernan KM, Blanchard A, Kotler L, Costa G (2012) Reagents, methods, and libraries for bead-based sequencing. United States Patent

Newman MM et al (2016) Changes in rhizosphere bacterial gene expression following glyphosate treatment. Sci Total Environ 553:32–41. https://doi.org/10.1016/j.scitotenv.2016.02.078

Nyrén P (2007) The history of pyrosequencing. Methods Mol Biol 373:1–14. https://doi.org/10.1385/1-59745-377-3:1

Ozsolak F (2012) Third-generation sequencing techniques and applications to drug discovery. Expert Opin Drug Discov 7:231–243. https://doi.org/10.1517/17460441.2012.660145

Ozsolak F et al (2009) Direct RNA sequencing. Nature 461:814–818. https://doi.org/10.1038/nature08390

Philippot L, Raaijmakers JM, Lemanceau P, van der Putten WH (2013) Going back to the roots: the microbial ecology of the rhizosphere. Nat Rev Microbiol 11:789–799. https://doi.org/10.1038/nrmicro3109

Prashar P, Kapoor N, Sachdeva S (2014) Rhizosphere: its structure, bacterial diversity and significance. Rev Environ Sci Biotechnol 13:63–77. https://doi.org/10.1007/s11157-013-9317-z

Ren N, Wang Y, Ye Y, Zhao Y, Huang Y, Fu W, Chu X (2020) Effects of continuous nitrogen fertilizer application on the diversity and composition of rhizosphere soil bacteria. Front Microbiol 11:1948. https://doi.org/10.3389/fmicb.2020.01948

Rhoads A, Au KF (2015) PacBio sequencing and its applications genomics. Proteom Bioinform 13:278–289. https://doi.org/10.1016/j.gpb.2015.08.002

Rothberg JM et al (2011) An integrated semiconductor device enabling non-optical genome sequencing. Nature 475:348–352. https://doi.org/10.1038/nature10242

Roumpeka DD, Wallace RJ, Escalettes F, Fotheringham I, Watson M (2017) A review of bioinformatics tools for bio-prospecting from metagenomic sequence data. Front Genet 8:23. https://doi.org/10.3389/fgene.2017.00023

Turner TR et al (2013) Comparative metatranscriptomics reveals kingdom level changes in the rhizosphere microbiome of plants. ISME J 7:2248–2258. https://doi.org/10.1038/ismej.2013.119

van Dijk EL, Jaszczyszyn Y, Thermes C (2014) Library preparation methods for next-generation sequencing: tone down the bias. Exp Cell Res 322:12–20. https://doi.org/10.1016/j.yexcr.2014.01.008

White RA III, Rivas-Ubach A, Borkum MI, Köberl M, Bilbao A, Colby SM, Hoyt DW, Bingol K, Kim YM, Wendler JP, Hixson KK (2017) The state of rhizospheric science in the era of multi-omics: a practical guide to omics technologies. Rhizosphere 3:212–221. https://doi.org/10.1016/j.rhisph.2017.05.003

White RA, Borkum MI, Rivas-Ubach A, Bilbao A, Wendler JP, Colby SM, Köberl M, Jansson C (2017) From data to knowledge: the future of multi-omics data analysis for the rhizosphere. Rhizosphere 3:222–229. https://doi.org/10.1016/j.rhisph.2017.05.001

Yim B et al (2020) Rhizosphere microbial communities associated to rose replant disease: links to plant growth and root metabolites. Horticult Res 7:144. https://doi.org/10.1038/s41438-020-00365-2

Zhuang X, McPhee KE, Coram TE, Peever TL, Chilvers MI (2012) Rapid transcriptome characterization and parsing of sequences in a non-model host-pathogen interaction; pea-Sclerotinia sclerotiorum. BMC Genomics 13:668. https://doi.org/10.1186/1471-2164-13-668

Chapter 13
Rhizoengineering: A Strategy to Enhance Soil and Crop Productivity

Kulandaivelu Velmourougane and D. Blaise

Abstract The persistence, survival, and availability of the applied agri-inputs (water, fertilizers, and other soil/plant amendments) including microbial inoculants in the rhizosphere of crop plants have become a major problem in agriculture. Rhizosphere engineering is an innovative approach through which the soil biophysical properties are modified to influence plant–microbiome–soil interactions to enhance soil and crop productivity through higher input use efficiency. The basic components of rhizosphere engineering include soil, plant, and microbes, which could be modified to optimize water and nutrient transport as well as microbial activity at the root–soil interface. Though genetic modification of crop plants and microbial engineering has taken back seat because of consumer awareness on human and environmental health, rhizosphere modification through agronomic approaches is the only hope at present to improve soil and crop productivity in an eco-friendly and sustainable manner. Natural way of modification of crop rhizosphere is expected to make soil healthy by avoiding indiscriminate use of plant protection chemicals and fertilizers. Hence, rhizoengineering approach should be advocated to farming community through eco-friendly farm amendments, instead of engineering crops or microorganisms.

Keywords Rhizoengineering · Microbiome engineering · Root exudates · Microbial biofilms · Plant and soil health

13.1 Introduction

The key challenges in agriculture are related to food security, land degradation, and crop yield. Since the microbes are critical drivers of soil functions and agricultural crop productivity, their significance is well recognized. A foremost challenge to

K. Velmourougane (✉) · D. Blaise
Division of Crop Production, ICAR—Central Institute for Cotton Research, Nagpur, Maharashtra, India

R. N. Pudake et al. (eds.), *Omics Science for Rhizosphere Biology*, Rhizosphere Biology, https://doi.org/10.1007/978-981-16-0889-6_13

agriculture at present climate change scenario is to make the maximum and efficient use of natural and external agri-inputs (water, fertilizers, and other amendment) resources by the crop plants, to enhance soil and crop productivity to feed the ever growing population (Lynch and Brown 2012). However, inefficient use of these resources makes the crops plants suffer, resulting in low agricultural production. Shortage of water (abiotic factor—drought) and infestation from pests and diseases (biotic factors) makes the crops vulnerable to express its potential productivity. Hence, worldwide attention has been diverted towards developing crop plants tolerant to biotic and abiotic factors to meet the targeted yield and quality. Thus, increasing biotic/abiotic tolerance in crop plants to improve water and nutrient use efficiency has become a prime objective in agricultural research. Though these factors have been achieved to a smaller extent by genetic modification of important crops, still, there is a long way to go, due to public acceptance of genetically modified crops especially on its effect on humans and environment. Hence, there is an urgent need for eco-friendly and alternative strategies to water and nutrient use efficiency in crops in a holistic manner. One such approach of late put forward to enhance crop productivity is the modification of crop rhizosphere (Rhizoengineering), which is expected to improve input use efficiency in crop plants, through modifications of soil physico-chemical and biological properties (Ahmed et al. 2018; Dessaux et al. 2016; Sposito 2013). In the present review, we discuss the significance of rhizoengineering in agriculture in enhancing soil and crop productivity.

13.2 Significance of Rhizosphere Biology in Agriculture

The rhizosphere is defined as a small area ranging from millimeter to centimeters adjacent to the roots (Hiltner 1904). In general, the rhizospheric region has three main zones: endorhizosphere, rhizoplane, and ectorhizosphere (Berendsen et al. 2012). The rhizosphere is considered as an active zone of biological activities, greatly influenced by several factors including the host, its metabolites (as root exudates or mucigels), and microbial groups (Darrah 1993; Mendes et al. 2013). The rhizosphere is considered to be a hotspot of microbial interactions (Bakker et al. 2013; Bonkowski et al. 2009; Jacoby et al. 2017; Raaijmakers et al. 2009), which acts as an important interface between plants–soil–microbe interactions, which provides important ecosystem services including carbon and water cycling, nutrient trapping, crop production, and carbon uptake and storage (Adl 2016; Lareen et al. 2016). Several metabolites are shown to be released from the plant roots as exudates, which include amino acids, organic acids, flavonols, glucosinolates, indole compounds, fatty acids, polysaccharides, and proteins (Li et al. 2014; Nguyen 2009). The rhizosphere not only represents the biologically active zone in the soils, but also it act as preventive microbial buffer zone that protects host plants against several pathogenic infections through production of several antimicrobial metabolites and imparting natural resistance (Kamalnath et al. 2019). Several studies on soil properties such as soil structure, soil aggregation, and water allocation around the roots have shown that they are physico-chemically and biologically different from its counterpart bulk soil (Ahmed et al. 2016a, b; Carminati et al. 2010; Watt et al. 1994; Young 1995). The root exudation or the presence of mucigels or mucilage plays a major role in differentiating the rhizosphere from the bulk soils. Recent studies have shown the role of mucilage in manipulating rhizosphere hydraulic properties and regulating root water uptake in crop plants (Ahmed et al. 2018). The root mucilage have been shown to influence the rhizospheric attributes including soil hydraulic properties, nutrient transformations, and mobilization to plants, through interactions between roots and surrounding soil particles (Ahmed et al. 2014; Carminati et al. 2011; McCully 1995). Thus managing the hydraulic properties of the rhizosphere can help in improvement of plants adaptation to drought situations (Ahmed et al. 2018).

13.3 Rhizoengineering

The rhizosphere is the area of soil (a few millimeters wide) that encircle a plant root where the biological and chemical activity in are influenced by compounds exuded by the root (rhizodeposits) and by associated microorganisms. Mucilage secreted from plant roots and extracellular polysaccharide (EPS) secreted by soil microorganisms changes the physical properties of the soil solution. Rhizosphere also acts as a hotspot for communication between the plant and the microbiome. "Rhizosphere engineering" is an innovative approach where soil biophysical properties are modified to positively influence plant–microbiome–soil interactions (Dessaux et al. 2016). The basic components of rhizosphere engineering include soil, plant, and microbes. The rhizosphere attributes could be altered to improve water and nutrient transport to plant as well as positive microbial activity at the root–soil interface. Higher mucilage exudations by plants and subsequent action of microbes increase soil aggregation properties, which increase CEC, BS, and hydraulic conductivity. Higher rhizodeposits by plants increase carbon sequestration in the rhizosphere, which increase organic carbon content of soil, apart from enhancing soil nutrient availability and plant uptake (Ahkami et al. 2017). The mucilage a component of plant root exudates is reported to play an important role in maintaining the hydraulic relationship between the soil and roots, especially during water-deficit conditions (Carminati et al. 2010, 2016; Read et al. 2003). Thus, modifications in mucilage swelling, rhizoligands may affect the hydraulic conductivity of the root–soil interface. The opening and closing of stomata during drying and wetting cycles were shown to be highly influenced by physical and chemical signals surrounding soil and plant systems (Dodd et al. 2015). Rhizoligands are reported to decrease the mucilage swelling after drying, thus limit the flow of mucilage far from the plant roots (Ahmed et al. 2018). The rhizoligands application in plants is shown to suppress the mucilage swelling, which results in a reduced hydraulic conductivity, lowering the transpiration in plants (Ahmed et al. 2018). Though the rhizoligand treatments were reported to increase the wettability of the rhizosphere, their effects on stomatal opening was not much faster as expected (Ahmed et al. 2018). They predicted that the applied rhizoligands might have induced some embolism that limited the recovery of transpiration, through entering the xylem and or they would have affected the plant hormone abscisic acid (ABA). Several data suggests that ABA functioning is one of the mechanisms involved in the control of stomatal conductance as the soil dries (Blackman and Davies 1985; Davies 1991; Dodd 2009; Dodd et al. 2015). The above facts suggest that the rewetting of rhizosphere by rhizoligand influences plant roots greatly and manages water and ABA transport to the xylem and ultimately to the shoots, where it briefly inhibits transpiration (Ahmed et al. 2018). The progress in noninvasive analytical methods and techniques like X-ray computed tomography, nuclear magnetic resonance, two-dimensional light transmission imaging, and neutron radiography are being used to improve the better understanding of rhizospheric phenomenon (Garrigues et al. 2006; Koebernick et al. 2014; Moradi et al. 2013; Pohlmeier et al. 2015). However, compared with other imaging methods, neutron radiography has been proven to be most useful method for studying root–soil water relations (Ahmed et al. 2016a, b; Carminati 2013; Carminati et al. 2010; Dara et al. 2015; Esser et al. 2010; Kroener et al. 2015).

13.4 Rhizoengineering and Its Types

The persistence and survival of applied agri-inputs including microbial inoculants to the crop plants have become a major problem in agriculture, due to limited survival and innate competition from resident microorganisms (Weller 1988). Hence, it is therefore crucial to develop methods to extend the survival and persistence of the bioinoculants in the rhizosphere of crops (O'Connell et al. 1996). This modifications in the rhizosphere is possible through either modifying the plants to release exudates supportive to the inoculated microbes or through exogenous addition of supplements, which supports the growth and proliferation of introduced microbes (Colbert et al. 1993a, b; Di Cello et al. 1997). The microbial populations were also reported to be highly dependent upon soil (Latour et al. 1996, 1999; Øvreås and Torsvik 1998; Parke 1991) and plant exudates (Graystone et al. 1998; Lemanceau et al. 1995; Miethling et al. 2000). The basic components of rhizosphere engineering includes soil, plants, and microbes (Dessaux et al. 2016). For better soil and plant productivity, all these three components can be engineered.

13.4.1 Rhizosphere Modification Through Plant Engineering

Conventional plant breeding methods and advanced plant molecular breeding methods including genetic engineering are promising ways to modify the plant genetic setup to enhance plant yield or quality attributes. Subsequently, these modifications in the plants, also lead to differences in plant root exudation, which changes the soil quality and subsequent biological (Ryan et al. 2009). Understanding the mechanism of photosynthate distribution in plants is key to alter its allocation between root and shoots (Su et al. 2015). Hence, there is a scope to modify the physico-chemical and biological properties (microbial community structure) in the rhizosphere through rhizoengineering. Since genes controlling plant exudation have been identified in several crops; it is practically possible to modify the rhizosphere for desired functional attributes. *Arabidopsis* vacuolar H^+-pyrophosphatase gene AVP1 was overexpressed in rice and tomato, and transgenic plants has shown 50% greater citrate and malate efflux than wild-types when treated with $AlPO_4$. This resulted in enhanced the crop resistance to Al^{3+} stress and also improved the ability to use insoluble phosphorus (Yang et al. 2007). Plants engineered to produce specific bacterial growth substrates have been shown to support the selective group of microbes in their rhizosphere (Guyon 1993; Oger et al. 1997; Savka and Farrand 1997). Primarily these plant produced substances found to be opines (Dessaux et al. 1992), a family of compounds derived from amino acids and/or sugars and specifically detected in the *Agrobacterium* infected tissues (Dessaux et al. 1992). Plants such as lotus and Solanum have been engineered through *Agrobacterium rhizogenes* mediated transformation to produce opine (Petit et al. 1987).

Since the microorganism has enormous potential related to enhancing soil and crop productivity, plant microbiome engineering offers us a great scope to enhance crop yields and quality. Transfer of bacterial N_2 fixing genes to cereals has offered a great possibility to meet the plant nitrogen requirements (Bageshwar et al. 2017; Fox et al. 2016; Geddes et al. 2015, 2019; Lugtenberg and Kamilova 2009; Mondy et al. 2014; Pankievicz et al. 2015). Transgenic plants that produced opine molecules, through gene transfer from *Agrobacterium*, were shown to enrich their rhizosphere with bacteria able to catabolize opines (Mondy et al. 2014; Oger et al. 1997; Savka et al. 2013). However, efforts to engineer rhizopine-producing plants (rhizopines, a compound recognized as chemical signal in plant–microbial interactions) in the past were unsuccessful (Savka et al. 2013). Rhizobia were reported to synthesize and utilize rhizopines as carbon and nitrogen sources (Murphy et al. 1987, 1993). Even though the role of rhizopines remains to be explained, earlier studies suggested that they may be excreted into the rhizosphere (Gordon et al. 1996; Murphy et al. 1995). Integration of synthetic signaling (rhizopine, scyllo-inosamine) networks between plants and bacteria, helps in targeted regulation of rhizospheric bacterial gene expression for achieving useful functions to plants (Geddes et al. 2019). Further, the signaling molecules integration in plants is important, as they serve as biocontainment strategies for detection of genetically modified bacterial inoculants getaway into the environment (Chan et al. 2016).

13.4.2 Rhizosphere Modification Through Microbiome Engineering

Microorganisms either through their beneficial or harmful interactions greatly influence or modify the crop rhizosphere biology. Microbiomes of crop plants either from rhizosphere or phyllosphere can be chosen for their plant beneficial traits such as nutrient bioavailability, biocontrol potential, production of growth hormones, and biotic/abiotic stress alleviator. However, the selected microbial inoculants' survival and persistence in the crop rhizosphere or phyllosphere has become a big challenge in agriculture. Hence, there is a need to modify or engineer the crop plants rhizosphere or phyllosphere to accommodate these exogenously applied beneficial microbial inoculants in the targeted niche. The main challenges include minimizing the parasitism and competition between the selected bioinoculants, while maximizing beneficial effects and cooperation (Foster and Bell 2012; Großkopf and Soyer 2014). The external environmental factors and the innate soil physico-chemical attributes also play an important role in proliferation of applied bioinoculants. Several microbial genera are known to colonize the crop plants rhizosphere and offer their ecological benefits in terms of plant growth promotion, biocontrol agent, nutrient solubilizers/mobilizers, etc. (Dong and Zhang 2014; Kim and Timmusk 2013; Medema et al. 2011). However, there is a drawback in engineering and development of synthetic microbial communities of some plant beneficial microorganisms, which have large genomes and mobile elements (Köberl et al. 2015). The most widely engineered bacterium is *Bacillus* spp, as it is relatively easy to engineer (Dong and Zhang 2014), has complete genome sequences (Sharma and Satyanarayana 2013), and possess multifunctional beneficial role in plants, including biocontrol (Arkhipova et al. 2005; Kim and Timmusk 2013; Köberl et al. 2013, 2015). Further, attempts were also made to engineer *Pseudomonas*, *Rhizobium*, and *Bradyrhizobium* for their plant beneficial roles (Großkopf and Soyer 2014).

Before designing a desired microbial community for an engineered rhizosphere, several crucial elements has to be considered, such as colonization efficiency and survival capabilities of the introduced microbes, competence of the introduced microbes to the native microorganisms, attachment and production of biofilms of its survival under undesirable situations, compatibility to the host plant biology including root exudation and rhizodeposits, multifunctional role of introduced microbes on plant growth and development, etc. (Bashan et al. 2014; Yang et al. 2009). In general, microorganisms which are compatible to host plant metabolites and have additional mechanisms for stronger attachment to roots are other plant parts have higher probability for proliferations in the plant systems (Bashan et al. 2014; Yang et al. 2009). Further, the microbes which have higher population density and antimicrobial mechanisms against the potential competitors including pathogens survive better in the hostile environment (Haas and Défago 2005). Additionally, the microbes should have mechanisms to tolerate or use the plant protection chemicals (seed treating chemicals, herbicides, fertilizers, and pesticides) applied to the host plants (O'Callaghan 2016). Hence, for microbiome engineering, the microbes chosen should possess all those above-mentioned attributes to qualify for specific plants or environment.

13.4.3 Rhizoengineering Through Soil Engineering

To improve soil overall quality, it can be amended or managed in such a way that its physical and chemical properties changes. The change in soil structure is gradually recognized as a form of soil degradation (Chan et al. 2003; Bhattacharyya et al. 2013) and is often associated with the land use and crop management operations. Soil structure influences various factors like soil water movement and its retention, erosion, crusting, nutrient recycling, root penetration, and ultimately crop yield. Different soil and crop management strategies can modify the soil quality or health, thus affecting crop rhizosphere and its functioning (Bhattacharyya et al. 2014; Ray et al. 2014; Velmourougane et al. 2017d). The enhancement in soil organic carbon, soil aggregation and hydraulic conductivity, which plays a major role in plant root proliferation and rhizosphere biological activity can be made through several soil and crop management practices or amendments including addition of fertilizers, management of grazing animals, allowing native vegetation, inclusion of cover crops, legumes and grasses, bioinoculation with beneficial microbes and irrigation (Bronick and Lal 2005; Nalayini et al. 2013; Raychaudhuri et al. 2014; Sidhu et al. 2014). Chemical compounds and microbes in the rhizosphere have stronger cementing effects on soil particles, resulting in greater soil aggregate stability in the rhizosphere than those in non-rhizosphere soil (Caravaca et al. 2002; Six et al. 2004, 2006). Crop management strategies such as growing a series of different crops and cover crops has found to improve soil aggregate dynamics (Lal and Jarecki 2003), which differs with crop chemical composition (Martens 2000), rooting structure and its ability to change the chemical and biological attributes of the soil (Castro Filho et al. 2002; Chan et al. 2003). The cover crops in agriculture has been reported to be helpful in increase of soil carbon content, reduction in erosion, and improve cation exchange capacity, aggregate stability, water infiltration, and nutrients recycling. Thus, it influence the crop rhizosphere biology and its functions (Bronick and Lal 2005). Residues of cover crops was also reported to increase microbial biomass, soil respiration, N mineralization, and shift in microbial community, thus influencing root chemistry and microbial activity (Schutter and Dick 2002; Velmourougane and Sahu 2013). The inclusion of leguminous trees in cropping systems also reported to reduce soil erosion and improve its productivity (Buresh and Tian 1998; Craswell et al. 1998).

The quantity and quality of fertilizers was also reported to affect soil properties and plant growth and development. Fertilizer applications though supplies nutrient requirements of the crop plants, under some situations, they may also decrease organic carbon content, reduce aggregation of soil and reduce microbial diversity compared to organic manured soils (Halvorson et al. 2002; Singh et al. 2014). Heavy dose of fertilizer are also reported to modify the soil pH and the electrolyte concentrations in soil, which can have adverse effects on soil structure and subsequent plant physiology and microbial activity affecting the crop rhizosphere (Haynes and Naidu 1998). Hence, optimum fertilizer application is always recommended to sustain normal plant growth and soil health, which supports normal microbial functions. Manuring has been reported to improve soil structure, macro-aggregation, porosity, aggregate stability, and organic carbon, which results in decreased bulk density and supports higher soil biological activity and plant functions (Hao and Chang 2002; Kay 1998). Earthworm populations in manured soil is higher, and this increased population results in increased soils pores through their boring activity, and improving in aggregate stability of soil through mucilage production (Six et al. 2004).,

Among soil management practices, tillage is found to influence soil structure and aggregate stability, thus affecting soil health and microbial activity. It disrupts the soil aggregates and leads to compaction of soil, and ultimately affect plant and microbial communities (Plante and McGill 2002). Differences in tillage intensity also have an impact on microbial diversity (Jackson et al. 2003). No tillage systems are reported to have more stable soil aggregates and organic carbon (Castro Filho et al. 2002). Reduced tillage is also reported to enhance higher macropores that affects water movement and its availability, helping plants to establish good growth and development (Benno 2001; Logan et al. 1991). Mulching and addition of composts to soils is also reported to plant growth and soil health (Caravaca et al. 2002; He et al. 2019; Leifheit et al. 2014; Rillig et al. 2015). Further, mulching practices are shown to increase the soil organic carbon reservoir (Duiker and Lal 1999; Jacinthe et al. 2002), which modifies soil temperature and moisture regimes and influences soil biology.

It has been known that microbial communities that are linked with different crop plants and their genotypes varies in terms of composition, activity, and nutrient content (Graystone et al. 1998; Schloter et al. 2003; Srivastava et al. 2014; Velmourougane and Sahu 2013; Velmourougane and Blaise 2014). The cropping systems, their rotation and types affects the soil biota (Orr et al. 2011; Velmourougane et al. 2014). Use of organic manure that are originated from leguminous green manure crops was reported to encourage more soil microflora as compared to field where chemical fertilizers are added, thus ultimately influencing plant growth (Bolton et al. 1985). Diverse cropping systems have been corelated to the increased microbial activity and diversity (Moore et al. 2000). Cultivating the different crops on the rotation basis is reported to increase soil carbon sequestration in comparison with mono-cropping system; and more intensive cropping rotations are also found to be responsible for increase in microbial activity (Six et al. 2006). Various soil management and cultural practices also impact soil microbial diversity and their activities (Velmourougane 2016; Wu et al. 2008). As compared to different soil type, crop management operation and type of crop cultivation have more control on soil biota (Fromm et al. 1993; Ibekwe et al. 2002; Velmourougane et al. 2014). Enzymes like urease, dehydrogenase, glucosidase, phosphatase, arylsulfatase, etc. that are present in soil are found to be involved in soil nutrient conversion, are also affected by soil attributes and plant types (Blaise and Velmourougane 2014;

Dorodnikov et al. 2009; Fliessbach et al. 2007; Kumar et al. 2015; Srivastava et al. 2014; Velmourougane et al. 2013, 2014; Velmourougane and Blaise 2014).

13.5 Factors Influencing Rhizosphere Functioning

13.5.1 Role of Rhizodeposits in Rhizosphere Biology and Functioning

Global climate change, together with increasing temperatures and global weather patterns such as rise in levels of atmospheric CO_2, were also shown to affect rhizosphere ecology, in several direct and indirect ways (Lobell and Field 2007). However, our understanding on the interactions between plant microbiome, root exudation, and plant growth and reproduction, remains limited (Bai et al. 2015). The rhizosphere biology is strongly influenced by plant metabolism including its secretion of photosynthate as root exudates (Bais et al. 2006; Estabrook and Yoder 1998). Plant root exudates play a vital role in influencing the interactions between soil, plants, and microbes, which in turn plays a major role in nutrient transformation and subsequent plant uptake (Xu et al. 2014). Several factors including type of plant species, varieties, developmental stages, and other biotic and abiotic factors regulate the quantity and quality of root exudates (Xu et al. 2014). Root exudates account for around 5–21% of total photosynthetically fixed carbon; however, they are the prime factors which influence community composition of the rhizosphere (Vandenkoornhuyse et al. 2015). The root exudates comprise the major portion of the rhizodeposits and are generally synthesized in the meristematic zone of root tips (Bais et al. 2006; Vicré et al. 2005; Watson et al. 2015) into the rhizosphere, which includes sloughed-off root cells and tissues, mucilages, root exudates, soluble lysates, volatile compounds, sugars, organic acids, phenolic compounds, alcohols, polypeptides, amino acids, proteins, plant cells, etc. (Nguyen 2009; Tian et al. 2020; Vicré et al. 2005; Watson et al. 2015). In general, rhizodeposits are reported to be rich in carbon and nitrogen compounds, and thus support growth of several rhizomicrobiome (Bais et al. 2006; Dennis et al. 2010).

Plant roots are reported to exude up to 20% of fixed carbon and 15% of nitrogen (Meharg and Killham 1988; El Zahar Haichar et al. 2016). The composition of rhizodeposits also varies spatially and temporally including the differences in plant species, plant physiological status, and upon biotic and abiotic stresses (Carvalhais et al. 2013; Chaparro et al. 2014; Edwards et al. 2015; Hirsch et al. 2013; Liu et al. 2019). Several metabolites including amino acids exuded from the plant roots were shown to influence the rhizosphere microbial community composition and their functioning (Hao et al. 2010). Amino acids excreted through roots also serve as nutrient sources for microbial growth (Hao et al. 2010). Phenolic acid exudates from the roots were also reported to act as inhibitors, attractants, or signaling molecules (Lanoue et al. 2010; Mandal et al. 2010). The root exudates are also shown to modulate a neutral, positive, or negative interactions among the soil microbes including soil-borne pathogens (Ling et al. 2013). The plant-derived carbon excretion in soils act as an nutrient source for different groups of microorganisms, leading to their rapid growth and community shifts based on the characteristics of the exudates (Jones et al. 2009; Steinauer et al. 2016), which in turn induces their enzymes mechanisms and transform nutrients in fixed form to plant available (Bonkowski and Clarholm 2012; Ekelund et al. 2009; Koller et al. 2013). The rhizodeposits also serve as and provide a physical barrier between the root cells and the pathogenic microorganism (Hirsch et al. 2013; Nguyen 2009). After being released into the rhizosphere, the rhizodeposits are reported to synthesis and release several secondary metabolites including antimicrobial compounds, which form first-line of defense against phytopathogens (Koroney et al. 2016). Apart from rhizodeposits, plant mucilages are also reported to release into the rhizosphere on the event of pathogens infections (Koroney et al. 2016), wherein they function as a lubricant, chelator, humectant, aggregator, as well as a carbon source (Hirsch et al. 2013; Nguyen 2009). The proteins and extracellular DNAs components of the mucilage are also reported to be involved in conferring defense against phytopathogens (Basu et al. 2006; Weiller et al. 2017). Exudates may be excess plant products (Preece et al. 2018; Preece and Peñuelas 2016), but they can also contain signaling and chemo-attractant molecules. The plant root exudates are also reported to function as microbial attractants/repellents, through production of different chemo-attractants, which attract or repel several groups of microbes towards the plants (Weisskopf et al. 2006, 2011). In some cases, plants in association with the interacting microbes are also reported to produce several compounds including volatiles which help plants to alleviate biotic or abiotic stresses (Sharifi et al. 2018; Zhalnina et al. 2018).

13.5.2 Soil Microbiome and Rhizosphere Functioning

The rhizosphere microbiome plays an effective role in influencing rhizosphere and its functioning, which aids in better plant growth and development through production of growth hormones, nutrient acquisition, imparting tolerance to biotic and abiotic stresses through soil–plant interactions by modifying the nutrient transformation and their mobilization into plant systems (Lareen et al. 2016; Lu et al. 2018a, b; Panke-Buisse et al. 2015; Soussi et al. 2016; Wagner et al. 2014). The beneficial microbial associations with crop plants were reported to promote root growth, which results in a larger root surface, which helps in greater water acquisition and nutrient uptake by the plants, which in turn increase specific metabolites in soil through root exudation benefit the growth of associated microbes. The quantity, type, and composition of root exudates were shown to modify the microbial dynamics and its composition in rhizosphere as well as the soils (Bais et al. 2006; Chaparro et al. 2014). Further, the quality of the root exudates released from the plants decides the variations in microbial community composition to be accommodated in its rhizosphere (Baudoin et al. 2002). However, studies have also demonstrated that these rhizospheric microbial compositions are significantly influenced by the developmental stages of the plants (Li et al. 2014; Okubo et al. 2014; Yuan et al. 2015). Several plant-associated microorganisms, including plant growth-promoting rhizobacteria (PGPR) provide several nutrients and growth factors to crop plants, which modifies the rhizosphere environment (Carvalhais et al. 2013; Karandashov and Bucher 2005; Ling et al. 2014; Prasad et al. 2015), helping in their growth promotion (do Amaral et al. 2016; Peiffer et al. 2013; Wintermans et al. 2016). Several plant-associated microorganisms including PGPR and endophytes also help the host plant in prevention of pathogen colonization in the rhizosphere and rhizoplane (Bloemberg and Lugtenberg 2001; Lau and Lennon 2012; Prasad et al. 2015; Van Der Ent et al. 2009) Similarly, the plant's root exudation can regulate the rhizosphere environment through its exudate composition, thereby modifying the microbial community structure of the rhizosphere (Chen et al. 2016). At least 21% of carbon fixed through photosynthesis was reported to enter the soils system, by way of root exudations, influencing the microbial community composition in the rhizosphere (Marschner 2012). Photoassimilates of cereals transferred to soil as root exudates was shown to play a crucial role in the plant–microbe interactions (El Zahar Haichar et al. 2014). Though the community structure of the rhizosphere microbiome was affected by several biotic and abiotic factors such soil physical and chemical properties, plant types and stages of development, rhizosphere attributes, and root exudation play a major role (Pérez-Jaramillo et al. 2016; Schreiter et al. 2014). The role of plant types and its root exudation on dynamics of the rhizosphere microbiome have been studied in several crops (De la Cruz-Barrón et al. 2017; Donn et al. 2015; Knox et al. 2014; Marasco et al. 2013; Marques et al. 2014; Smalla et al. 2001; Zarraonaindia et al. 2015). Recent studies also propose that the microbiota present around the roots may contribute to phenotypic plasticity, which has significant role in our understanding of plant phenology in a changing climate and also for improving crop production (Chen et al. 2017; Singh et al. 2010).

The synthesis of plant hormones by PGPR was also reported to regulate initiation of lateral root primordial, root elongation, and root architecture (Aloni et al. 2006; Dobbelaere et al. 2001; Patten and Glick 2002). The interactions between of protists and bacteria also reported to influence microbial plant hormone production, plant productivity, root diameter, root architecture and nutrient uptake efficiency in several plant species including garden cress, rice , *Plantago lanceolata* L., and *A. thaliana* (Jousset 2017; Koller et al. 2013; Kreuzer et al. 2006; Krome et al. 2009, 2010). Root gene expression was also shown to be regulated by presence of pathogenic bacteria (Chen et al. 2014), PGPR (Camilios-Neto et al. 2014; Do Amaral et al. 2014), and mycorrhiza in the rhizosphere (Dhawi et al. 2015; Gupta et al. 2017). Decrease in plant stress levels also shown to cause protist-induced shifts in microbial communities in the rhizosphere (Kuppardt et al. 2018).

13.5.3 Soil and Crop Management

The significance of rhizosphere processes and root–soil interactions in influencing soil organic matter decomposition has been increasingly recognized (Dijkstra and Cheng 2007a; Fontaine et al. 2007; Wallenstein and Weintraub 2008). Both abiotic and biotic factors, such as soil nutrient status (Liljeroth et al. 1994), soil moisture (Dijkstra and Cheng 2007b), CO_2 concentration (Carney et al. 2007), light intensity (Kuzyakov and Cheng 2001), plant phenology (Cheng et al. 2003), biomass (Dijkstra et al. 2006), and rhizodeposition (Dijkstra and Cheng 2007a) have been reported to manipulate rhizosphere biology and its functioning.

In some of the crop plants, root growth is reported to create soil compaction, which reduces the porosity of the rhizosphere (Aravena et al. 2011). Roots are also shown to shrink when they are dry, creating space between soil and roots. Repeated drying and wetting cycles in the rhizosphere also reported to enhance soil aggregation (Carminati et al. 2009). Several root attributes were shown to involved in the extraction of water and nutrients from the soil, including root architecture (Wasson et al. 2012), root depth (Tron et al. 2015), internal axial and radial conductivity and the conductance of the roots to the shoot base (Lobet et al. 2014), the ability of roots to reach the subsoil (Lynch and Wojciechowski 2015), and root-shoot signaling (Huber et al. 2014). Plants were also shown to modify the rhizosphere, to enhance their water uptake (Hinsinger et al. 2009; York et al. 2016). Although, the mucilage secreted by the plants was reported to absorb large volumes of water (McCully and Boyer 1997), they also shown to have a small fraction of amphiphilic components such as lipids (Read et al. 2003). Under water-deficit conditions, the hydrophilic ligands of mucilage may bind to each other or to the soil particles and may leave the hydrophobic end towards the air-filled phase of the pore space, resulting into water repellency in the rhizosphere (Ahmed et al. 2016a, b). Zarebanadkouki et al. (2016) has reported that the rhizosphere water repellency is temporarily limited to the root water uptake following soil drying and rewetting. Rhizosphere water repellency has been reported in several crops, including barley (Hallett et al. 2003), lupines (Moradi et al. 2012; Zarebanadkouki et al. 2016), beans and wheat (Zickenrott et al. 2016), and maize (Ahmed et al. 2015, 2016a, b). It has been shown that lowering the root hydraulic conductance increased the collar xylem potential during lengthy periods of soil drying (Couvreur et al. 2014). Similarly, lowering the root hydraulic conductance was reported to induce an early closure of the stomata, resulting in a reduced transpiration and water consumption, which can improve plant performance (Tardieu et al. 2017). In recent years, use of rhizoligands as an additive to enhance the rhizosphere wettability especially in water-repellent soils is also proposed (Dekker et al. 2005; Kostka 2000; Kostka et al. 1997). Rhizoligands are reported to have hydrophilic and hydrophobic functional groups (Ahmed et al. 2018). Simovic et al. (1999) reported that, at specific concentrations, nonionic surfactants decrease the swelling and increase the viscosity of gels containing hydrophobic components.

13.6 Role of Microbial Biofilms and in Rhizoengineering

Microbial biofilms are aggregation of their cells in a self-produced polymeric matrix attached to biotic and abiotic surfaces (Costerton et al. 1978). Microbial cells form single, dual, or multispecies biofilms depending on the ecological conditions in their habitat. Biofilms are generally formed by microorganisms to overcome stress in the environment, or nutrient limitation, change in cultural conditions, competition, presence of antibiotics, disinfectants, etc. (Karatan and Watnick 2009). Apart from being a stress avoidance strategy, biofilm style of growth also confers a reproductive fitness advantage, making them better adapted to endure challenges, as compared to the free floating cells (planktonic cells) (Davey and O'Toole 2000). Formation of biofilm has been considered as a developmental process (Monds and O'Toole 2009), as it is assisted by microbial attributes such as cell-to-cell communication (Lopez et al. 2010), cell differentiation (López and Kolter 2010) and pattern formation (Asally et al. 2012). Microbial organelles including flagella, fimbriae, pili, and other microbial components such as lipopolysaccharides and membrane proteins are reported to be involved in biofilm formation (Hinsa et al. 2003). Biofilms are recognized to have various application in many areas like health, agriculture, food, biopolymers, oil refinery, metal extraction, mineral exploration, biofuel cells, bioremediation, etc. However, microbial biofilm application in agriculture, especially in crop production and protection has gained importance, recently (Velmourougane et al. 2017a).

Biofilm formation on the plant roots is a valuable tool of rhizospheric microorganisms that protect them from being separated from the roots due to various natural processes occurring in the soil (Velmourougane et al. 2017a). Secretion of EPS by bacteria have been also reported to help the microbes to colonize the roots in an effective way by formation of biofilms that protect them from desiccation (Sandhya et al. 2009, 2010a; Velmourougane et al. 2017c, 2019a; Velmourougane and Prasanna 2017). Many bacterial EPS have the property of trapping water in its polymeric matrix, which safeguards bacteria from drying under drought stress by maintaining a wet microenvironment and controlling the diffusion of organic carbon sources from microbes to plants and contrariwise (Sandhya et al. 2009, 2010a, b; Vardharajula et al. 2011). It was reported to have beneficial changes in the soil structure, root-associated soil/ root ratio, and the macroaggregate properties resulting in better plant growth, by increasing uptake of water and nutrients through extended roots . The EPS composition was reported to change during stress conditions producing more glucose, rhamnose, mannose, and trehalose that enhance EPS property to retain more water (Tewari and Arora 2014). Promotion of in situ biofilm production or using the native biofilms are promising technologies for future agriculture, due to their huge potential to provide multiple benefits via single inoculant. It has potential to improve plant growth, nutrient mobilization, biotic/abiotic stress tolerance, etc. Recent research on biofilms containing multiple species has found to be a more attractive option in agriculture and other industries as they produce new types of polysaccharides with different composition or other bioactive compounds, compared to single species biofilm (Velmourougane et al. 2017, 2017c). The enhancement in soil physical assets such as improved soil aggregation, reduction in bulk density, better water and nutrients holding in rhizospheric region, can be achieved due to the polysaccharides released by microbial biofilms. Some microbes were also reported to modulate antioxidants enzymes (peroxidases, polyphenol oxidases, catalase, PAL, PEP carboxylase, phenols etc.), compatible osmolytes, and polyamines in crop plants to endure the drought stress (Ghosh et al. 2017; Sandhya et al. 2010a, b; Sen et al. 2018). Foliar application of these biofilm-produced microbial polysaccharides can be explored as tool for protecting crops from the damages of biotic and abiotic stresses (Velmourougane et al. 2017a, b, 2019a). The use of microbial biofilms can be a novel option, in lieu of their preponderance, versatility in function and resilience to environmental stressors. Microbial biofilms, by virtue of their resilience to environmental perturbations and stress can also be a promising option due to their dynamic nature and ability to persist in unfavorable environments (Bharti et al. 2017). This opens the avenue of using microbial biofilm-based agents as sustainable approach in future crop cultivation and management.

Microbial biofilms (cyanobacterial or fungal-based bacterial biofilms) have been successfully developed, and inoculation in rice, wheat, maize, cotton, legume crops has improved plant growth attributes, soil nutrient availability, and caused the expression of plant defense system (Babu et al. 2015; Triveni et al. 2013; Velmourougane et al. 2017a). Recently, the use of biofilm developed by *Azotobacter chroococcum*) and *Trichoderma viride* has have shown higher polysaccharide production when grown together as biofilms (Velmourougane et al. 2017). Further, our group characterized the qualitative aspects and modulation of exopolymeric substances (EPS) produced during this biofilm development (Velmourougane et al. 2017a; Velmourougane and Prasanna 2017), and found that increased production of hydrolytic enzymes (β-1, 3-glucanases, and chitosanase) by biofilm as compared to individual inoculation. Bacterial (*Az*) and fungal (*Tv*) partners in biofilms were reported to change their gene expression (Velmourougane et al. 2019b).

Microbial biofilm-based formulation is expected to assist in engineering the root and rhizosphere, which helps in retention of higher moisture in the rhizosphere, thereby improve root architecture, enhance soil aggregation and hydraulics leading to improved soil moisture and nutrient availability under moisture-deficit situations. Further, the microbial biofilms are also an eco-friendly alternative to chemical rhizoligands/hydrogels/osmoprotectants (synthetic polymers) to enhance soil hydration, soil functional diversity, and subsequent plant water/nutrient use efficiency in crops rhizosphere, leading to better soil health and productivity. From an environmental perspective, the use of rhizoengineering using biofilm is expected to reduce the usage of plant growth regulators and osmoprotectants, which are environmental polluters in long-term usage.

13.7 Conclusion and Future Outlook

Rhizoengineering is a recent innovative approach to modify the crop plants rhizosphere to derive maximum benefits in terms of soil and crop productivity. Though exogenous application of agri-inputs including bioinoculants to crops have proved beneficial effects, most of these functions are found to be transient in nature. However, modification of rhizosphere through crop and soil management interventions was proved to be advantageous in several crops. Though genetic modification of crop plants and microbial engineering has taken back seat because of consumer awareness on human and environmental health, rhizosphere modification through agronomic approaches is the only hope at present to improve soil and crop productivity in an eco-friendly and sustainable manner. Inclusion of cover crops and following crop rotations and right type of nutrients in farming systems were shown to enhance soil health to support microbial groups, and their beneficial interactions with the crop plants. In recent years, application of microbial inoculants in the form of microbial biofilms is also shown to modify rhizosphere attributes in terms of enhancement in microbial colonization and associated benefits. Further, understanding of soil–plant–microbial interactions is still in infant stage to be applied in rhizoengineering approaches in agriculture. However, overall rhizosphere engineering is reported to help plants to tolerate several biotic and abiotic stresses under climate change situations. With the recent advancement in omics techniques, it is expected to enhance our knowledge on rhizosphere engineering and its functionality in agriculture. Natural way of modification of crop rhizosphere is also expected to make soil healthy by avoiding indiscriminate use of plant protection chemicals and fertilizers. Hence, rhizoengineering approach should be advocated to farming community through eco-friendly farm amendments, instead of engineering crops or microorganisms.

References

Adl S (2016) Rhizosphere, food security, and climate change: a critical role for plant-soil research. Rhizosphere 1:1–3

Ahkami AH, Allen White R, Handakumbura PP, Jansson C (2017) Rhizosphere engineering: enhancing sustainable plant ecosystem productivity. Rhizosphere 3:233–243

Ahmed MA, Kroener E, Holz M, Zarebanadkouki M, Carminati A (2014) Mucilage exudation facilitates root water uptake in dry soils. Funct Plant Biol 41:1129–1137

Ahmed MA, Holz M, Woche SK, Bachmann J, Carminati A (2015) Effect of soil drying on mucilage exudation and its water repellency: a new method to collect mucilage. J Plant Nutr Soil Sci 178:821–824

Ahmed MA, Kroener E, Benard P, Zarebanadkouki M, Kaestner A, Carminati A (2016a) Drying of mucilage causes water repellency in the rhizosphere of maize: measurements and modelling. Plant Soil 407:161–171

Ahmed MA, Zarebanadkouki M, Kaestner A, Carminati A (2016b) Measurements of water uptake of maize roots: the key function of lateral roots. Plant Soil 398:59–77

Ahmed MA, Zarebanadkouki M, Ahmadi K, Kroener E, Kostka S, Kaestner A, Carminati A (2018) Engineering rhizosphere hydraulics: pathways to improve plant adaptation to drought. Vadose Zo J 17:160090

Aloni R, Aloni E, Langhans M, Ullrich CI (2006) Role of cytokinin and auxin in shaping root architecture: regulating vascular differentiation, lateral root initiation, root apical dominance and root gravitropism. Ann Bot 97:883–893

Aravena JE, Berli M, Ghezzehei TA, Tyler SW (2011) Effects of root-induced compaction on rhizosphere hydraulic properties—X-ray microtomography imaging and numerical simulations. Environ Sci Technol 45:425–431

Arkhipova TN, Veselov SU, Melentiev AI, Martynenko EV, Kudoyarova GR (2005) Ability of bacterium Bacillus subtilis to produce cytokinins and to influence the growth and endogenous hormone content of lettuce plants. Plant Soil 272:201–209

Asally M, Kittisopikul M, Rué P, Du Y, Hu Z, Çağatay T, Robinson AB, Lu H, Garcia-Ojalvo J, Süel GM (2012) Localized cell death focuses mechanical forces during 3D patterning in a biofilm. Proc Natl Acad Sci U S A 109:18891–18896

Babu S, Bidyarani N, Chopra P, Monga D, Kumar R, Prasanna R, Kranthi S, Saxena AK (2015) Evaluating microbe-plant interactions and varietal differences for enhancing biocontrol efficacy in root rot disease challenged cotton crop. Eur J Plant Pathol 142:345–362

Bageshwar UK, Srivastava M, Pardha-Saradhi PP, Paul S, Gothandapani S, Jaat RS, Shankar P, Yadav R, Biswas DR, Kumar PA et al (2017) An environment friendly engineered Azotobacter can replace substantial amount of urea fertilizer and yet sustain same wheat yield. Appl Environ Microbiol 83:AEM–00590

Bai Y, Müller DB, Srinivas G, Garrido-Oter R, Potthoff E, Rott M, Dombrowski N, Münch PC, Spaepen S, Remus-Emsermann M, Hüttel B, McHardy AC, Vorholt JA, Schulze-Lefert P (2015) Functional overlap of the Arabidopsis leaf and root microbiota. Nature 528:364–369

Bais HP, Weir TL, Perry LG, Gilroy S, Vivanco JM (2006) The role of root exudates in rhizosphere interactions with plants and other organisms. Annu Rev Plant Biol 57:233–266

Bakker PAHM, Berendsen RL, Doornbos RF, Wintermans PCA, Pieterse CMJ (2013) The rhizosphere revisited: root microbiomics. Front Plant Sci 4:165

Bashan Y, de-Bashan LE, Prabhu SR, Hernandez JP (2014) Advances in plant growth-promoting bacterial inoculant technology: formulations and practical perspectives (1998-2013). Plant Soil 378:1–33

Basu U, Francis JL, Whittal RM, Stephens JL, Wang Y, Zaiane OR, Goebel R, Muench DG, Good AG, Taylor GJ (2006) Extracellular proteomes of Arabidopsis thaliana and Brassica napus roots: analysis and comparison by MudPIT and LC-MS/MS. Plant Soil 286:357–376

Baudoin E, Benizri E, Guckert A (2002) Impact of growth stage on the bacterial community structure along maize roots, as determined by metabolic and genetic fingerprinting. Appl Soil Ecol 19:135–145

Benno PW (2001) The tillage effect in sustaining soil functions. J Plant Nutr Soil Sci 164:345–350

Berendsen RL, Pieterse CMJ, Bakker PAHM (2012) The rhizosphere microbiome and plant health. Trends Plant Sci 17:478–486

Bharti A, Velmourougane K, Prasanna R (2017) Phototrophic biofilms: diversity, ecology and applications. J Appl Phycol 29:2729–2744

Bhattacharyya T, Pal DK, Mandal C et al (2013) Soils of India: their historical perspective, classification and recent advances. Curr Sci 104:1308–1323

Bhattacharyya T et al (2014) Soil information system: use and potentials in humid and semi-arid tropics. Curr Sci 107:1550–1564

Blackman PG, Davies WJ (1985) Root to shoot communication in maize plants of the effects of soil drying. J Exp Bot 36:39–48

Blaise D, Velmourougane K (2014) Dynamics of soil dehydrogenase and alkaline phosphatase activities under organically and conventionally grown cotton on rainfed vertisols of Central India. Cotton Res J 6:54–57

Bloemberg GV, Lugtenberg B (2001) Molecular basis of plant growth promotion and biocontrol by rhizobacteria. Curr Opin Plant Biol 4:343–350
Bolton H, Elliott LF, Papendick RI, Bezdicek DF (1985) Soil microbial biomass and selected soil enzyme activities: effect of fertilization and cropping practices. Soil Biol Biochem 17:297–302
Bonkowski M, Clarholm M (2012) Stimulation of plant growth through interactions of bacteria and protozoa: testing the auxiliary microbial loop hypothesis. Acta Protozool 51:237–247
Bonkowski M, Villenave C, Griffiths B (2009) Rhizosphere fauna: the functional and structural diversity of intimate interactions of soil fauna with plant roots. Plant Soil 321:213–233
Bronick CJ, Lal R (2005) Soil structure and management: a review. Geoderma 124:3–22
Buresh RJ, Tian G 1998 Soil improvement by trees in sub-Saharan Africa, pp 51–76
Camilios-Neto D, Bonato P, Wassem R, Tadra-Sfeir MZ, Brusamarello-Santos LCC, Valdameri G, Donatti L, Faoro H, Weiss VA, Chubatsu LS, Pedrosa FO, Souza EM (2014) Dual RNA-seq transcriptional analysis of wheat roots colonized by Azospirillum brasilense reveals up-regulation of nutrient acquisition and cell cycle genes. BMC Genomics 15:378
Caravaca F, Hernández T, García C, Roldán A (2002) Improvement of rhizosphere aggregate stability of afforested semiarid plant species subjected to mycorrhizal inoculation and compost addition. Geoderma 108:133–144
Carminati A (2013) Rhizosphere wettability decreases with root age: a problem or a strategy to increase water uptake of young roots? Front Plant Sci 4:298
Carminati A, Vetterlein D, Weller U, Vogel H-J, Oswald SE (2009) When roots lose contact. Vadose Zo J 8:805–809
Carminati A, Moradi AB, Vetterlein D, Vontobel P, Lehmann E, Weller U, Vogel HJ, Oswald SE (2010) Dynamics of soil water content in the rhizosphere. Plant Soil 332:163–176
Carminati A, Schneider CL, Moradi AB, Zarebanadkouki M, Vetterlein D, Vogel H-J, Hildebrandt A, Weller U, Schüler L, Oswald SE (2011) How the rhizosphere may favor water availability to roots. Vadose Zo J 10:988–998
Carminati A, Kroener E, Ahmed MA, Zarebanadkouki M, Holz M, Ghezzehei T (2016) Water for carbon, carbon for water. Vadose Zo J 15:vzj2015.04.0060
Carney KM, Hungate BA, Drake BG, Megonigal JP (2007) Altered soil microbial community at elevated CO2 leads to loss of soil carbon. Proc Natl Acad Sci U S A 104:4990–4995
Carvalhais LC, Dennis PG, Fan B, Fedoseyenko D, Kierul K, Becker A, von Wiren N, Borriss R (2013) Linking plant nutritional status to plant-microbe interactions. PLoS One 8:e68555
Castro Filho C, Lourenço A, Guimarães MDF, Fonseca ICB (2002) Aggregate stability under different soil management systems in a red latosol in the state of Parana, Brazil. Soil Tillage Res 65:45–51
Chan KY, Heenan DP, So HB (2003) Sequestration of carbon and changes in soil quality under conservation tillage on light-textured soils in Australia: a review. Aust J Exp Agric 43:325–334
Chan CTY, Lee JW, Cameron DE, Bashor CJ, Collins JJ (2016) 'Deadman' and 'Passcode' microbial kill switches for bacterial containment. Nat Chem Biol 12:82–86
Chaparro JM, Badri DV, Vivanco JM (2014) Rhizosphere microbiome assemblage is affected by plant development. ISME J 8:790–803
Chen YN, Ren XP, Zhou XJ, Huang L, Huang JQ, Yan LY, Lei Y, Qi Y, Wei WH, Jiang HF (2014) Alteration of gene expression profile in the roots of wild diploid Arachis duranensis inoculated with Ralstonia solanacearum. Plant Pathol 63:803–811
Chen ZJ, Tian YH, Zhang Y, Song BR, Li HC, Chen ZH (2016) Effects of root organic exudates on rhizosphere microbes and nutrient removal in the constructed wetlands. Ecol Eng 92:243–250
Chen S, Li X, Lavoie M, Jin Y, Xu J, Fu Z, Qian H (2017) Diclofop-methyl affects microbial rhizosphere community and induces systemic acquired resistance in rice. J Environ Sci (China) 51:352–360
Cheng W, Johnson DW, Fu S (2003) Rhizosphere effects on decomposition: controls of plant species, phenology, and fertilization. Soil Sci Soc Am J 67:1418–1427

Colbert SF, Isakeit T, Ferri M, Weinhold AR, Hendson M, Schroth MN (1993a) Use of an exotic carbon source to selectively increase metabolic activity and growth of Pseudomonas putida in soil. Appl Environ Microbiol 59:2056–2063
Colbert SF, Schroth MN, Weinhold AR, Hendson M (1993b) Enhancement of population densities of Pseudomonas putida PpG7 in agricultural ecosystems by selective feeding with the carbon source salicylate. Appl Environ Microbiol 59:2064–2070
Costerton JW, Geesey GG, Cheng KJ (1978) How bacteria stick. Sci Am 238:86–95
Couvreur V, Vanderborght J, Draye X, Javaux M (2014) Dynamic aspects of soil water availability for isohydric plants: focus on root hydraulic resistances. Water Resour Res 50:8891–8906
Craswell ET, Sajjapongse A, Howlett DJB, Dowling AJ (1998) Agroforestry in the management of sloping lands in Asia and the Pacific, pp 121–137
Dara A, Moradi BA, Vontobel P, Oswald SE (2015) Mapping compensating root water uptake in heterogeneous soil conditions via neutron radiography. Plant Soil 397:273–287
Darrah PR (1993) The rhizosphere and plant nutrition: a quantitative approach. Plant Nutr 3–22
Davey ME, O'Toole GA (2000) Microbial biofilms: from ecology to molecular genetics. Microbiol Mol Biol Rev 64:847–867
Davies W (1991) Root signals and the regulation of growth and development of plants in drying soil. Annu Rev Plant Physiol Plant Mol Biol 42:55–76
De la Cruz-Barrón M, Cruz-Mendoza A, Navarro-Noya YE, Ruiz-Valdiviezo VM, Ortíz-Gutiérrez D, Ramírez-Villanueva DA, Luna-Guido M, Thierfelder C, Wall PC, Verhulst N, Govaerts B, Dendooven L (2017) The bacterial community structure and dynamics of carbon and nitrogen when Maize (*Zea mays* L.) and its neutral detergent fibre were added to soil from Zimbabwe with contrasting management practices. Microb Ecol 73:135–152
Dekker LW, Oostindie K, Kostka SJ, Ritsema CJ (2005) Effects of surfactant treatments on the wettability of a water repellent grass-covered dune sand. Aust J Soil Res 43:383–395
Dennis PG, Miller AJ, Hirsch PR (2010) Are root exudates more important than other sources of rhizodeposits in structuring rhizosphere bacterial communities? FEMS Microbiol Ecol 72:313–327
Dessaux Y, Petit A, Tempe J (1992) Opines in Agrobacterium biology. In: Molecular signals in plant-microbe communications, pp 109–136
Dessaux Y, Grandclément C, Faure D (2016) Engineering the rhizosphere. Trends Plant Sci 21:266–278
Dhawi F, Datta R, Ramakrishna W (2015) Mycorrhiza and PGPB modulate maize biomass, nutrient uptake and metabolic pathways in maize grown in mining-impacted soil. Plant Physiol Biochem 97:390–399
Di Cello F, Bevivino A, Chiarini L, Fani R, Paffetti D, Tabacchioni S, Dalmastri C (1997) Biodiversity of a Burkholderia cepacia population isolated from the maize rhizosphere at different plant growth stages. Appl Environ Microbiol 63:4485–4493
Dijkstra FA, Cheng W (2007a) Interactions between soil and tree roots accelerate long-term soil carbon decomposition. Ecol Lett 10:1046–1053
Dijkstra FA, Cheng W (2007b) Moisture modulates rhizosphere effects on C decomposition in two different soil types. Soil Biol Biochem 39:2264–2274
Dijkstra FA, Cheng W, Johnson DW (2006) Plant biomass influences rhizosphere priming effects on soil organic matter decomposition in two differently managed soils. Soil Biol Biochem 38:2519–2526
Do Amaral FP, Bueno JCF, Hermes VS, Arisi ACM (2014) Gene expression analysis of maize seedlings (DKB240 variety) inoculated with plant growth promoting bacterium Herbaspirillum seropedicae. Symbiosis 62:41–50
do Amaral FP, Pankievicz VCS, Arisi ACM, de Souza EM, Pedrosa F, Stacey G (2016) Differential growth responses of Brachypodium distachyon genotypes to inoculation with plant growth promoting rhizobacteria. Plant Mol Biol 90:689–697
Dobbelaere S, Croonenborghs A, Thys A, Ptacek D, Vanderleyden J, Dutto P, Labandera-Gonzalez C, Caballero-Mellado J, Aguirre JF, Kapulnik Y, Brener S, Burdman S, Kadouri D,

Sarig S, Okon Y (2001) Responses of agronomically important crops to inoculation with Azospirillum. Aust J Plant Physiol 28:871–879
Dodd IC (2009) Rhizosphere manipulations to maximize 'crop per drop' during deficit irrigation. J Exp Bot 60:2454–2459
Dodd IC, Puértolas J, Huber K, Pérez-Pérez JG, Wright HR, Blackwell MSA (2015) The importance of soil drying and re-wetting in crop phytohormonal and nutritional responses to deficit irrigation. J Exp Bot 66:2239–2252
Dong HN, Zhang DW (2014) Current development in genetic engineering strategies of Bacillus species. Microb Cell Factories 13:63
Donn S, Kirkegaard JA, Perera G, Richardson AE, Watt M (2015) Evolution of bacterial communities in the wheat crop rhizosphere. Environ Microbiol 17:610–621
Dorodnikov M, Blagodatskaya E, Blagodatsky S, Marhan S, Fangmeier A, Kuzyakov Y (2009) Stimulation of microbial extracellular enzyme activities by elevated CO2 depends on soil aggregate size. Glob Chang Biol 15:1603–1614
Duiker SW, Lal R (1999) Crop residue and tillage effects on carbon sequestration in a Luvisol in central Ohio. Soil Tillage Res 52:73–81
Edwards J, Johnson C, Santos-Medellín C, Lurie E, Podishetty NK, Bhatnagar S, Eisen JA, Sundaresan V, Jeffery LD (2015) Structure, variation, and assembly of the root-associated microbiomes of rice. Proc Natl Acad Sci U S A 112:E911–E920
Ekelund F, Saj S, Vestergård M, Bertaux J, Mikola J (2009) The 'soil microbial loop' is not always needed to explain protozoan stimulation of plants. Soil Biol Biochem 41:2336–2342
El Zahar Haichar F, Santaella C, Heulin T, Achouak W (2014) Root exudates mediated interactions belowground. Soil Biol Biochem 77:69–80
El Zahar Haichar F, Heulin T, Guyonnet JP, Achouak W (2016) Stable isotope probing of carbon flow in the plant holobiont. Curr Opin Biotechnol 41:9–13
Esser HG, Carminati A, Vontobel P, Lehmann EH, Oswald SE (2010) Neutron radiography and tomography of water distribution in the root zone. J Plant Nutr Soil Sci 173:757–764
Estabrook EM, Yoder JI (1998) Plant-plant communications: rhizosphere signaling between parasitic angiosperms and their hosts. Plant Physiol 116:1–7
Fliessbach A, Mader PDA, Oberholzer HR, Gunst L (2007) Soil organic matter and biological soil quality indicators after 21 years of organic and conventional farming. Agric Ecosyst Environ 118:273–284. ST-Soil organic matter and biological s
Fontaine S, Barot S, Barré P, Bdioui N, Mary B, Rumpel C (2007) Stability of organic carbon in deep soil layers controlled by fresh carbon supply. Nature 450:277–280
Foster KR, Bell T (2012) Competition, not cooperation, dominates interactions among culturable microbial species. Curr Biol 22:1845–1850
Fox AR, Soto G, Valverde C, Russo D, Lagares A, Zorreguieta Á, Alleva K, Pascuan C, Frare R, Mercado-Blanco J, Dixon R, Ayub ND (2016) Major cereal crops benefit from biological nitrogen fixation when inoculated with the nitrogen-fixing bacterium Pseudomonas protegens Pf-5 X940. Environ Microbiol 18:3522–3534
Fromm H, Winter K, Filser J, Hantschel R, Beese F (1993) The influence of soil type and cultivation system on the spatial distributions of the soil fauna and microorganisms and their interactions. Geoderma 60:109–118
Garrigues E, Doussan C, Pierret A (2006) Water uptake by plant roots: I—Formation and propagation of a water extraction front in mature root systems as evidenced by 2D light transmission imaging. Plant Soil 283:83–98
Geddes BA, Ryu MH, Mus F, Garcia Costas A, Peters JW, Voigt CA, Poole P (2015) Use of plant colonizing bacteria as chassis for transfer of N2-fixation to cereals. Curr Opin Biotechnol 32:216–222
Geddes BA, Paramasivan P, Joffrin A, Thompson AL, Christensen K, Jorrin B, Brett P, Conway SJ, Oldroyd GED, Poole PS (2019) Engineering transkingdom signalling in plants to control gene expression in rhizosphere bacteria. Nat Commun 10:3430

Ghosh D, Sen S, Mohapatra S (2017) Modulation of proline metabolic gene expression in Arabidopsis thaliana under water-stressed conditions by a drought-mitigating Pseudomonas putida strain. Ann Microbiol 67:655–668
Gordon DM, Ryder MH, Heinrich K, Murphy PJ (1996) An experimental test of the rhizopine concept in Rhizobium meliloti. Appl Environ Microbiol 62:3991–3996
Graystone SJ, Wang S, Campbell CD, Edwards AC (1998) Selective influence of plant species on microbial diversity in the rhizosphere. Soil Biol Biochem 30:369–378
Großkopf T, Soyer OS (2014) Synthetic microbial communities. Curr Opin Microbiol 18:72–77
Gupta R, Singh A, Srivastava M, Singh V, Gupta MM, Pandey R (2017) Microbial modulation of bacoside A biosynthetic pathway and systemic defense mechanism in Bacopa monnieri under Meloidogyne incognita stress. Sci Rep 7:41867
Guyon P (1993) Transformed plants producing opines specifically promote growth of opine-degrading Agrobacteria. Mol Plant-Microbe Interact 6:92–98
Haas D, Défago G (2005) Biological control of soil-borne pathogens by fluorescent pseudomonads. Nat Rev Microbiol 3:307–319
Hallett PD, Gordon DC, Bengough AG (2003) Plant influence on rhizosphere hydraulic properties: direct measurements using a miniaturized infiltrometer. New Phytol 157:597–603
Halvorson AD, Wienhold BJ, Black AL (2002) Tillage, nitrogen, and cropping system effects on soil carbon sequestration. Soil Sci Soc Am J 66:906–912
Hao X, Chang C (2002) Effect of 25 annual cattle manure applications on soluble and exchangeable cations in soil. Soil Sci 167:126–134
Hao WY, Ren LX, Ran W, Shen QR (2010) Allelopathic effects of root exudates from watermelon and rice plants on Fusarium oxysporum f.sp. niveum. Plant Soil 336:485–497
Haynes RJ, Naidu R (1998) Influence of lime, fertilizer and manure applications on soil organic matter content and soil physical conditions: a review. Nutr Cycl Agroecosyst 51:123–137
He Y, Zhou X, Cheng W, Zhou L, Zhang G, Zhou G, Liu R, Shao J, Zhu K, Cheng W (2019) Linking improvement of soil structure to soil carbon storage following invasion by a C4 Plant Spartina alterniflora. Ecosystems 22:859–872
Hiltner L (1904) Über neurer erfahrungen und probleme auf dem gebiete der bodenbakteriologie unter besonderer berücksichtigung der gründüngung und brache. Arbeiten der Deitschen. Landwirtsch. Gesellschaft 98:59–78
Hinsa SM, Espinosa-Urgel M, Ramos JL, O'Toole GA (2003) Transition from reversible to irreversible attachment during biofilm formation by Pseudomonas fluorescens WCS365 requires an ABC transporter and a large secreted protein. Mol Microbiol 49:905–918
Hinsinger P, Bengough AG, Vetterlein D, Young IM (2009) Rhizosphere: biophysics, biogeochemistry and ecological relevance. Plant Soil 321:117–152
Hirsch PR, Miller AJ, Dennis PG (2013) Do root exudates exert more influence on rhizosphere bacterial community structure than other rhizodeposits? In: Molecular microbial ecology of the rhizosphere, pp 229–242
Huber K, Vanderborght J, Javaux M, Schröder N, Dodd IC, Vereecken H (2014) Modelling the impact of heterogeneous rootzone water distribution on the regulation of transpiration by hormone transport and/or hydraulic pressures. Plant Soil 384:93–112
Ibekwe AM, Kennedy AC, Frohne PS, Papiernik SK, Yang CH, Crowley DE (2002) Microbial diversity along a transect of agronomic zones. FEMS Microbiol Ecol 39:183–191
Jacinthe PA, Lal R, Kimble JM (2002) Carbon budget and seasonal carbon dioxide emission from a central Ohio Luvisol as influenced by wheat residue amendment. Soil Tillage Res 67:147–157
Jackson LE, Calderon FJ, Steenwerth KL, Scow KM, Rolston DE (2003) Responses of soil microbial processes and community structure to tillage events and implications for soil quality. Geoderma 114:305–317
Jacoby R, Peukert M, Succurro A, Koprivova A, Kopriva S (2017) The role of soil microorganisms in plant mineral nutrition—current knowledge and future directions. Front Plant Sci 8:1617

Jones DL, Nguyen C, Finlay RD (2009) Carbon flow in the rhizosphere: carbon trading at the soil-root interface. Plant Soil 321:5–33

Jousset A (2017) Application of protists to improve plant growth in sustainable agriculture. In: Rhizotrophs: plant growth promotion to bioremediation, pp 263–273

Kamalnath M, Rao MS, Umamaheswari R (2019) Rhizophere engineering with beneficial microbes for growth enhancement and nematode disease complex management in gherkin (Cucumis anguria L.). Sci Hortic 257:108681

Karandashov V, Bucher M (2005) Symbiotic phosphate transport in arbuscular mycorrhizas. Trends Plant Sci 10:22–29

Karatan E, Watnick P (2009) Signals, regulatory networks, and materials that build and break bacterial biofilms. Microbiol Mol Biol Rev 73:310–347

Kay BD (1998) Soil structure and organic carbon: a review. In: Soil processes and the carbon cycle, p 609

Kim SB, Timmusk S (2013) A simplified method for gene knockout and direct screening of recombinant clones for application in Paenibacillus polymyxa. PLoS One 8:e68092

Knox OGG, Gupta VVSR, Lardner R (2014) Field evaluation of the effects of cotton variety and GM status on rhizosphere microbial diversity and function in Australian soils. Soil Res 52:203–215

Köberl M, Ramadan EM, Adam M, Cardinale M, Hallmann J, Heuer H, Smalla K, Berg G (2013) Bacillus and streptomyces were selected as broad-spectrum antagonists against soilborne pathogens from arid areas in egypt. FEMS Microbiol Lett 342:168–178

Köberl M, White RA, Erschen S, El-Arabi TF, Jansson JK, Berg G (2015) Draft genome sequence of Streptomyces sp. strain Wb2n-11, a desert isolate with broadspectrum antagonism against soilborne phytopathogens. Genome Announc 3(4):e00860–e00815

Koebernick N, Weller U, Huber K, Schlüter S, Vogel H-J, Jahn R, Vereecken H, Vetterlein D (2014) In Situ visualization and quantification of three-dimensional root system architecture and growth using X-ray computed tomography. Vadose Zo J 13:vzj2014.03.0024

Koller R, Scheu S, Bonkowski M, Robin C (2013) Protozoa stimulate N uptake and growth of arbuscular mycorrhizal plants. Soil Biol Biochem 65:204–210

Koroney AS, Plasson C, Pawlak B, Sidikou R, Driouich A, Menu-Bouaouiche L, Vicré-Gibouin M (2016) Root exudate of solanum tuberosum is enriched in galactose-containing molecules and impacts the growth of pectobacterium atrosepticum. Ann Bot 118:797–808

Kostka SJ (2000) Amelioration of water repellency in highly managed soils and the enhancement of turfgrass performance through the systematic application of surfactants. J Hydrol 231–232:359–368

Kostka SJ, Cisar JL, Short JR, Mane S (1997) Evaluation of soil surfactants for the management of soil water repellency in turfgrass. Int Turfgrass Soc Res J 8:485–494

Kreuzer K, Adamczyk J, Iijima M, Wagner M, Scheu S, Bonkowski M (2006) Grazing of a common species of soil protozoa (Acanthamoeba castellanii) affects rhizosphere bacterial community composition and root architecture of rice (Oryza sativa L.). Soil Biol Biochem 38:1665–1672

Kroener E, Ahmed MA, Carminati A (2015) Roots at the percolation threshold. Phys Rev E 91:042706

Krome K, Rosenberg K, Bonkowski M, Scheu S (2009) Grazing of protozoa on rhizosphere bacteria alters growth and reproduction of Arabidopsis thaliana. Soil Biol Biochem 41:1866–1873

Krome K, Rosenberg K, Dickler C, Kreuzer K, Ludwig-Müller J, Ullrich-Eberius C, Scheu S, Bonkowski M (2010) Soil bacteria and protozoa affect root branching via effects on the auxin and cytokinin balance in plants. Plant Soil 328:191–201

Kumar A et al (2015) Urease activity and its kinetics in selected bench mark soils of Indo-Gangetic plains. Proc Natl Acad Sci India Sec B Biol Sci 85:407–413

Kuppardt A, Fester T, Härtig C, Chatzinotas A (2018) Rhizosphere protists change metabolite profiles in Zea mays. Front Microbiol 9:857

Kuzyakov Y, Cheng W (2001) Photosynthesis controls of rhizosphere respiration and organic matter decomposition. Soil Biol Biochem 33:1915–1925
Lal R, Jarecki MK (2003) Crop management for soil carbon sequestration. CRC Crit Rev Plant Sci 22:471–502
Lanoue A, Burlat V, Henkes GJ, Koch I, Schurr U, Röse USR (2010) De novo biosynthesis of defense root exudates in response to Fusarium attack in barley. New Phytol 185:577–588
Lareen A, Burton F, Schäfer P (2016) Plant root-microbe communication in shaping root microbiomes. Plant Mol Biol 90:575–587
Latour X, Corberand T, Laguerre G, Allard F, Lemanceau P (1996) The composition of fluorescent pseudomonad populations associated with roots is influenced by plant and soil type. Appl Environ Microbiol 62:2449–2456
Latour X, Philippot L, Corberand T, Lemanceau P (1999) The establishment of an introduced community of fluorescent pseudomonads in the soil and in the rhizosphere is affected by the soil type. FEMS Microbiol Ecol 30:163–170
Lau JA, Lennon JT (2012) Rapid responses of soil microorganisms improve plant fitness in novel environments. Proc Natl Acad Sci U S A 109:14058–14062
Leifheit EF, Veresoglou SD, Lehmann A, Morris EK, Rillig MC (2014) Multiple factors influence the role of arbuscular mycorrhizal fungi in soil aggregation-a meta-analysis. Plant Soil 374:523–537
Lemanceau P, Corberand T, Gardan L, Latour X, Laguerre G, Boeufgras JM, Alabouvette C (1995) Effect of two plant species, flax (Linum usitatissinum L.) and Tomato (Lycopersicon esculentum Mill.), on the diversity of soilborne populations of fluorescent pseudomonads. Appl Environ Microbiol 61:1004–1012
Li X, Rui J, Mao Y, Yannarell A, Mackie R (2014) Dynamics of the bacterial community structure in the rhizosphere of a maize cultivar. Soil Biol Biochem 68:392–401
Liljeroth E, Kuikman P, Van Veen JA (1994) Carbon translocation to the rhizosphere of maize and wheat and influence on the turnover of native soil organic matter at different soil nitrogen levels. Plant Soil 161:233–240
Ling N, Zhang W, Wang D, Mao J, Huang Q, Guo S, Shen Q (2013) Root exudates from grafted-root watermelon showed a certain contribution in inhibiting Fusarium oxysporum f. sp. niveum. PLoS One 8(5):e63383
Ling N, Deng K, Song Y, Wu Y, Zhao J, Raza W, Huang Q, Shen Q (2014) Variation of rhizosphere bacterial community in watermelon continuous mono-cropping soil by long-term application of a novel bioorganic fertilizer. Microbiol Res 169:570–578
Liu F, Hewezi T, Lebeis SL, Pantalone V, Grewal PS, Staton ME (2019) Soil indigenous microbiome and plant genotypes cooperatively modify soybean rhizosphere microbiome assembly. BMC Microbiol 19:201
Lobell DB, Field CB (2007) Global scale climate-crop yield relationships and the impacts of recent warming. Environ Res Lett 2:2–9
Lobet G, Couvreur V, Meunier F, Javaux M, Draye X (2014) Plant water uptake in drying soils. Plant Physiol 164:1619–1627
Logan TJ, Lal R, Dick WA (1991) Tillage systems and soil properties in North America. Soil Tillage Res 20:241–270
López D, Kolter R (2010) Extracellular signals that define distinct and coexisting cell fates in Bacillus subtilis. FEMS Microbiol Rev 34:134–149
Lopez D, Vlamakis H, Kolter R (2010) Biofilms. CSH Perspect Biol 2:1
Lu T, Ke M, Lavoie M, Jin Y, Fan X, Zhang Z, Fu Z, Sun L, Gillings M, Peñuelas J, Qian H, Zhu YG (2018a) Rhizosphere microorganisms can influence the timing of plant flowering. Microbiome 6:231
Lu T, Ke M, Peijnenburg WJGM, Zhu Y, Zhang M, Sun L, Fu Z, Qian H (2018b) Investigation of rhizospheric microbial communities in wheat, barley, and two rice varieties at the seedling stage. J Agric Food Chem 66:2645–2653

Lugtenberg B, Kamilova F (2009) Plant-growth-promoting rhizobacteria. Annu Rev Microbiol 63:541–556

Lynch JP, Brown KM (2012) New roots for agriculture: exploiting the root phenome. Philos Trans R Soc B Biol Sci 367:1598–1604

Lynch JP, Wojciechowski T (2015) Opportunities and challenges in the subsoil: pathways to deeper rooted crops. J Exp Bot 66:2199–2210

Mandal SM, Chakraborty D, Dey S (2010) Phenolic acids act as signaling molecules in plant-microbe symbioses. Plant Signal Behav 5:359–368

Marasco R, Rolli E, Vigani G, Borin S, Sorlini C, Ouzari H, Zocchi G, Daffonchio D (2013) Are drought-resistance promoting bacteria cross-compatible with different plant models? Plant Signal Behav 8(10):e26741

Marques JM, da Silva TF, Vollu RE, Blank AF, Ding GC, Seldin L, Smalla K (2014) Plant age and genotype affect the bacterial community composition in the tuber rhizosphere of field-grown sweet potato plants. FEMS Microbiol Ecol 88:424–435

Marschner H (2012) Mineral nutrition of higher plants, 3rd edn. Academic, London

Martens DA (2000) Plant residue biochemistry regulates soil carbon cycling and carbon sequestration. Soil Biol Biochem 32:361–369

McCully M (1995) How do real roots work? Some new views of root structure. Plant Physiol 109:1–6

McCully ME, Boyer JS (1997) The expansion of maize root-cap mucilage during hydration. 3. Changes in water potential and water content. Physiol Plant 99:169–177

Medema MH, Breitling R, Takano E (2011) Synthetic biology in streptomyces bacteria. Methods Enzymol 497:485–502

Meharg AA, Killham KA (1988) Comparison of carbon flow from pre-labelled and pulse-labelled plants. Plant Soil 112:225–231

Mendes R, Garbeva P, Raaijmakers JM (2013) The rhizosphere microbiome: significance of plant beneficial, plant pathogenic, and human pathogenic microorganisms. FEMS Microbiol Rev 37:634–663

Miethling R, Wieland G, Backhaus H, Tebbe CC (2000) Variation of microbial rhizosphere communities in response to crop species, soil origin, and inoculation with Sinorhizobium meliloti L33. Microb Ecol 40:43–56

Monds RD, O'Toole GA (2009) The developmental model of microbial biofilms: ten years of a paradigm up for review. Trends Microbiol 17:73–87

Mondy S, Lenglet A, Beury-Cirou A, Libanga C, Ratet P, Faure D, Dessaux Y (2014) An increasing opine carbon bias in artificial exudation systems and genetically modified plant rhizospheres leads to an increasing reshaping of bacterial populations. Mol Ecol 23:4846–4861

Moore JM, Klose S, Tabatabai MA (2000) Soil microbial biomass carbon and nitrogen as affected by cropping systems. Biol Fertil Soils 31:200–210

Moradi AB, Carminati A, Lamparter A, Woche SK, Bachmann J, Vetterlein D, Vogel H-J, Oswald SE (2012) Is the rhizosphere temporarily water repellent? Vadose Zo J 11:vzj2011.0120

Moradi AB, Hopmans JW, Oswald SE, Menon M, Carminati A, Lehmann E, et al (2013) Applications of neutron imaging in soil-water-root systems. In: Anderson SH, Hopmans JW (eds) Soil-water-root processes: advances in tomography and imaging. SSSA Spec. Publ. 61. SSSA, Madison, WI, p 113–136

Murphy PJ, Heycke N, Banfalvi Z (1987) Genes for the catabolism and synthesis of an opine-like compound in Rhizobium meliloti are closely linked and on the Sym plasmid. Proc Natl Acad Sci U S A 84:493–497

Murphy PJ, Trenz SP, Grzemski W, De Bruijn FJ, Schell J (1993) The Rhizobium meliloti rhizopine mos locus is a mosaic structure facilitating its symbiotic regulation. J Bacteriol 175:5193–5204

Murphy PJ, Wexler W, Grzemski W, Rao JP, Gordon D (1995) Rhizopines-their role in symbiosis and competition. Soil Biol Biochem 27:525–529

Nalayini P, Sankaranarayanan K, Velmourougane K (2013) Herbigation in cotton (Gossypium spp): effects on weed control, soil microflora and succeeding greengram (Vigna radiata). Indian J Agric Sci 83:1144–1148

Nguyen C (2009) Rhizodeposition of organic C by plant: mechanisms and controls. In: Sustainable agriculture, pp 97–123

O'Callaghan M (2016) Microbial inoculation of seed for improved crop performance: issues and opportunities. Appl Microbiol Biotechnol 100:5729–5746

O'Connell KP, Goodman RM, Handelsman J (1996) Engineering the rhizosphere: expressing a bias. Trends Biotechnol 14:83–88

Oger P, Petit A, Dessaux Y (1997) Genetically engineered plants producing opines alter their biological environment. Nat Biotechnol 15:369–372

Okubo T, Tokida T, Ikeda S, Bao Z, Tago K, Hayatsu M, Nakamura H, Sakai H, Usui Y, Hayashi K, Hasegawa T, Minamisawa K (2014) Effects of elevated carbon dioxide, elevated temperature, and rice growth stage on the community structure of rice root-associated bacteria. Microbes Environ 29:184–190

Orr CH, James A, Leifert C, Cooper JM, Cummings SP (2011) Diversity and activity of free-living nitrogen-fixing bacteria and total bacteria in organic and conventionally managed soils. Appl Environ Microbiol 77:911–919

Øvreås L, Torsvik V (1998) Microbial diversity and community structure in two different agricultural soil communities. Microb Ecol 36:303–315

Panke-Buisse K, Poole AC, Goodrich JK, Ley RE, Kao-Kniffin J (2015) Selection on soil microbiomes reveals reproducible impacts on plant function. ISME J 9:980–989

Pankievicz VCS, Do Amaral FP, Santos KFDN, Agtuca B, Xu Y, Schueller MJ, Arisi ACM, Steffens MBR, De Souza EM, Pedrosa FO, Stacey G, Ferrieri RA (2015) Robust biological nitrogen fixation in a model grass-bacterial association. Plant J 81:907–919

Parke JL (1991) Root colonization by indigenous and introduced microorganisms. In: The rhizosphere and plant growth, p 33–42

Patten CL, Glick BR (2002) Role of Pseudomonas putida indoleacetic acid in development of the host plant root system. Appl Environ Microbiol 68:3795–3801

Peiffer JA, Spor A, Koren O, Jin Z, Tringe SG, Dangl JL, Buckler ES, Ley RE (2013) Diversity and heritability of the maize rhizosphere microbiome under field conditions. Proc Natl Acad Sci U S A 110:6548–6553

Pérez-Jaramillo JE, Mendes R, Raaijmakers JM (2016) Impact of plant domestication on rhizosphere microbiome assembly and functions. Plant Mol Biol 90:635–644

Petit A, Stougaard J, Kühle A, Marcker KA, Tempé J (1987) Transformation and regeneration of the legume Lotus corniculatus: a system for molecular studies of symbiotic nitrogen fixation. MGG Mol Gen Genet 207:245–250

Plante AF, McGill WB (2002) Soil aggregate dynamics and the retention of organic matter in laboratory-incubated soil with differing simulated tillage frequencies. Soil Tillage Res 66:79–92

Pohlmeier A, Haber-Pohlmeier S, Javaux M, Vereecken H (2015) Magnetic resonance imaging techniques for visualization of root growth and root water uptake processes, pp 137–156

Prasad R, Kumar M, Varma A (2015) In: Egamberdieva D, Shrivastava S, Varma A (eds) Plant-growth-promoting rhizobacteria (PGPR) and medicinal plants. Springer International Publishing, pp 247–260

Preece C, Peñuelas J (2016) Rhizodeposition under drought and consequences for soil communities and ecosystem resilience. Plant Soil 409

Preece C, Farré-Armengol G, Llusià J, Peñuelas J (2018) Thirsty tree roots exude more carbon. Tree Physiol 38:690–695

Raaijmakers JM, Paulitz TC, Steinberg C, Alabouvette C, Moënne-Loccoz Y (2009) The rhizosphere: a playground and battlefield for soilborne pathogens and beneficial microorganisms. Plant Soil 321:341–361

Ray SK et al (2014) Soil and land quality indicators of the Indo-Gangetic Plains of India. Curr Sci 107:1470–1486

Raychaudhuri M et al (2014) Soil physical quality of the Indo-Gangetic Plains and black soil region. Curr Sci 107:1440–1451

Read DB, Bengough AG, Gregory PJ, Crawford JW, Robinson D, Scrimgeour CM, Young IM, Zhang K, Zhang X (2003) Plant roots release phospholipid surfactants that modify the physical and chemical properties of soil. New Phytol 157:315–326

Rillig MC, Aguilar-Trigueros AC, Bergmann J, Verbruggen E, Veresoglou SD, Lehmann A (2015) Plant root and mycorrhizal fungal traits for understanding soil aggregation. New Phytol 205:1385–1388

Ryan PR, Dessaux Y, Thomashow LS, Weller DM (2009) Rhizosphere engineering and management for sustainable agriculture. Plant Soil 321:363–383

Sandhya V, Skz A, Grover M, Reddy G, Venkateswarlu B (2009) Alleviation of drought stress effects in sunflower seedlings by the exopolysaccharides producing Pseudomonas putida strain GAP-p45. Biol Fertil Soils 46:17–26

Sandhya V, Ali SZ, Grover M, Reddy G, Venkateswarlu B (2010a) Effect of plant growth promoting Pseudomonas spp. on compatible solutes, antioxidant status and plant growth of maize under drought stress. Plant Growth Regul 62:21–30

Sandhya V, Ali SZ, Venkateswarlu B, Reddy G, Grover M (2010b) Effect of osmotic stress on plant growth promoting pseudomonas spp. Arch Microbiol 192:867–876

Savka MA, Farrand SK (1997) Modification of rhizobacterial populations by engineering bacterium utilization of a novel plant-produced resource. Nat Biotechnol 15:363–368

Savka MA, Dessaux Y, Mcspadden Gardener BB, Mondy S, Kohler PRA, de Bruijn FJ, Rossbach S (2013) The 'biased rhizosphere' concept and advances in the omics era to study bacterial competitiveness and persistence in the phytosphere. Mol Microb Ecol Rhizosph 2:1145–1161

Schloter M, Dilly O, Munch JC (2003) Indicators for evaluating soil quality. Agric Ecosyst Environ 98:255–262

Schreiter S, Ding GC, Heuer H, Neumann G, Sandmann M, Grosch R, Kropf S, Smalla K (2014) Effect of the soil type on the microbiome in the rhizosphere of field-grown lettuce. Front Microbiol 5:144

Schutter ME, Dick RP (2002) Microbial community profiles and activities among aggregates of winter fallow and cover-cropped soil. Soil Sci Soc Am J 66:142

Sen S, Ghosh D, Mohapatra S (2018) Modulation of polyamine biosynthesis in Arabidopsis thaliana by a drought mitigating Pseudomonas putida strain. Plant Physiol Biochem 129:180–188

Sharifi R, Lee SM, Ryu CM (2018) Microbe-induced plant volatiles. New Phytol 220:684–691

Sharma A, Satyanarayana T (2013) Comparative genomics of Bacillus species and its relevance in industrial microbiology. Genomics Insights 6:25–36

Sidhu GS et al (2014) Impact of management levels and land-use changes on soil properties in rice-wheat cropping system of the Indo-Gangetic Plains. Curr Sci 107:1487–1501

Simovic S, Tamburic S, Milic-Askrabic J, Rajic D (1999) An investigation into interactions between polyacrylic polymers and a non-ionic surfactant: an emulsion preformulation study. Int J Pharm 184:207–217

Singh BK, Bardgett RD, Smith P, Reay DS (2010) Microorganisms and climate change: terrestrial feedbacks and mitigation options. Nat Rev Microbiol 8:779–790

Singh J, Velmourougane K, Kranthi S, Shanty V (2014) Evaluation of organic and inorganic cultivation systems of cotton on bollworm, soil microbial population and yield. Cotton Res J 5:117–119

Six J, Bossuyt H, Degryze S, Denef K (2004) A history of research on the link between (micro)-aggregates, soil biota, and soil organic matter dynamics. Soil Tillage Res 79:7–31

Six J, Frey SD, Thiet RK, Batten KM (2006) Bacterial and fungal contributions to carbon sequestration in agroecosystems. Soil Sci Soc Am J 70:555–569

Smalla K, Wieland G, Buchner A, Zock A, Parzy J, Kaiser S, Roskot N, Heuer H, Berg G (2001) Bulk and rhizosphere soil bacterial communities studied by denaturing gradient gel

electrophoresis: plant-dependent enrichment and seasonal shifts revealed. Appl Environ Microbiol 67:4742–4751
Soussi A, Ferjani R, Marasco R, Guesmi A, Cherif H, Rolli E, Mapelli F, Ouzari HI, Daffonchio D, Cherif A (2016) Plant-associated microbiomes in arid lands: diversity, ecology and biotechnological potential. Plant Soil 405:357–370
Sposito G (2013) Green water and global food security. Vadose Zo J 12:vzj2013.02.0041
Srivastava AK et al (2014) Impacts of agro-climates and land use systems on culturable microbial population in soils of the Indo-Gangetic Plains, India. Curr Sci 107:1464–1469
Steinauer K, Chatzinotas A, Eisenhauer N (2016) Root exudate cocktails: the link between plant diversity and soil microorganisms? Ecol Evol 6:7387–7396
Su J, Hu C, Yan X, Jin Y, Chen Z, Guan Q, Wang Y, Zhong D, Jansson C, Wang F, Schnürer A, Sun C (2015) Expression of barley SUSIBA2 transcription factor yields high-starch low-methane rice. Nature 523:602–606
Tardieu F, Draye X, Javaux M (2017) Root water uptake and ideotypes of the root system: whole-plant controls matter. Vadose Zo J 16:vzj2017.05.0107
Tewari S, Arora NK (2014) Multifunctional exopolysaccharides from Pseudomonas aeruginosa PF23 involved in plant growth stimulation, biocontrol and stress amelioration in sunflower under saline conditions. Curr Microbiol 69:484–494
Tian T, Reverdy A, She Q, Sun B, Chai Y (2020) The role of rhizodeposits in shaping rhizomicrobiome. Environ Microbiol Rep 12:160–172
Triveni S, Prasanna R, Shukla L, Saxena AK (2013) Evaluating the biochemical traits of novel Trichoderma-based biofilms for use as plant growth-promoting inoculants. Ann Microbiol 63:1147–1156
Tron S, Bodner G, Laio F, Ridolfi L, Leitner D (2015) Can diversity in root architecture explain plant water use efficiency? A modeling study. Ecol Modell 312:200–210
Van Der Ent S, Van Hulten M, Pozo MJ, Czechowski T, Udvardi MK, Pieterse CMJ, Ton J (2009) Priming of plant innate immunity by rhizobacteria and β-aminobutyric acid: differences and similarities in regulation. New Phytol 183:419–431
Vandenkoornhuyse P, Quaiser A, Duhamel M, Le Van A, Dufresne A (2015) The importance of the microbiome of the plant holobiont. New Phytol 206:1196–1206
Vardharajula S, Ali SZ, Grover M, Reddy G, Bandi V (2011) Drought-tolerant plant growth promoting bacillus spp.: effect on growth, osmolytes, and antioxidant status of maize under drought stress. J Plant Interact 6:1–14
Velmourougane K (2016) Impact of organic and conventional systems of coffee farming on soil properties and culturable microbial diversity. Scientifica (Cairo) 2016:3604026
Velmourougane K, Blaise D (2014) Transgenic cotton and its impact on microbial diversity. In: Maheshwari DK (ed) Bacterial diversity in sustainable agriculture. Springer International Publishing, Basel, pp 191–204
Velmourougane K, Prasanna R (2017) Influence of L-amino acids on aggregation and biofilm formation in Azotobacter chroococcum and Trichoderma viride. J Appl Microbiol 123:977–991
Velmourougane K, Sahu A (2013) Impact of transgenic cottons expressing cry1Ac on soil biological attributes. Plant Soil Environ 59:108–114
Velmourougane K, Venugopalan MV, Bhattacharyya T, Sarkar D, Pal DK, Sahu A, Ray SK, Nair KM, Prasad J, Singh RS (2013) Soil dehydrogenase activity in agro-ecological sub regions of black soil regions in India. Geoderma 197–198:186–192
Velmourougane K et al (2014) Impacts of bioclimates, cropping systems, land use and management on the cultural microbial population in black soil regions of India. Curr Sci 107:1452–1463
Velmourougane K, Prasanna R, Saxena AK, Singh SB, Chawla G, Kaushik R, Ramakrishnan B, Nain L (2017) Modulation of growth media influences aggregation and biofilm formation between Azotobacter chroococcum and Trichoderma viride. Appl Biochem Microbiol 53:546–556
Velmourougane K, Prasanna R, Saxena AK (2017a) Agriculturally important microbial biofilms: present status and future prospects. J Basic Microbiol 57:548–573

Velmourougane K, Prasanna R, Singh S, Chawla G, Kumar A, Saxena AK (2017b) Modulating rhizosphere colonisation, plant growth, soil nutrient availability and plant defense enzyme activity through Trichoderma viride-Azotobacter chroococcum biofilm inoculation in chickpea. Plant Soil 421:157–174

Velmourougane K, Prasanna R, Singh SB, Kumar R, Saha S (2017c) Sequence of inoculation influences the nature of extracellular polymeric substances and biofilm formation in Azotobacter chroococcum and Trichoderma viride. FEMS Microbiol Ecol 93

Velmourougane K, Saxena G, Prasanna R (2017d) Plant-microbe interactions in the rhizosphere: mechanisms and their ecological benefits. In: Plant-microbe interactions in agro-ecological perspectives, pp 193–219

Velmourougane K, Prasanna R, Chawla G, Nain L, Kumar A, Saxena AK (2019a) Trichoderma–Azotobacter biofilm inoculation improves soil nutrient availability and plant growth in wheat and cotton. J Basic Microbiol 59:632–644

Velmourougane K, Prasanna R, Supriya P, Ramakrishnan B, Thapa S, Saxena AK (2019b) Transcriptome profiling provides insights into regulatory factors involved in Trichoderma viride-Azotobacter chroococcum biofilm formation. Microbiol Res 227:126292

Vicré M, Santaella C, Blanchet S, Gateau A, Driouich A (2005) Root border-like cells of Arabidopsis. Microscopical characterization and role in the interaction with rhizobacteria. Plant Physiol 138:998–1008

Wagner MR, Lundberg DS, Coleman-Derr D, Tringe SG, Dangl JL, Mitchell-Olds T (2014) Natural soil microbes alter flowering phenology and the intensity of selection on flowering time in a wild Arabidopsis relative. Ecol Lett 17:717–726

Wallenstein MD, Weintraub MN (2008) Emerging tools for measuring and modeling the in situ activity of soil extracellular enzymes. Soil Biol Biochem 40:2098–2106

Wasson AP, Richards RA, Chatrath R, Misra SC, Prasad SVS, Rebetzke GJ, Kirkegaard JA, Christopher J, Watt M (2012) Traits and selection strategies to improve root systems and water uptake in water-limited wheat crops. J Exp Bot 63:3485–3498

Watson BS, Bedair MF, Urbanczyk-Wochniak E, Huhman DV, Yang DS, Allen SN, Li W, Tang Y, Sumner LW (2015) Integrated metabolomics and transcriptomics reveal enhanced specialized metabolism in Medicago truncatula root border cells. Plant Physiol 167:1699–1716

Watt M, McCully ME, Canny MJ (1994) Formation and stabilization of rhizosheaths of Zea mays L. Effect of soil water content. Plant Physiol 106:179–186

Weiller F, Moore JP, Young P, Driouich A, Vivier MA (2017) The brassicaceae species heliophila coronopifolia produces root border-like cells that protect the root tip and secrete defensin peptides. Ann Bot 119:803–813

Weisskopf L, Abou-Mansour E, Fromin N, Tomasi N, Santelia D, Edelkott I, Neumann G, Aragno M, Tabacchi R, Martinoia E (2006) White lupin has developed a complex strategy to limit microbial degradation of secreted citrate required for phosphate acquisition. Plant Cell Environ 29:919–927

Weisskopf L, Heller S, Eberl L (2011) Burkholderia species are major inhabitants of white lupin cluster roots. Appl Environ Microbiol 77:7715–7720

Weller D (1988) Biological control of soilborne plant pathogens in the rhizosphere with bacteria. Annu Rev Phytopathol 26:379–407

Wintermans PCA, Bakker PAHM, Pieterse CMJ (2016) Natural genetic variation in Arabidopsis for responsiveness to plant growth-promoting rhizobacteria. Plant Mol Biol 90:623–634

Wu T, Chellemi DO, Graham JH, Martin KJ, Rosskopf EN (2008) Comparison of soil bacterial communities under diverse agricultural land management and crop production practices. Microb Ecol 55:293–310

Xu W, Liu D, Wu F, Liu S (2014) Root exudates of wheat are involved in suppression of Fusarium wilt in watermelon in watermelon-wheat companion cropping. Eur J Plant Pathol 141:209–216

Yang H, Knapp J, Koirala P, Rajagopal D, Peer WA, Silbart LK, Murphy A, Gaxiola RA (2007) Enhanced phosphorus nutrition in monocots and dicots over-expressing a phosphorus-responsive type I H+-pyrophosphatase. Plant Biotechnol J 5:735–745

Yang J, Kloepper JW, Ryu CM (2009) Rhizosphere bacteria help plants tolerate abiotic stress. Trends Plant Sci 14:1–4
York LM, Carminati A, Mooney SJ, Ritz K, Bennett MJ (2016) The holistic rhizosphere: integrating zones, processes, and semantics in the soil influenced by roots. J Exp Bot 67:3629–3643
Young IM (1995) Variation in moisture contents between bulk soil and the rhizosheath of wheat (Triticum aestivum L. cv. Wembley). New Phytol 130:135–139
Yuan J, Chaparro JM, Manter DK, Zhang R, Vivanco JM, Shen Q (2015) Roots from distinct plant developmental stages are capable of rapidly selecting their own microbiome without the influence of environmental and soil edaphic factors. Soil Biol Biochem 89:206–209
Zarebanadkouki M, Ahmed MA, Carminati A (2016) Hydraulic conductivity of the root-soil interface of lupin in sandy soil after drying and rewetting. Plant Soil 398:267–280
Zarraonaindia I, Owens SM, Weisenhorn P, West K, Hampton-Marcell J, Lax S, Bokulich NA, Mills DA, Martin G, Taghavi S, van der Lelie D, Gilbert JA (2015) The soil microbiome influences grapevine-associated microbiota. MBio 6
Zhalnina K, Louie KB, Hao Z, Mansoori N, Da Rocha UN, Shi S, Cho H, Karaoz U, Loqué D, Bowen BP, Firestone MK, Northen TR, Brodie EL (2018) Dynamic root exudate chemistry and microbial substrate preferences drive patterns in rhizosphere microbial community assembly. Nat Microbiol 3:470–480
Zickenrott IM, Woche SK, Bachmann J, Ahmed MA, Vetterlein D (2016) An efficient method for the collection of root mucilage from different plant species—A case study on the effect of mucilage on soil water repellency. J Plant Nutr Soil Sci 179:294–302

Chapter 14
Endospheric Microbiome-Assisted Alteration in the Metabolomic Profiling of Host towards Abiotic Stress Mitigation

Srishti Kar, Priya Jaiswal, Sankalp Misra, and Puneet Singh Chauhan

Abstract Abiotic stress such as salinity, low or high temperatures, alkalinity, drought, and other environmental extremes may be defined as a negative effect of non-living factors on a living thing thus inhibiting plant function. Abiotic stress tends to negatively impact growth, development, seed quality, and yield of the crop and other plants. To overcome this problem, various researches are being carried out in genetic engineering to develop plant varieties that are tolerant against abiotic stress. An alternative strategy has also been observed in the present scenario, where microbial endophytes play a key role in plant survival under abiotic stress. Endophytes that live internally in plant tissues for a part of their life cycle are known to regulate homeostasis in plants during stressed environmental conditions. This potential of endophytes to promote plant growth during abiotic stress has been explored with several in vitro studies. Several mechanisms that are employed by endophytes to overcome abiotic stress include accumulation of stress responsible molecules, secondary metabolites, increased production of phytohormones, and production of antioxidant enzymes. The tools of omics can be used further to provide detailed insight into how endophytic diversity influences the metabolomics of hosts during abiotic stresses. This chapter mainly emphasizes on the endophyte microbiome and its role in altering the mechanisms of a host to mitigate abiotic stress.

Keywords Abiotic stress · Biotic stress · Phytohormones · Secondary metabolites

S. Kar
Microbial Technology Division, Council of Scientific and Industrial Research-National Botanical Research Institute (CSIR-NBRI), Rana Pratap Marg, Lucknow, India

P. Jaiswal · S. Misra · P. S. Chauhan (✉)
Microbial Technology Division, Council of Scientific and Industrial Research-National Botanical Research Institute (CSIR-NBRI), Rana Pratap Marg, Lucknow, India

Academy of Scientific and Innovative Research (AcSIR), Ghaziabad, India
e-mail: puneet@nbri.res.in

R. N. Pudake et al. (eds.), *Omics Science for Rhizosphere Biology*, Rhizosphere Biology, https://doi.org/10.1007/978-981-16-0889-6_14

14.1 Introduction

The word endophyte refers to something that is present inside the plant. Thus, in literal terms, endophyte could refer simply to the location of the organism. However, it has been this term is not very useful in case of the absence of a mutualistic relationship. The usage of this term is as broad as its literal definition and spectrum of potential hosts and inhabitants, e.g., bacteria (Kobayashi and Palumbo 2000), fungi (Stone et al. 2000), plants (Marler et al. 1999), and insects in plants (Feller 1995), but also for algae within algae (Peters 1991). There have been various definitions for the term endophyte, but the "endophytes" can most commonly be defined as those organisms whose "...infections are inconspicuous, the infected host tissues are at least transiently symptomless, and the microbial colonization can be demonstrated to be internal..." (Stone et al. 2000). Petrini's characterization of endophytic interactions as not "causing apparent harm" (Petrini 1991), which presumably refers to an absence of macroscopically visible symptoms further adds validity to the description of the definition of endophytes. The term "endophyte" can be used to describe both bacteria and fungi that can be detected at a particular moment within the tissues of apparently healthy plant hosts (Schulz and Boyle 2005).

The plant without endophytes can be contemplated as the unconventional one to that found in nature (Partida-Martinez and Heil 2011). Endophytes lacking plants are basically prone to stress circumstances related to the environment and often lack in the potential to hold back the pathogens (Timmusk et al. 2011). The endophytes origin is quite complex to be interpreted, as it possesses a multifaceted association with the host plant and also because of the host's natural surroundings and their unique ability to multiply. However, endophytes nativity has been hypothesized by two theories one being the endogenous theory other being the exogenous theory. **Endogenous hypothesis** concluded that the nativity of endophytes appeared previously from the mitochondria and chloroplast of the plant, and hence possess a commensurate genetic background to the host (Wen 2004). **Exogenous hypothesis** proposes that endophytes are from the external environment of the plant and were inculcated into the host from the various surface, induced channels, or root wounds (Li 2005).

It has been known that host plant roots are colonized by variants of the microbial community including bacteria, algae, fungi, and actinomycetes (Mishra et al. 2017; Misra et al. 2019; Dixit et al. 2020; Prashar et al. 2014; Saharan and Nehra 2011). The transmission of endophytes takes place across the seeds between the soil rhizosphere. Spread is quick between the endo-rhizosphere through the lateral root junction incited by nematodes or microbial phytopathogens (Chi et al. 2005). Also, there is another route present for the bacterial endophytes to enter their host plant roots which is through the available spaces present between the epidermal cells and root hairs (Hardoim et al. 2008). The most common endophytic fungi isolated and identified from numerous plants are *Penicillium* sp., *Alternaria infectoria*, *Colletotrichum musae, Aspergillus* sp., *Colletotrichum gloeosporioides*, *Nigrospora sphaerica, Nigrospora oryzae*, *Phomopsis* sp., *Guignardia* sp., *Rhizoctonia* sp.,

Cordana musae, *Phialocephala sphaeroides*, *P. chrysogenum*, *A. alternata* (Fouda et al. 2015;Wilson et al. 2004).

14.2 Endophytes and Their Potential Roles

Endophytes and plants exhibit a mutualistic relation, where endophytes tend to colonize in plant tissues internally, obtaining protection and nutrition from them and exhibits a favour to the host plant in return by providing induced fitness with the help of production of certain metabolites. Along with these secreted functional metabolites, there are several other mechanisms that play a potential role in keeping the plant healthy and briefly discussed below.

14.2.1 As Plant Growth Promoters

The plants that are infected with endophytes often tend to grow faster than non-infected ones. Hence, it has been interpreted that this effect exists due to the ability of endophytes to produce phytohormones such as cytokines, indole-3-acetic acid (IAA), and other plant growth-promoting substances. Also, this can be partly considered that the host's uptake of nutritional elements such as nitrogen and phosphorus can be readily induced by endophytes potential functions (Papik et al. 2020).

Microbes involved in plant growth promotion are found to be associated with many plant species that have beneficial effects involving extensive growth of plant and less prone to diseases that are caused by various categories of plant–pathogen including nematodes, fungi, viruses, and bacteria (Bisht et al. 2020; Dixit et al. 2016). The primary role of plant growth promotion is carried out by various beneficial activities such as phosphate solubilization, nutrient uptake, and release of indole-3-acetic acid (IAA), gibberellins, cytokinins, siderophores, and antagonism to phytopathogens (Misra et al. 2017). Plant growth-promoting microbes can also activate physical and chemical changes by a unique process termed as induced systemic resistance helping in the protection of plants.

"Induced resistance" may be explained as a unique process where endophytes enhance plant defences against various pathogens. It is a form of resistance that is initiated by several chemical and biological agents thus helping plants to evolve against the pathogen. Various bacterial factors such as salicylic acid, siderophores, lipopolysaccharides, N-acyl- homoserine lactones, antibiotics, jasmonic acid, and volatiles (e.g., acetoin) are responsible for the induction of ISR (Bordiec et al. 2011). Induced systemic resistance involves the protection of plants by various defence mechanisms against pathogens and herbivorous insects. Endophytic bacteria have been considered to enhance induced systemic resistance through the induction of salicylic acid, plant hormones such as jasmonic acid (JA) and ethylene (ET) also possess essential roles in regulation of signalling pathways involved in initiation of ISR (Pieterse et al. 2012). *Pseudomonas fluorescens* 89B-61 is known to be the first reported endophytic bacterium that imparted ISR to protect cucumber plants against cucumber anthracnose (Kloepper and Ryu 2006).

14.2.2 Endophytes in Mitigation of Abiotic Stress

Agricultural productivity is hugely governed by environmental variations. Alterations in environmental conditions is one of the major causes of increasing abiotic stresses. Salinity, alkalinity, drought, nutrient deficiency, high and low temperatures, heavy metal toxicity, and pesticide stress are the major categories of abiotic stresses faced by plants. In field conditions, crops are subjected to a combination of several abiotic stresses. To cope up with these variations, plants synchronize with their symbiotic microbial partners. Endophytes that live in close association with their host often contribute to the better survival and growth of their partners. Endophytes employ several mechanisms to improve their hosts endurance to abiotic stresses (Table 14.1).

Plants have developed several morphophysiological, biochemical, and molecular responses to cope with drought stress, and endophytic bacteria via their close associations are capable of altering these processes and improve tolerance in plants. Tolerance to abiotic stress is regulated by changes in osmolytes, relative water content, generation of antioxidants, and regulation of phytohormones.

Osmolyte regulation is an effective mechanism to maintain hydration in cells. Osmotic adjustment is maintained by the active accumulation of osmoprotectants also known as compatible solutes. These compatible solutes are highly soluble, electrically neutral and pose low toxicity to cells at high concentrations. Sugars, glycine, betaine, inorganic ions (e.g., calcium), organic acids (e.g., malate), and proline are the major constituents of osmoprotectants. The term osmoprotectants has been denoted for these solutes as they act as protectants for cellular organelles, enzymes, proteins, cell membranes, and genetic materials during stress conditions like drought (Ullah et al. 2017). Various research works have illustrated the role of endophytes in osmotic regulation. For example, endophytic bacteria, *Sphingomonas* sp. LK11 isolated from the leaves of *Tephrosia apollinea* was found to enhance the drought tolerance in soybean plants. In an in vitro drought-induced stress, soybean plants treated with *Sphingomonas* LK11 produced higher amounts of sugars and amino acids (proline glycine and glutamate) which improved the osmoregulation (Asaf et al. 2017). *Arthrobacter* sp. and *Bacillus* sp. isolated from pepper seeds (*Capsicum annuum* L.) were found to increase proline content under in vitro osmotic stress experiment (Sziderics et al. 2007).

Table 14.1 Potential endophytes and their beneficial uses as plant growth promoters and as effective biocontrol agents for sustainable agriculture

Endophyte	Source of isolation	Application	Citation and year
Aspergillus fumigatus sp. LH02	Soyabean plant roots	Gibberellin (GA) production under salt stress	Khan et al. (2011)
Aureobasidium sp. BSS6 and *Preussia* sp. BSL10	*Boswellia sacra*	Plant growth promotion through production of IAA	Khan et al. (2016a)
Azospirillum brasilense	–	Ameliorates the *Arabidopsis thaliana* of reaction to drought by enhancing of abscisic acid (ABA) levels	Cohen et al. (2015)
Azospirillum lipoferum	–	Produces ABA, IAA, and gibberellins and phytohormones. Its inoculation to what plant has shown to alleviate drought stress	Creus et al. (2004)
Bacillus amyloliquefaciens	Rice seeds	Produced ABA under salt-stressed conditions	Shahzad et al. (2017)
Bacillus licheniformis	*Helianthus annuus* roots	Produced ABA in minimal media	Cohen et al. (2009)
Bacillus subtilis	Leaves of *Speranskia tuberculate (*bail)	Effective biocontrol agent with antagonistic action against pathogen causes rotting in tomatoes	Wang et al. (2009)
Paecilomyces formosus	Cucumber plants roots	Gibberellin (GA) production under saline conditions	Khan et al. (2012)
Pantoea alhagi	*Alhagi sparsifolia* leaves	Improves the drought tolerance mechanism of wheat by the production of IAA	Chen et al. (2017)
Penicillium resedanum	Drought-stressed *Capsicum annuum* plants	Gibberellin (GA) production under abiotic stress	Khan et al. (2014)
Pseudomnonas fluoroescens	*Helianthus annuus* roots	Produced ABA in minimal media	Cohen et al. (2009)
Streptomyces capillispiralis Ca-1	Recombinant endophytic strain	Uses synthesized copper nanoparticles for biocontrolling these insects	Hassan et al. (2018)
Yarrowia lipolytica	*Euphorbia milli* L. spines	Produces high indole-3-acetic acid (IAA), indole-3-acetamide (IAM) under salinity stress	Jan et al. (2019b)

14.2.2.1 Regulation of Relative Water Content

Slatyer in 1967 coined the term "relative water content" (RWC) which is the most probable measurement of plant water status relative to its extreme water holding capacity and is an indicator of the water balance of a plant. Plant species possessing higher relative water content have better potential to survive during drought conditions. With continuously changing environmental conditions and declining water tables plants have adopted some strategies to maintain high water potential to mitigate drought stress. Plant growth-promoting microbes are known to act as stress protecting agents. Endophytes are naturally associated with their host plants and interact via several mechanisms during abiotic stresses. Several endophytic strains have been isolated which have shown to improve relative water content under drought stress. *Burkholderia phytofirmans* is capable of colonizing maize seeds, and it has shown to improve the relative water content under drought stress. Another endophyte—*Pseudomonas azotoformans* isolated from the leaves of *Alyssum serpyllifolium* improves relative water content of *Trifolium arvense* under drought stress (Ma et al. 2017).

14.2.2.2 Antioxidant Enzymes

Reactive oxygen species are produced during natural metabolic processes and play significant roles in maintaining cellular homeostasis and signal transduction. Higher concentrations of ROS are detrimental for living organisms and cause oxidative stress. ROS comprises of free radicals like hydroxyl radical ($^{\bullet}OH$), superoxide anion ($O2{\bullet}-$), and non-radical molecules like singlet oxygen ($^{1}O_2$), hydrogen peroxide (H_2O_2). ROS family is involved in several physiological phenomena they act as secondary messengers and induce oxidative damages under several environmental stress conditions like cold, drought, salinity, heavy metals, and extreme temperatures. The generation and elimination of ROS are important for maintaining cellular homeostasis. Generation of reactive oxygen species increases during abiotic stresses which can cause oxidative damage to proteins, lipids, membranes, and other macromolecules. To combat the oxidative damage caused by high levels of ROS, plants utilize their endogenous antioxidant defence system. This antioxidant defence system comprises of enzymatic and nonenzymatic components. Enzymatic components comprise catalases, superoxide dismutases (SOD), glutathione reductases (GR), ascorbate peroxidases (APX), guaiacol peroxidase (GPX), dehydroascorbate reductases (DHAR), and mono-dehydroascorbate reductases (MDHAR) and nonenzymic components comprise of glutathione (γ-glutamyl-cysteinyl-glycine, GSH), cellular redox buffers ascorbate (AsA), carotenoids, tocopherols, and phenolic compounds.

Microorganisms also employ antioxidant enzymes like plants to combat unfavourable conditions. However, the generation of ROS can be minimized by employing endophytes under drought stress (Khan et al. 2016b). They isolated seven endophytic bacteria, i.e., *Acinetobacter calcoaceticus, Rhizobium tropici, Burkholderia vietnameinsis, Sphingomonas yanoikuya, Enterobacter asburiae, Burkholderia* sp.*, Rhanella sp., Pseudomonas sp.,* and *Curtobacterium sp.* from willow and poplar stem and examined their potential for drought tolerance in poplar plants. The treated plants exhibited decreased ROS activity when exposed to drought conditions. Xu et al. (2017) studied the potential protective activity of endophytic fungus *Piriformospora indica* against the detrimental effects of drought stress. Maize and finger millets plants inoculated with *P. indica* produced higher levels of antioxidant enzymes thereby lowering the membrane damage (Xu et al. 2017; Tyagi et al. 2017). *Burkholderia phytofirmans* and *Enterobacter sp.* alleviated drought stress in maize plants by producing enzymes like peroxidase, catalase, and superoxide dismutase (Naveed et al. 2014).

14.2.2.3 Production of Phytohormones

Phytohormones help plants to survive or escape abiotic stresses. Plants along with their symbiotic partners, i.e., endophytes modulate the production localization and distribution of phytohormones to protect themselves under stressed conditions. Plant growth-promoting bacteria synthesize phytohormones like auxin, abscisic acid, ethylene causing physiological, biochemical, and molecular changes in plants and improve tolerance to environmental stresses. Auxin is the most abundant phytohormone found in plants and it predominantly exists as IAA. Auxin has an important role in the regulation of plant growth, seed germination, cell division, and cell elongation. IAA is the most abundant phytohormone produced by rhizospheric bacteria and endophytes. Endophytic bacteria influence the auxin levels by interacting with IAA transporters in plants (Sukumar et al. 2013). Inoculation of plants with endophytic bacteria showed an increase in the lateral roots and root hair formation resulting in increased surface area and higher uptake of moisture and minerals from the soil (Egamberdieva et al. 2017).

Phytohormone abscisic acid is involved in plant responses to several abiotic stress (extreme temperature, drought, and salinity) responses and adaptation (Vysotskaya et al. 2009). Under drought stress, ABA improves the root system to enhance optimal water and nutrient acquisition by increasing root length and density (Shahzad et al. 2017). Abscisic acid sustains the cell turgor potential in plants by maximum utilization of soil moisture which leads to better drought tolerance. This increased tolerance is achieved by the upregulation of the antioxidants and the accumulation of compatible osmolytes which maintains the relative water content of plants during drought conditions.

Ethylene is the only gaseous phytohormone that is responsible for plant growth and development. It also regulates senescence, and abscission, and fruit ripening (Glick 2014). Abiotic stress (drought) leads to increased production of ethylene which can become inhibitory for plant growth. To minimize the ethylene levels, several bacteria produce ACC deaminase which cleaves ACC into ammonia and α-ketobutyrate and reduces the ethylene levels. Regarding the role of ACC deaminase-producing endophytic bacteria in drought stress, Naveed et al. (2014) demonstrated that ACC deaminase-producing endophytic bacteria *Burkholderia phytofirmans* enhanced drought stress tolerance in maize. Salicylic acid (SA) was increased in heavy metal-treated plants inoculated with endophytic bacteria. These results suggest that both hormones may have synergistic effects on heavy metal stress, as observed in another study (Jan et al. 2019a) (Fig. 14.1).

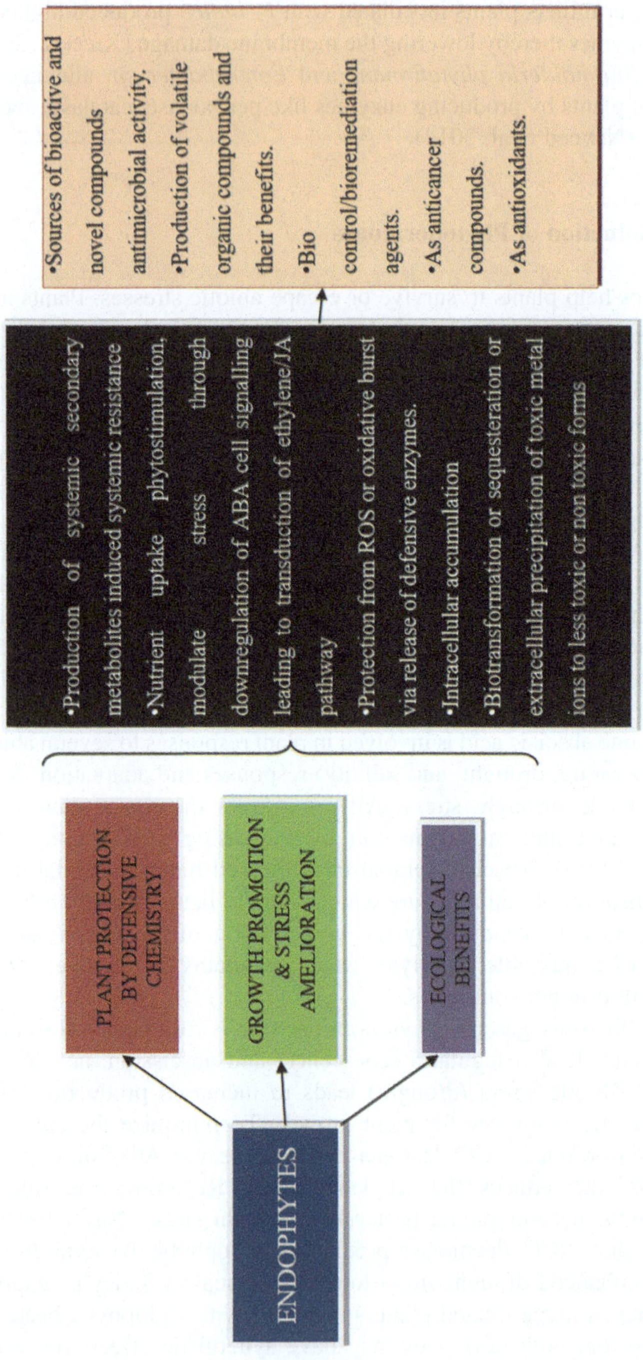

Fig 14.1 Endophytes and their multifarious nature

14.3 Endophytes and Sustainable Agriculture

Biocontrol: Endophytes can be used as potential biocontrol agents, in comparison to the controls by chemical agents such as pesticides and insecticides which is harmful to the whole trophic level. Numerous cases exist where it has been proven that they act as potential biocontrol agents (Kusari et al. 2013). Few are also listed in Table 14.2.

Bio-remediators: This has been clearly observed by various findings that endophytes possess an ability to bioremediate hazardous wastes and pollutants, i.e., they are able to eliminate them by breaking them down into less harmful substances or compounds and finally degrading them or volatizing them through a certain biological process. Mastretta et al. (2009) has reported that with the help of introduction of endophytes into *Nicotiana tabacum* plants have an increase in the plant biomass

Table 14.2 Naturally obtained secondary metabolites from endophytes and their potential functions giving a varied explanation about host endophyte interaction and why there is a need to decipher them (Tan and Zou 2001)

Group	Secondary metabolite	Endophyte	Function
Alkaloids	Peramine	*Neotyphodium coenophialum, N. lolli, Epichloe festucae, and E. typhina* present in the stem and leaf of tall fescue, ryegrass, and other grasses	Toxic to insects (extremely to Argentine stem weevil) without any harmful impact on mammals
Steroids	3β-hydroxyergosta-5-ene, 3-oxoergosta-4,6,8(14), 22-tetraene, 3β, 5-α-dihydroxy-6βacetoxyergosta-7, 22-diene and 3β, 5α-dihydroxy-6-β-phenylacetoxyergosta-7, 22-diene	*Colletotrichum* sp. of *Artemisia annua*	Antifungal against some crop pathogens *Gaeumannomyces graminis var. tritici, Rhizoctonia cerealis, Helminthosporium sativum, and Phytophthora capsici.*
Isocoumarin derivatives	(R)-Mellein	*Pezicula* spp.	Fungicidal, herbicidal, and algicidal
Quinones	Rugulosin	*Hormonema dematioides*, an endophytic fungus of balsam fir	Insecticide
Peptides	Cryptocandin	*Cryptosporiopsis* cf. *quercina* of redwood	Potent antifungal activity
Phenol and phenolic acids	2-Methoxy-4hydroxy-6-methoxymethylbenzaldehyde	Tree endophyte *Pezicula sp.*	Antifungal by the bioautography assay against phytopathogen *Cladosporium cucumerinum*

even under the presence of Cadmium (Cd). Many other studies have proved the useful effects of endophytes in the case of bioremediation process (Thijs et al. 2016).

14.4 Metabolomics Approach for Amelioration of Abiotic Stress

An organism's physical traits is generally considered to be the outcome of the amalgamation of various entwined, precise/non-precise and potent kind of interaction amidst varied elements involving RNA, metabolites, DNA, and proteins in accordance to the environment consisting of certain adverse conditions and/or advancing stages equivalent to nutrient availability, water and temperature or salinity. Therefore, a meticulous explanation of the physical characteristics other than the scanning of metabolites, proteins, and RNA transcripts is required by almost all genome-scale studies. Nevertheless, to obtain a translucent frame work for the traits acquired physically of a given organism only the sum of these three aspects is not enough rather an approach where elements can be characterized in a sequential manner, respectively, should be considered. There is a demanding need to involve all these various conditions as an individual or the classical approach lacks in the characterization of the evolving properties of an organism (Arbona et al. 2013). This kind of need is now a universal requirement and more helpful as well as informative to the motive where one is evaluating plant relationship dynamics in accordance with the environment. When such dynamics is being taken into consideration, it has been realized that the approach involving expression of proteins and definitive genes display only plants capabilities to respond in stressed conditions but the approach that involves metabolites includes or constitutes both protein and gene expression in addition to the environment dynamics making it a suitable and an universal approach amidst other omics technologies with fairly less amendments and more applications to a varied number of organisms. An area where metabolomics has been implemented and gained successful conclusions involves finding similar patterns that are able to confer stress especially abiotic stress in plants by deciphering their molecular physical traits (Arbona et al. 2013).

Therefore, the definition of metabolomics can be summarized as the recognition of metabolites that possess lower molecular weights of an organism at particular tissue, organ, and cell type or at a specific developmental level/levels (Arbona et al. 2009; Fiehn 2001). This summarization of metabolomics appears to be peculiar and tough to defy for a sole reason that these constitutes of high number of continuous molecules that possess varied chemical composition and structures. But still, broad categories of metabolites that are contained in plant kingdom and their overall analysis are yet uncharacterized. Such issue can be overcome by or can be deciphered by certain kind of techniques that include a specific combination of separation technique paired up with a detection device which is usually mass spectrometry (Arbona et al. 2013). The role of the separation technique is primarily to provide the selective pattern or a layer required by the varied categories of metabolites, as we can see in case of gas chromatography that has the primarily objective of selectivity for primary volatile metabolites, e.g., amino acid, tricarboxylic acid (TCA), or sugar intermediates after they have been derivatized (Arbona et al. 2009). Another technique, namely liquid chromatography which is considered quite adjustable can be used for a wide range of compounds to decipher secondary metabolites—a category of most important sort of metabolites in plants. This technique is carried out without the metabolites being derivatized means first through LC metabolites are screened and then derivatization takes place. In comparison to this capillary zone electrophoresis (CZE) is also possesses similar sort of objectives. But in case of CZE, there is a step ahead of separation of even ionic metabolites also. Various useful objectives are being served by the category of plant metabolomics, and some of which involves as follows:

(a) To decipher or to trail various compounds to their category which is involved in a specific kind of degradation or biosynthetic pathway.
(b) To know about the effects of the functioning of plant metabolism when the host plants have conferred stress or have been treated with such impactful conditions.
(c) Plant metabolomics also helps in various samples classification.

Activation of a specific metabolic pathway releases certain intermediates or precursors that pilot the release of certain bioactive molecules like signalling compound, an antioxidant, a cell structure biosynthesis intermediate, or even a storage compound. Once they are produced, they are controlled or levelled by various other elements such as plant hormones or signalling molecules. These regulatory elements are in fact considered to be not in relation to any kind of such pathways which can activate or deactivate varied steps involved in metabolism.

This above-explained mechanism can be clearly observed in the case of a very important process known as photosynthesis, which is also considered to be sensitive process to abiotic stress. Therefore, when the host plants undergo through abiotic stress, carbon assimilation, and the primary metabolism are largely affected. Mostly the concentration of essential metabolites like sugar alcohols, amino acids, and sugars is affected due to stress conditions. But the effect is not just because of conferring a particular stress condition, the effect is also considered to be the outcome of another complex regulatory network (Arbona et al. 2013).

14.5 Deciphering Host Endophyte Interaction Through Metabolomics

The interaction between host plants and endophytes is a complicated one as it revolves not only between these two but with whole microbiome from the outer surroundings. So, what actually happens is that the endophytes adapt to certain aspects in order to survive through unknown conditions, firstly they start their evolution through coexisting. To achieve this, they develop or inbuilt varieties of traits within themselves in different range. For example:

- Production of certain molecules that would help in inhibiting quorum sensing.
- Production of certain other compounds that play role in defence actions, such as enhancers to produce precursors that would catalyse certain important unravelled biosynthesis pathways effectively.
- Or by being epigenetic modulators (Kusari and Spiteller 2011; Jia et al. 2016; Mookherjee et al. 2018; Scherlach and Hertweck 2018).

This has been fairly concluded by several researches that the production of these molecules or compounds have an essential ecological role. The significant discovery was the endophytic fungus *Neotyphodium coenophialum* that found in tall fescue. It produces alkaloids that are toxic in nature and results in the disease known as fescue toxicosis in livestock, hence protecting their host plants against herbivores (Bacon et al. 1977).

Now a major question lies on the point that how can such biomolecules/compounds can be identified or unravelled? Metabolomics in combination with other analytical techniques can be deciphered. In this case, the most used technique is low chromatography mass spectrometry (LC-MS) which is generally used to identify complex peptides or non-volatile molecules from endophytes. With the help of metabolomics approach certain peculiar compounds or substrates involved in interlinked mechanisms can be decoded. Usually, the function of these interlinked compounds is in activating defence mechanism of the hosts. This is also an escape from the traditional old methods that have been used until now. These profiling methods are considered to be way more effective. In addition to this, these when linked with other present omics technologies are capable of decoding the whole process and give us a platform to redesign the beneficial activities in a much broader scale as in terms of whole ecology.

14.6 Metabolomic Profiling

Metabolite profiling with multiple variables and data based is considered to be a very effective and dedicated approach to classify samples where plant microbes are interrelated that may lead to recognition of various metabolites found in plants, i.e., basically, this technique is used to explain that irrespective of interaction between plant and endophytes there always exists varied kind of metabolites. So basically, the metabolomic profiling technique has been designed in such a pattern that the information of metabolite when deciphered; is in respect to the coexistence of metabolites present in both plant and microbe. For example, the technique GC-TOF-MS also known as gas chromatography coupled with time-of-flight mass spectrometry is said to decipher or to find a total number of microbes present in that particular area specifically defined as microbial biomass. Basically, the emphasis of this metabolomic tool is to unravel certain interesting topics related to endophytes, may it be in terms of their importance in relation to environment or their sustainable impact on plants (Christian et al. 2009; Barsch et al. 2006a, b; Andrea and Paul 2011). In recent study, in situ analysis was done in *Setaria viridis* roots colonized by beneficial endophytic bacteria with in situ laser ablation electrospray ionization mass spectrometry to investigate the metabolic changes in roots (Agtuca et al. 2020). The present omics world also includes studies that identify the genes and proteins potentially engaged in the plant–microbe interaction. Understanding how plant–microbe interaction happens allows researchers to ameliorate crop yield and reduce environmental stresses.

Hence, this whole procedure could be summarized into different steps which can be described as follows:

1. The quick study of metabolites at their early stages can be achieved with the help of techniques like 1H nuclear magnetic resonance and high resolution mass spectrometry (Emwas et al. 2019).
2. To properly execute, a technique known as metabolite fingerprinting shall be applied at precursor stages, i.e., to the cultivation media that is yielding the active metabolites at maximum levels. By doing this, we simply increase or enhance the production of active metabolites to optimum/desired levels.
3. Then comes the procedure of metabolite profiling which involves the deciphering of such extracts that can be processed further for fractionation.

Therefore, after this whole guided procedure of metabolomics, it becomes easier to obtain the potential metabolites that were present in certain combinations. Metabolomics as a whole in respect to endophytes can prove as a very diligent and effective technique to decipher various useful compounds or substances that are involved in various major process of plants or investigate novel leads that lead to the survival of plants in such extreme conditions. Metabolomics is a powerful facilitator in the discovery of natural products, which are considered as an excellent source for novel leads, and even more, as a means to highlight active targets (Kamal et al. 2017; Maciá-Vicente et al. 2018; Tawfike et al. 2019; Wei et al. 2020). Table 14.2 lists such various secondary metabolites from endophytes and their uses thus proving that metabolomics is a potent tool.

14.7 Conclusions

The metabolome study explains more accurately and precisely the phenotype of a given plant species by indicating the integration of the genetic background and the influence of the environmental conditions. In response to adverse abiotic stimuli, plants orchestrate an array of responses oriented to stress avoidance, defence, or resistance, depending on the particular stress tolerance. It would be an interesting section to explore the impact of endosymbionts on the host's gene expression, metabolism, and other physiological aspects essential in conferring resistance against biotic and abiotic stresses. A more intriguing and inexplicable issue with many endophytes that must be critically evaluated is their ability to produce host metabolites, which can be harnessed on a large scale for potential use in diverse areas. It is a much required need of the hour to obtain the biochemistry and physiology of endophytes up to genomic and metabolomic levels. To date, there are no databases exclusively available for endophytic microorganisms and their metabolites, which can be of great importance and provide solutions to many issues. Hence, with the help of omics technologies, it is important to decipher the role of metabolites and their possible mechanism of action to gain all possible benefits from the remarkable association.

Acknowledgements The authors are grateful to Director CSIR-National Botanical Research Institute for providing facilities and support during the study. This work is supported by the CSIR-Network project (MLP0048; OLP 0109) funded by Council of Scientific and Industrial Research, New Delhi, India.

References

Agtuca BJ et al (2020) In-situ metabolomic analysis of Setaria viridis roots colonized by beneficial endophytic bacteria. Mol Plant-Microbe Interact 33:272–283

Andrea P-A, Paul B (2011) Hidden fungi, emergent properties: endophytes and microbiomes. Annu Rev Phytopathol 49:291–315

Arbona V, Iglesias DJ, Talon M, Cadenas AG (2009) Plant phenotype demarcation using non targeted LC-MS and GC-MS metabolite profiling. J Agric Food Chem 57:7338–7347

Arbona V, Manzi M, de Ollas C, Cadenas AG (2013) Metabolomics as a tool to investigate abiotic stress tolerance in plants. Int J Mol Sci 14:4885–4911

Asaf S, Khan MA, Khan AL, Waqas M, Shahzad R, Kim AY, Kang SM, Lee IJ (2017) Bacterial endophytes from arid land plants regulate endogenous hormone content and promote growth in crop plants: an example of *Sphingomonas* sp. and *Serratia marcescens*. J Plant Interact 12 (1):31–38

Bacon C, Porter J, Robbins J, Luttrell E (1977) Epichloë typhina from toxic tall fescue grasses. Appl Environ Microbiol 34:576–581

Barsch A, Carvalho HG, Cullimore JV, Niehaus K (2006a) GC MS based metabolite profiling implies three interdependent ways of ammonium assimilation in *Medicago truncatula* root nodules. J Biotechnol 127:79–83

Barsch A, Tellstrom V, Patschkowski T, Kuster H, Niehaus K (2006b) Metabolite profiles of nodulated alfalfa plants indicate that distinct stages of nodule organogenesis are accompanied by global physiological adaptations. Mol Plant-Microbe Interact 19:998–1013

Bisht N, Mishra SK, Chauhan PS (2020) *Bacillus amyloliquefaciens* inoculation alters physiology of rice (*Oryza sativa* L. var. IR-36) through modulating carbohydrate metabolism to mitigate stress induced by nutrient starvation. Int J Biol Macromol 143:937–951

Bordiec S, Paquis S, Lacroix H, Dhondt S, Ait Barka E, Kauffmann S, Jeandet P, Mazeyrat-Gourbeyre F, Clément C, Baillieul F, Dorey S (2011) Comparative analysis of defence responses induced by the endophytic plant growth-promoting rhizobacterium *Burkholderia phytofirmans* strain PsJN and the non-host bacterium *Pseudomonas syringae* pv. *pisi* in grapevine cell suspensions. J Exp Bot 62(2):595–603

Chen C, Xin K, Liu H, Cheng J, Shen X, Wang Y, Zhang L (2017) Pantoea alhagi, a novel endophytic bacterium with ability to improve growth and drought tolerance in wheat. Sci Rep 7:41564

Chi F, Shen SH, Cheng HP, Jing YX, Yanni YG, Dazzo FB (2005) Ascending migration of endophytic rhizobia, from roots to leaves, inside rice plants and assessment of benefits to rice growth physiology. Appl Environ Microbiol 71(11):7271–7278

Christian S, Kristina U, Dietrich E, Wolfram W (2009) A metabolic signature of the beneficial interaction of the endophyte Paenibacillus sp. isolate and in vitro—grown poplar plants revealed by metabolomics. Mol Plant-Microbe Interact 22:1032–1037

Cohen AC, Travaglia CN, Bottini R, Piccoli PN (2009) Participation of abscisic acid and gibberellins produced by endophytic Azospirillum in the alleviation of drought effects in maize. Bot 87:455–462

Cohen AC, Bottini R, Pontin M, Berli FJ, Moreno D, Boccanlandro H, Travaglia CN, Piccoli PN (2015) Azospirillum brasilense ameliorates the response of Arabidopsis thaliana to drought mainly via enhancement of ABA levels. Physiol Plant 153(1):79–90

Creus CM, Sueldo RJ, Barassi CA (2004) Water relations and yield in Azospirillum-inoculated wheat exposed to drought in the field. Can J Bot 82:273–281

Dixit R, Agrawal L, Gupta S, Kumar M, Yadav S, Chauhan PS, Nautiyal CS (2016) Southern blight disease of tomato control by 1-aminocyclopropane-1-carboxylate (ACC) deaminase producing *Paenibacillus lentimorbus* B-30488. Plant Signal Behav 11:e1113363

Dixit VK, Misra S, Mishra SK, Tewari SK, Joshi N, Chauhan PS (2020) Characterization of plant growth-promoting alkalotolerant *Alcaligenes* and *Bacillus* strains for mitigating the alkaline stress in *Zea mays*. Antonie Van Leeuwenhoek 9:1–7

Egamberdieva D, Wirth SJ, Shurigin VV, Hashem A, Abd Allah EF (2017) Endophytic bacteria improve plant growth, symbiotic performance of chickpea (Cicer arietinum L.) and induce suppression of root rot caused by Fusarium solani under salt stress. Front Microbiol 8:1887

Emwas A-H et al (2019) NMR spectroscopy for metabolomics research. Metabolites 9:123

Feller IC (1995) Effects of nutrient enrichment on growth and herbivory of dwarf red mangrove (Rhizophora mangle). Ecol Monogr 65(4):477–505

Fiehn O (2001) Combining genomics, metabolome analysis, and biochemical modelling to understand metabolic networks. Comp Funct Genomics 2:155–168

Fouda AH, Hassan SE, Eid AM, Ewais EE (2015) Biotechnological applications of fungal endophytes associated with medicinal plant Asclepias sinaica (Bioss.). Ann Agric Sci 60 (1):95–104

Glick BR (2014) Bacteria with ACC deaminase can promote plant growth and help to feed the world. Microbiol Res 169(1):30–39

Hardoim PR, van O, verbeek LS, van E, lsas JD (2008) Properties of bacterialendophytes and their proposed role in plant growth. Trends Microbiol 16:463–471

Hassan SE, Salem SS, Fouda A, Awad MA, El-Gamal MS, Abdo AM (2018) New approach for antimicrobial activity and bio-control of various pathogens by biosynthesized copper nanoparticles using endophytic actinomycetes. J Radiat Res Appl Sci 11(3):262–270

Jan R, Khan MA, Asaf S, Lee IJ, Kim KM (2019a) Metal resistant endophytic bacteria reduces cadmium, nickel toxicity, and enhances expression of metal stress related genes with improved growth of *Oryza Sativa*, via regulating its antioxidant machinery and endogenous hormones. Plan Theory 8(10):363

Jan FG, Hamayun M, Hussain A, Jan G, Iqbal A, Khan A, Lee IJ (2019b) An endophytic isolate of the fungus *Yarrowia lipolytica* produces metabolites that ameliorate the negative impact of salt stress on the physiology of maize. BMC Microbiol 19(1):1

Jia M, Chen L, Xin HL, Zheng CJ, Rahman K, Han T (2016) A friendly relationship between endophytic fungi and medicinal plants: a systematic review. Front Microbiol 7:906. https://doi.org/10.3389/fmicb.2016.00906

Kamal N, Viegelmann CV, Clements CJ, Edrada-Ebel R (2017) Metabolomics-guided isolation of anti-trypanosomal metabolites from the endophytic fungus *Lasiodiplodia theobromae*. Planta Med 83(6):565–573

Khan AL, Hamayun M, Kim YH, Kang SM, Lee JH, Lee IJ (2011) Gibberellins producing endophytic Aspergillus fumigatus sp. LH02 influenced endogenous phytohormonal levels, isoflavonoids production and plant growth in salinity stress. Process Biochem 46(2):440–447

Khan AL, Hamayun M, Kang SM, Kim YH, Jung HY, Lee JH, Lee IJ (2012) Endophytic fungal association via gibberellins and indole acetic acid can improve plant growth under abiotic stress: an example of *Paecilomyces formosus* LHL10. BMC Microbiol 12(1):3

Khan AL, Waqas M, Kang SM, Al-Harrasi A, Hussain J, Al-Rawahi A, Al-Khiziri S, Ullah I, Ali L, Jung HY, Lee IJ (2014) Bacterial endophyte *Sphingomonas* sp. LK11 produces gibberellins and IAA and promotes tomato plant growth. J Microbiol 52(8):689–695

Khan AL et al (2016a) Endophytic fungi from Frankincense tree improves host growth and produces extracellular enzymes and indole acetic acid. PLoS One 11:e0158207

Khan Z et al (2016b) Growth enhancement and drought tolerance of hybrid poplar upon inoculation with endophyte consortia. Curr Plant Biol 6:38–47

Kloepper JW, Ryu CM (2006) Bacterial endophytes as elicitors of induced systemic resistance. In: Microbial root endophytes. Springer, Berlin, pp 33–52

Kobayashi DY, Palumbo JD (2000) Bacterial endophytes and their effects on plants and uses in agriculture. In: Bacon CW, White JF (eds) Microbial endophytes. Dekker, New York, pp 199–236

Kusari S, Spiteller M (2011) Are we ready for industrial production of bioactive plant secondary metabolites utilizing endophytes? Nat Prod Rep 28:1203–1207

Kusari P, Kusari S, Spiteller M, Kayser O (2013) Endophytic fungi harbored in Cannabis sativa L.: diversity and potential as biocontrol agents against host plant-specific phytopathogens. Fungal Div 60:137–151

Li WK (2005) Endophytes and natural medicines. Chin J Nat Med 3(4):193–199

Ma Y, Rajkumar M, Moreno A, Zhang C, Freitas H (2017) Serpentine endophytic bacterium Pseudomonas azotoformans ASS1 accelerates phytoremediation of soil metals under drought stress. Chemosphere 185:75–85

Maciá-Vicente JG et al (2018) Metabolomics-based chemotaxonomy of root endophytic fungi for natural products discovery. Environ Microbiol 20:1253–1270

Marler MJ, Zabinski CA, Callaway RM (1999) Mycorrhizae indirectly enhance competitive effects of an invasive forb on a native bunchgrass. Ecology 80(4):1180–1186

Mastretta C, Taghavi S, Van Der Lelie D, Mengoni A, Galardi F, Gonnelli C, Barac T, Boulet J, Weyens N, Vangronsveld J (2009) Endophytic bacteria from seeds of *Nicotiana tabacum* can reduce cadmium phytotoxicity. Int J Phytoremediation 11(3):251–267

Mishra SK, Khan MH, Misra S, Kant VK, Khare P, Srivastava S, Chauhan PS (2017) Characterization of *Pseudomonas*spp. and *Ochrobactrum* sp. isolated from volcanic soil. Antonie Van Leeuwenhoek 110:253–270

Misra S, Dixit VK, Khan MH, Mishra SK, Dviwedi G, Yadav S, Lehri A, Chauhan PS (2017) Exploitation of agro-climatic environment for selection of 1-aminocyclopropane-1-carboxylic acid (ACC) deaminase producing salt-tolerant indigenous plant growth promoting rhizobacteria. Microbiol Res 205:25–34

Misra S, Dixit VK, Mishra SK, Chauhan PS (2019) Demonstrating the potential of abiotic stress-tolerant *Jeotgalicoccushuakuii* NBRI 13E for plant growth promotion and salt stress amelioration. Ann Microbiol 69:419–434

Mookherjee A, Singh S, Maiti MK (2018) Quorum sensing inhibitors: can endophytes be prospective sources? Arch Microbiol 200:355–369

Naveed M, Mitter B, ReichenauerTG WK, Sessitsch A (2014) Increased drought stress resilience of maize through endophytic colonization by *Burkholderia phytofirmans PsJN* and *Enterobacter sp,* FD17. Environ Exp Bot 97:30–39

Papik J, Folkmanova M, Majorova MP, Suman J, Uhlik O (2020) The invisible life inside plants: deciphering the riddles of endophytic bacterial diversity. Biotech Adv 44:107614

Partida-Martinez LP, Heil M (2011) The microbe-free plant: fact or artifact? Front Plant Sci 2:100

Peters AF (1991) Field and culture studies of *Streblonema macrocystis* sp. nov.(Ectocapales, Phaeophyceae) from Chile, a sexual endophyte of giant kelp. Phycologia 30(4):365–377

Petrini O (1991) Fungal endophytes of tree leaves. In: Andrews J, Hirano S (eds) Microbial ecology of leaves. Springer, pp 179–197

Pieterse CM, Van der Does D, Zamioudis C, Leon-Reyes A, Van Wees SC (2012) Hormonal modulation of plant immunity. Annu Rev Cell Dev Biol 11:28

Prashar P, Kapoor N, Sachdeva S (2014) Rhizosphere: its structure, bacterial diversity and significance. Rev Environ Sci Biotechnol 13(1):63–77

Saharan BS, Nehra V (2011) Plant growth promoting rhizobacteria: a critical review. Life Sci Med Res 21(1):30

Scherlach K, Hertweck C (2018) Mediators of mutualistic microbe-microbe interactions. Nat Prod Rep 35(4):303–308

Schulz B, Boyle C (2005) The endophytic continuum. Mycol Res 109:661–687

Shahzad R, Khan AL, Bilal S, Waqas M, Kang SM, Lee IJ (2017) Inoculation of abscisic acid-producing endophytic bacteria enhances salinity stress tolerance in *Oryza sativa*. Environ Exp Bot 136:68–77

Stone JK, Bacon CW, White JF Jr (2000) An overview of endophytic microbes: endophytism defined. In: Microbial endophytes: a continuum of interactions with plants. Marcel Dekker Inc, New York, pp 3–29

Sukumar P, Legue V, Vayssieres A, Martin F, Tuskan GA, Kalluri UC (2013) Involvement of auxin pathways in modulating root architecture during beneficial plant microorganisms' interactions. Plant Cell Environ 36:909–919

Sziderics AH, Rasche F, Trognitz F, Sessitsch A, Wilhelm E (2007) Bacterial endophytes contribute to abiotic stress adaptation in pepper plants (*Capsicum annuum* L.). Can J Microbiol 53 (11):1195–1202

Tan RX, Zou WX (2001) Endophytes: a rich source of functional metabolites. Nat Prod Rep 18:448–459

Tawfike AF et al (2019) Isolation of anticancer and anti-trypanosome secondary metabolites from the endophytic fungus *Aspergillus flocculus* via bioactivity guided isolation and MS based metabolomics. J Chromatogr B 1106:71–83

Thijs S, Sillen W, Rineau F, Weyens N, Vangronsveld J (2016) Towards an enhanced understanding of plant–microbiome interactions to improve phytoremediation: engineering the metaorganism. Front Microbiol 7:341

Timmusk S, Paalme V, Pavlicek T, Bergquist J, Vangala A, Danilas T, Nevo E (2011) Bacterial distribution in the rhizosphere of wild barley under contrasting microclimates. PLoS One 6(3): e17968

Tyagi J, Varma A, Pudake RN (2017) Evaluation of comparative effects of arbuscular mycorrhiza (*Rhizophagus intraradices*) and endophyte (*Piriformospora indica*) association with finger millet (Eleusine coracana) under drought stress. Eur J Soil Biol 81:1

Ullah A, Sun H, Yang X, Zhang X (2017) Drought coping strategies in cotton: increased crop per drop. Plant Biotechnol J 15(3):271–284

Vysotskaya LB, Korobova AV, Veselov SY, Dodd IC, Kudoyarova GR (2009) ABA mediation of shoot cytokinin oxidase activity: assessing its impacts on cytokinin status and biomass allocation of nutrient-deprived durum wheat. Funct Plant Biol 36(1):66–72

Wang S, Hu T, Jiao Y, Wei J, Cao K (2009) Isolation and characterization of Bacillus subtilis EB-28, an endophytic bacterium strain displaying biocontrol activity against *Botrytis cinereal* Pers. Front Agric China 3(3):247–252

Wei J et al (2020) Comparative metabolomics revealed the potential antitumor characteristics of four endophytic fungi of Brassica rapa L. ACS Omega 5:5939–5950

Wen CY (2004) Recent advances and issues on the endophyte. Chinese J Ecol 23(2):86–91

Wilson BJ, Addy HD, Tsuneda A, Hambleton S, Currah RS (2004) *Phialocephala sphaeroides* sp. nov., a new species among the dark septate endophytes from a boreal wetland in Canada. Can J Bot 82:607–617

Xu L, Wang A, Wang J, Wei Q, Zhang W (2017) *Piriformospora indica* confers drought tolerance on *Zea mays* L. through enhancement of antioxidant activity and expression of drought-related genes. Crop J 5(3):251–258

The manufacturer's authorised representative in the EU is Springer Nature Customer Service Centre GmbH, Europaplatz 3, 69115 Heidelberg, Germany. If you have any concerns regarding our products, please contact ProductSafety@springernature.com

Printed and bound by CPI Group (UK) Ltd, Croydon, CR0 4YY
30/09/2026
02225332-0008